# Advances in
# MICROBIAL ECOLOGY

**Volume 1**

# ADVANCES IN MICROBIAL ECOLOGY

Sponsored by International Commission on Microbial Ecology,
a unit of International Association of Microbiological Societies
and the Division of Environmental Biology of the
International Union of Biological Societies

A Continuation Order Plan is available for this series. A continuation order will bring
delivery of each new volume immediately upon publication. Volumes are billed only upon
actual shipment. For further information please contact the publisher.

# Advances in
# MICROBIAL ECOLOGY

## Volume 1

Edited by

## M. Alexander

*Cornell University*
*Ithaca, New York*

PLENUM PRESS · NEW YORK AND LONDON

ISBN 0-306-38161-3

# Contributors

**Dennis J. Cosgrove**, CSIRO, Division of Plant Industry, Canberra, Australia.

**T. M. Fenchel**, Institute of Ecology and Genetics, University of Aarhus, DK–8000 Aarhus C, Denmark.

**D. D. Focht**, Department of Soil Science and Agricultural Engineering, University of California, Riverside, California, U.S.A.

**B. Barker Jørgensen**, Institute of Ecology and Genetics, University of Aarhus, DK–8000 Aarhus C, Denmark.

**J. W. M. la Rivière**, International Institute for Hydraulic and Environmental Engineering, Delft University of Technology, Delft, The Netherlands.

**H. Veldkamp**, Department of Microbiology, University of Groningen, Biological Centre, Haren (Gr.), The Netherlands.

**W. Verstraete**, Laboratory of General and Industrial Microbiology, State University of Ghent, Belgium.

v

# Preface

Microbial ecology has come of age. A long history of exploratory research and problem definition is admittedly behind us, and outstanding scientists from many regions of the world have contributed major findings to the development of the discipline. The early literature is indeed vast and impressive, yet it is still fair to say that the field is just beginning to develop in a mature fashion. The major microbial habitats and the chief inhabitants have been delineated and are only now being adequately explored; the processes that are brought about by the microscopic residents are at the present time being investigated in definitive ways; new concepts are being formulated; and novel, exciting, and useful techniques have been devised.

In recognition of the emergence of microbial ecology as a rapidly growing and independent discipline, the International Commission on Microbial Ecology—a unit of the International Association of Microbiological Societies and the Division of Environmental Biology of the International Union of Biological Societies—has established the *Advances in Microbial Ecology* and has appointed an editor and editorial board for the series. The purposes of this review series are to present in one group of volumes an overview of the information that is now spread in an enormous array of technical journals, to have authors in various areas of microbial ecology summarize and present views in depth of their own specialties, and to help promote the continued growth of microbial ecology. It is planned that reviews will deal with diverse ecosystems and processes, including both basic and applied problems and all major categories of microorganisms—bacteria, fungi, algae, protozoa, and viruses.

The editorial board, the composition of which is given below, welcomes suggestions, comments, and criticisms. We are endeavoring to serve a useful function for our colleagues, our discipline, and future generations of microbial ecologists, and we hope that *Advances in Microbial Ecology* will play a pivotal role in the future of this field.

The editorial board expresses its gratitude to Dr. N. E. Gibbons, former secretary-general of the International Association of Microbiological Societies, for his assistance and guidance in arranging for the establishment of this new series of monographs.

M. Alexander, Editor
T. Rosswall
M. Shilo
H. Veldkamp

# Contents

Chapter 1
**Detritus Food Chains of Aquatic Ecosystems: The Role of Bacteria**
T. M. Fenchel and B. Barker Jørgensen

Chapter 2
Ecological Studies with the Chemostat
H. Veldkamp

Chapter 3
Microbial Transformations in the Phosphorus Cycle
Dennis J. Cosgrove

# Chapter 4
# Biochemical Ecology of Nitrification and Denitrification
D. D. Focht and W. Verstraete

# Chapter 5
# Microbial Ecology of Liquid Waste Treatment
J. W. M. la Rivière

# Detritus Food Chains of Aquatic Ecosystems: The Role of Bacteria

## T. M. FENCHEL AND B. BARKER JØRGENSEN

## 1. Introduction

The purpose of this review is to present and discuss current knowledge on the role played by bacteria in the carbon cycle of ecosystems. The ecological experience of the authors explains why our examples are mainly derived from aquatic systems; many of the discussed principles, however, apply equally well to terrestrial systems. In fact the one environmental requirement common to all microorganisms is the presence of free water, and in a sense all microorganisms are aquatic organisms.

In contrast to soil biologists, students of aquatic ecosystems began to appreciate the quantitative importance of bacteria in the transformation of matter only relatively recently, although some early studies (e.g., ZoBell and Feltham, 1937–38, 1942; ZoBell and Grant, 1943; ZoBell and Landon, 1937) emphasized the importance of bacteria in marine food chains. However, while marine ecologists early in this century were aware of the importance of detritus food chains in the sea (e.g., Boysen-Jensen, 1914), in general only little attention was given to bacteria. Within the last 15 years, the interest in aquatic microbial ecology has increased drastically as witnessed by a large number of textbooks and symposia on the topic (e.g., Cairns, 1971; Colwell and Morita, 1974; Melchiorri-Santolini and Hopton, 1972; Oppenheimer, 1963; Rodina, 1972; Rosswall, 1973; Wood, 1965, 1967). Reviewing this literature reveals a large (and warranted) emphasis on

**T. M. FENCHEL and B. BARKER JØRGENSEN** • Institute of Ecology and Genetics, University of Aarhus, DK–8000 Aarhus C, Denmark.

methodological problems, in particular with respect to the quantification of numbers and activity of microorganisms in nature, and many problems remain to be solved before a complete understanding of the quantitative importance of bacteria will be achieved. It is understandable that many biologists, and microbiologists make no exception, like their favorite organisms to play an important role in nature, and they like to be able to demonstrate this by quantitative means. It is admittedly an important task of microbial ecology to provide these data. In the present review, however, we will also emphasize qualitative aspects. In particular we will discuss properties of ecosystems and of bacteria which explain why and how the most simple types of life known play a key role in the function of ecosystems.

When discussing the role of bacteria in food chains, we will first of all observe that bacteria (with the exception of photosynthetic forms) nearly always exploit dead organic material as a source of energy and carbon (or inorganic compounds, e.g., $H_2S$, are used as source of energy, but the potential energy of these compounds derives from the incomplete de-composition of organic compounds). To be sure, parasitic bacteria which attack living plant tissue or predatory bacteria such as *Bdellovibrio* exist and take part in food chains in nature, but their contribution to the carbon cycle is trivial. In the first section we will therefore explore why a large part of the organic production and in particular of the primary production passes detritus food chains, i.e., becomes available to consumers in the form of dead organic material rather than being utilized by herbi-vorous animals. In following sections we will discuss why bacteria dominate as primary decomposers of detritus and will describe the colonization and growth of bacteria on dissolved or particulate detritus, respectively. Among the bacteria, a diversity in metabolic pathways and in potential sub-strates is found, unparallelled in eucaryotic organisms. This is an important aspect of the role of bacteria in nature and will be discussed in detail in a separate chapter. We will then discuss various qualitative and quantitative aspects of the exploitation of bacteria by animals, the effect of grazing on bacterial populations, and symbiotic relationships between bacteria and animals. Finally, in the last section we will summarize what is known about the quantitative importance of bacteria for the energy flow of aquatic systems.

## 2. Detritus Food Chains

### 2.1. The Definition and Origin of Detritus

The term *detritus* is ambiguous in the literature. We will here follow Wetzel *et al.* (1972), who define detritus as *non-predatory losses of organic carbon from any trophic level (includes egestion, excretion, secretion, etc.) or inputs from sources external to the ecosystem that enter and cycle in the system (allochthonous organic carbon).* The term *detritus food chain* can thus be defined as *any route by which chemical energy contained within detrital organic carbon becomes available to the biota.* These definitions differ from the usual meaning of the term by including dissolved organic material as detritus. Furthermore, the definition of Wetzel *et al.* (1972) excludes living components as a part of detritus. To the zoologist who studies detritus-feeding animals which ingest the whole complex of dead and living organic material indiscriminately, it is natural to consider it all as detritus. In the present context, where we will also discuss the interrelationships between the dead organic material and the associated microflora, it is useful to consider detritus as only consisting of dead organic material. Wetzel *et al.* (1972) finally coined the term *detrital electron flux* to infer that the potential chemical energy of detritus may be transferred by inorganic compounds (e.g., $H_2S$, $NO_2^-$) in a detrital food chain.

Although a whole continuum of particle sizes from small organic molecules to particles several millimeters long are found in nature, dissolved detritus is usually defined arbitrarily as organic material passing filters with pore sizes of 0.2 or 0.45 $\mu m$; thus the smallest detrital particles are just visible under the light microscope.

In the oceans dissolved organic matter, particulate dead organic matter, and living biomass are found in the proportions $100:10:2$ (Parsons, 1963), and similar values are known from lakes (e.g., Wetzel *et al.*, 1972). The concentration of dissolved detritus in seawater is usually within the range 1–5 mg/liter but may exceed 40 mg/liter in some inshore waters. In lakes, values between 1 and 50 mg/liter are usually reported (Jørgensen, 1966; Ogura, 1975; Wetzel *et al.*, 1972).

Particulate detritus usually constitutes the largest part of suspended particles in natural waters; in oceanic waters it is found in concentrations between 0.02 and 6 mg/liter. In inshore waters and in freshwater systems a much wider range, i.e., 0.1 to 40 mg/liter has been recorded (Jørgensen, 1966). Large amounts of particulate detritus are also found on and in sediments which may contain from less than 1% up to 100% organic material. Sometimes the distinction between suspended and sedimented detritus is not very sharp since cycles of sedimentation and resuspension may take place (Steele and Baird, 1972).

Dissolved detritus derives from several sources. Thus healthy and growing microphytes as well as macrophytes secrete a part of the assimilated carbon as dissolved organic matter. Wetzel *et al.* (1972) found that in a hard-water lake the phytoplankton, epiphytic algae, epipelic algae, and macrophytes release 5.7, 5, 5, and 4%, respectively, of the photosynthetically fixed carbon. These are annual averages as different environmental factors influence the amount of released organic matter. Riley (1970) concluded on the basis of the literature that as much as 10–30% of marine phyto-plankton production is lost as dissolved organic matter, and Brylinski (quoted from Mann, 1972) found that of six species of marine macrophytes, none released less than 4% of their primary production of dissolved organic matter. Other important sources of dissolved detritus derive from the autolysis of dead plant material (Mann, 1972; Otsuki and Wetzel, 1974) and by loss during feeding by or secretion from animals. Thus secretion may amount to more than 33% of the assimilated food (e.g., Field, 1972; Hargrave, 1971; Johannes and Satomi, 1967; Kofoed, 1975b; Webb and Johannes, 1967). Active bacteria also exude dissolved organic matter, e.g., in connection with the decomposition of particulate material (Otsuki and Hanya, 1972).

Particulate detritus also derives from different sources. Animals which have avoided predators will eventually contribute to the pool of dead particulate organic material as will, for example, chitin from crustacean exuviae. Many herbivorous (and detritivorous) animals have a low efficiency of assimilation, and their fecal pellets may constitute a significant contribution to particulate detritus (Frankenberg and Smith, 1967; Newell, 1965). By far the greatest contribution of detritus derives, however, directly from the primary producers. It is a fact that only a fraction of the primary production due to macrophytes is consumed by herbivores, and the rest consequently enters detritus food chains.

There exist different estimates for the fraction of primary production consumed by herbivores in different plant communities or stands of particular macrophyte species. Jørgensen (1966) estimated that 80% of the phytoplankton production may be utilized by herbivores. Teal (1962) found that about 7% of the production of *Spartina* is grazed, and values for mangroves of 5.1% (Heald, 1969, quoted from Mann, 1972) and kelp beds of less than 10% (Mann, 1972) are reported. It should be noted that as plant size decreases an increasing amount of the production is consumed by grazers, but even in the case of phytoplankton a substantial fraction enters detritus food chains. The communities of macrophytes, such as the boreal kelp beds or seagrass meadows, are only to a very limited extent, or not at all, grazed by animals (seagrasses especially by brent geese); yet they are often by far the most important primary producers in shallow water habitats (Mann, 1972; McRoy, 1970). In the tropics, seagrasses are utilized directly by some herbivores, i.e., sea urchins and especially vertebrates such as turtles, fish, and manatees, but here also the larger part ends up as particulate detritus (Bertram and Bertram, 1968; Randall, 1965; Wood et al., 1969).

## 2.2. Reasons for the Quantitative Importance of Detritus Food Chains

We may here consider it as a fact of plant physiology that a sub-stantial part of the primary production may be lost as dissolved organic matter. It is less obvious why such a comparatively small fraction of the produced living plant tissue is consumed by herbivores. A teleological argument is that if there were no mechanisms preventing the biomass of larger plants, which have a high ratio of biomass to productivity, to be utilized rapidly by smaller herbivores with a much higher population growth potential, then the existence of macrophytes would not be possible. Thus, as plant size increases a decreasing fraction of the productivity is consumed by herbivores, a relation which is also found in terrestrial ecosystems (Wiegert and Owen, 1971). Macrophytes have by natural selection aquired mechanisms which minimize herbivore exploitation, and for existing plants this evolutionary trend must have kept pace with a parallel evolution of herbivores to become more efficient in the exploitation of their food. These mechanisms in part also explain why detritus is primarily degraded by bacteria rather than by animals.

The most important mechanism involved in limiting herbivory is related to the chemical composition of plant tissue. Although plant material to a large extent consists of structural polysaccharides or other structural polymers, e.g., lignin, the ability to hydrolyze these compounds is very rare among animals, and when ingested, these compounds usually pass unchanged through the digestive system. Hylleberg Kristensen (1972) quantified the activity of digestive carbohydrases in 22 shallow-water, marine invertebrates. Only one case of a moderate alginase activity was found; otherwise, the significant hydrolysis of structural polysaccharides characteristic of different marine macroalgae and vascular plants was not detected. Similar results were obtained by Nielsen (1966) in a study of digestive enzymes of invertebrates living in wrack strings along beaches.

Another important characteristic of macrophytes is a low content of nutrients, in particular nitrogen and phosphorus. Russell-Hunter (1970) generalized that all animals of all trophic levels (with the exception of ruminants, and probably there are other exceptions, too) require a diet with a $C:N$ ratio of less than 17. Data in the literature on the nitrogen contents of aquatic macrophytes are not quite consistent due to seasonal variations (cf. Harrison and Mann, 1975a). However, typical values for the $C:N$ ratio of large brown algae (*Fucus, Laminaria*) are around 30 (16–68); those for large, multicellular green algae are in the range 10–60; those of rhodophyceans are around 20; and for *Zostera* they are between 17 and 70. For comparison, typical values for microalgae are 6.5 (growing diatoms), 6 (chlorophyceans), 6.3 (cyanophyceans), and 11 (peridineans). Bacteria have $C:N$ ratios of around 5.7 (Mann, 1972; Spector, 1956; Vinogradov, 1953). Similarly, the $C:P$ ratios are very high in macrophytes ($\sim 200$) as compared with microalgae ($\sim 70$) or bacteria ($\sim 27$) (Spector, 1956; Vinogradov, 1953).

Thus herbivores may be resource limited by the availability of nutrients in their food rather than by organic carbon. This, in conjunction with the fact that animals in general cannot digest structural plant compounds, explains why the highest proportion of the production of macrophytes is channeled through detritus food chains.

The many examples of special adaptations found in terrestrial plants (toxins, aromatic resins, thorns, and other mechanical features) which limit herbivory are not as widespread (or studied) in aquatic plants. Blue-green algae are often digested with a very low efficiency compared with other small algae or are passed unharmed through the gut of herbivorous animals (Hargrave, 1970a; Kofoed, 1975a), and this could be due to the mucous sheath of cyanophyceans.

## 3. The Primary Decomposition of Detritus

### 3.1. Factors Influencing the Role of Bacteria

Four properties of bacteria can essentially explain their dominating role as primary decomposers in ecosystems. First, they can utilize dissolved organic substrates at low concentrations and also assimilate dissolved inorganic nutrients such as nitrate and phosphate and at the same time decompose nutrient-poor plant tissue. These properties are in part due to the small size and thus the large surface to volume ratio of bacteria. Bacteria also have enzyme systems that can hydrolyze structural plant compounds efficiently; and finally, many forms have developed efficient systems of anaerobic metabolism.

From the viewpoint of the bacteriologist, the distinction between the decomposition of dissolved and particulate detritus made in the following sections might seem somewhat artificial since bacteria only assimilate dissolved substrates and solid substrates are first hydrolyzed by extracellular enzymes before being assimilated. However, when discussing the fate of detritus in ecosystems, the distinction is relevant.

### 3.2. Dissolved Organic Matter

As previously discussed the great majority of all organic matter in aquatic habitats is in a dissolved form. We must assume that from 10 to 30% of all primary production enters the pool of dissolved organic matter and that this material is mineralized or rendered available to higher levels of the food chains as particulate organic matter mainly through the activity of bacteria. Assuming an average content in the oceans of between 0.5 and 2.5 mg dissolved organic matter per liter and that 10% of the primary production (i.e., $1.3 \times 10^9$ tons) enters this pool annually, Jørgensen (1966) estimated a turnover time of between 500 and 2500 yr. However, the pool of dissolved organic matter constitutes a heterogeneous assemblage of molecules, and a large part of the input has a very rapid turnover, in particular in the surface layers and in coastal waters. Thus, Ogura (1975) found in coastal seawater a concentration of dissolved organic material of about 2.7 mg/liter of which 24% was of low molecular weight (< 500). The mineralization of the dissolved material through microbial activity was initially rapid (0.01–0.09/day), and mostly the low molecular fraction was broken down; later the mineralization rate decreased to levels between 0.001 and 0.009/day. There is evidence to show that a large part of the carbon flow is in the form of low molecular weight compounds such as glucose and amino acids, which occur in very low concentrations (typically of the magnitude of 1 and 10 μg/liter for glucose and total amino acids, respectively) but have a very rapid turnover.

Our knowledge of the turnover and microbial utilization of such compounds is mainly due to experiments with [14]C-labeled substrates. The main difficulty of this approach is that natural substrate levels are often so low (e.g., for glucose) as to be difficult to quantify. Furthermore, since [14]C-labeled organic tracers have a relatively low specific radioactivity, the addition of the labeled compounds results in a significant increase in the concentration relative to the naturally occurring substrate level. Wright and Hobbie (1965) attempted to solve this problem by assuming that the microbial substrate uptake can be described by Michaelis–Menten kinetics. By measuring the [14]C substrate uptake at different levels of substrate addition, the maximum uptake velocity ($V_{max}$), which is taken as a measure of "heterotrophic potential," is found. Furthermore, by analyzing the data graphically with a modification of a Lineweaver–Burk plot and by extrapolating the data, the true turnover time, $T$, is found. If the true substrate concentration is known, the true uptake velocity, $V$, can then be calculated. The method is, as mentioned above, dependent on the assumption that the substrate uptake follows Michaelis–Menten kinetics in natural waters. Natural waters contain a heterogeneous assemblage of bacteria, and as shown by Williams (1973), unless all the bacteria have identical kinetic constants ($V_{max}$ and $K_s$) this assemblage will not behave according to the Michaelis–Menten equation; hence, an estimate of $T$ by a Lineweaver–Burk plot may lead to serious errors. It is therefore a tacit assumption of the Wright–Hobbie method that under given conditions (substrate concentrations), bacteria with given kinetic constants will be selected for, so that the total population is homogeneous in this respect.

There is in fact no reason to believe this to be so, or in other words, simple chemostat kinetics does not describe nature. First of all, different bacteria may utilize more than one substrate with different efficiencies, and there is nothing in theoretical population biology which prohibits two bacteria, of which one has a high affinity for substrate $A$ and a low affinity for substrate $B$ and the other a high affinity for substrate $B$ and a low affinity for substrate $A$, to coexist under a wide combination of substrate concentrations of $A$ and $B$. Furthermore, bacteria associated with detrital particles or occurring in organic aggregates will experience a higher substrate concentration than isolated bacteria so that even the planktonic environment is not homogeneous. This problem becomes acute when the method is applied to highly heterogeneous systems such as sediments (e.g., Hall *et al.*, 1972); here the results of the Wright–Hobbie method hardly yield more information than to show that microbial activity is much higher in sediments than in the overlying water. Thus, while the method is accurate and reproducible, it does not yield results which are easy to interpret.

Still, the method has given the best understanding so far achieved of the dynamics of bacterial uptake of dissolved organic matter in natural waters (e.g., Crawford *et al.*, 1974; Hobbie, 1967; Hobbie and Crawford, 1969; Gordon *et al.*, 1973). These studies have revealed a rapid uptake of low molecular weight compounds and high variations in parameter values among different water bodies and seasonally within one habitat. Thus, a span in $V_{max}$ of four orders of magnitude and a span of turnover times ranging from less than 0.5 hr to $10^4$ hr have been found for glucose and acetate when comparing different aquatic habitats ranging from polluted ponds and productive estuaries at one end of the scale to oligotrophic arctic lakes at the other end. For a Swedish lake, turnover times ranging from 10 hr in the summer to more than 1000 hr in the winter were demonstrated for glucose (Hobbie, 1971). The peaks of maximum turnover and $V_{max}$ are clearly correlated with peaks in phytoplankton productivity. Assuming substrate levels of glucose and acetate to be around 2.5 and 5 $\mu$g/liter respectively, in the mentioned lake, the flux of these two compounds through bacteria was estimated to be 8 g carbon/m$^2$/yr. Since there are many other compounds (amino acids, carbohydrates, fatty acids) which are taken up, this figure shows that bacterial production based on dissolved organic matter must be considerable.

Studies on the uptake of different amino acids in an estuary (Crawford *et al.*, 1974) indicate that bacterial production based on the uptake of amino acids alone constitutes about 10% of the algal production during the summer and thus is a significant contribution of particulate organic matter to the system. In this study, the actual uptake velocities were only small fractions of $V_{max}$ (1–15%), and turnover times varied seasonally as well as for different amino acids from 0.7 to more than 200 hr. The percentage of assimilated substrate which is respired varies from 10 to 20% for glucose and leucin, for example, and to around 50% for aspartic acid.

Glycolic acid is an important constituent of algal excretion of dissolved organic matter and therefore a potentially important bacterial substrate. A recent study (Wright and Shah, 1975) has shown that glycolic acid has turnover times varying from 7 to 2000 hr, the most rapid turnover being found in the surface layers. Two-thirds of 141 bacterial strains isolated from seawater were capable of taking up and metabolizing glycolic acid, but the percentage respired (70%) is unusually high for this substrate. Gordon et al. (1973), in a study of the uptake velocities of different organic acids in a lake, showed a clear correlation between bacterial densities and acid uptake.

These studies have all demonstrated the significance and dynamic nature of the bacterial uptake of dissolved organic matter in aquatic environments, although as already discussed the quantitative estimates arrived at must be considered with some reservation.

ZoBell (1943) showed that microbial activity in seawater is enhanced by the presence of solid surfaces. When seawater is stored in bottles, microbial activity (measured as oxygen uptake in ZoBell's experiments) increases, and this effect can be further stimulated by adding inert particles, e.g., sand. The results of these experiments were explained as the effect of adsorption and thus increased concentration of dilute nutrients on surfaces, where the substrates can be utilized more efficiently by bacteria. The experiments do show that more easily decomposable substrates often occur at limiting, low levels in natural waters. Jannasch and Pritchard (1972) studied this effect in batch and continuous cultures using defined bacterial strains and the addition of natural silt or particles consisting of chitin, clay, or cellulose. It could be shown that substrates at low concentrations were utilized more efficiently in the presence of inert particles and that bacteria adhere to such particles at low substrate levels but not in the absence of the substrate or when it is present in high concentrations (Fig. 1). It could also be shown that the outcome of competition for the same substrate between two strains of bacteria in chemostats may be altered by the presence of inert particles. Field observations and experimental studies (e.g., Pearl, 1974; Seki, 1972) have also shown the importance of this effect in freshwater as well as in the sea. Planktonic bacteria are to a large extent found in aggregates in conjunction with suspended detrital particles, diatom frustules, etc., in which filamentous bacteria play a large role together with rods and cocci, and the aggregates are apparently held together by bacterial capsule material. The enrichment of the water with dissolved organic matter enhances the formation of such aggregates.

This adherence of bacteria to particles may be of considerable significance for filter-feeding animals by enriching particles of a given size range with bacterial biomass. An example of this was given by Fenchel et al. (1975). The shallow-water burying amphipod Corophium volutator collects sediment or suspended particles with the gnathopods, which act like a filter. The ingested particles have a size range of 4–60 μm which is in accordance

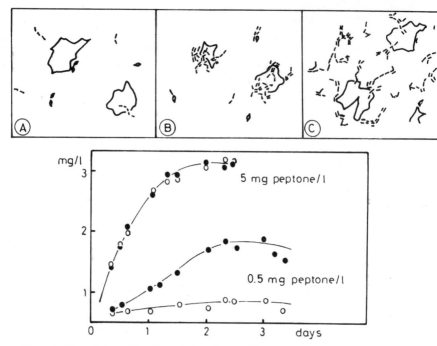

**Figure 1.** The relation of *Bacillus subtilis* cells to chitin particles in cultures with 0 (A), 0.5 (B), and 5 (C) mg peptone/liter. Below: the oxygen consumption of *Achromobacter* sp. in batch cultures with two different substrate concentrations and with (filled circles) and without (open circles) the presence of chitin particles. Redrawn from Jannasch and Pritchard (1972).

with the spacing of the bristles on the gnathopods. When $^{14}$C-labeled bacteria are added to the sediment or water in dishes with buried amphipods, the latter become radioactive by ingesting the bacteria. However, if clay and silt particles ($<62$ $\mu$m), which constitute about 5% of the natural sediment, have been removed from the sediment, the amphipods do not become radioactive since there are no particles to which bacteria can adhere and then be filtered and ingested (Fig. 2).

We cannot conclude a discussion on the bacterial utilization of dissolved organic matter without touching on the question of whether this resource is also being utilized directly by higher (i.e., eucaryotic) organisms. With respect to animals, much of the literature has been reviewed by Jørgensen (1966, 1976). A large part of this literature is inconclusive since unnaturally high substrate concentrations (often two or three orders of magnitude higher than natural concentrations) have been used in experiments, and the simultaneous excretion of the tested substrates has often been neglected. The main conclusion, however, is that while many aquatic invertebrates do have the ability to assimilate dissolved organic matter,

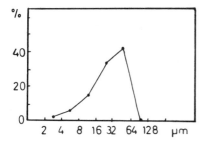 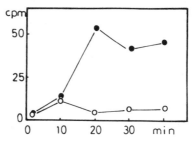

**Figure 2.** The particle size distribution (volume-%) of ingested particles of *Corophium volutator* (left) and the radioactivity per individual of animals feeding on [14]C-labeled bacteria in sediments with (filled circles) and without (open circles) the presence of mineral particles <62 μm. Redrawn from Fenchel *et al.* (1975).

animals do not depend on this resource for their metabolic needs, and considering the low substrate levels of natural waters, animal uptake of dissolved organic matter can at most play a trivial role. All free-living animals (including protozoa) are capable of feeding on particulate matter and have developed various kinds of often sophisticated and complex mechanisms for collecting and ingesting particulate food ranging in size from a bacterium to a large fish. On the other hand, parasitic forms which actually depend on the uptake of dissolved organic matter (e.g., cestodes, astomatid ciliates) have lost such features as a mouth, cytostome, intestinal tract, etc. This is to our opinion convincing evidence that free-living animals do not normally depend on dissolved organic matter. The ciliate *Tetrahymena* feeds on bacteria in nature; in the laboratory, it can be grown axenically in peptone solutions at concentrations from about 1 g/liter, and the uptake is saturated at concentrations at about 10 g/liter. Even at these high concentrations, however, growth is slow compared to cultures fed with bacteria, and the cell yield of peptone-grown cultures is only about 9% compared with about 41% for bacteria-based cultures (Sleigh, 1973). *Paramecium* cannot maintain itself even in high concentrations of dissolved substrates unless inert particles which adsorb the nutrients and which can be ingested by the ciliates are present (Reilly, 1964). For comparison, it can be mentioned that bacteria have been isolated from lake water with $K_s$ values of 0.007 mg/liter for glucose (Wright and Hobbie, 1965), whereas Jørgensen (1976) reports values between 0.94 and 12.6 mg/liter for some marine invertebrates. It is therefore clear that free-living protozoa and metazoa cannot compete with bacteria with respect to high substrate affinity, and under most natural conditions they are dependent on more concentrated food in the form of particulate matter.

More recently, critical experiments have shown that a number of marine invertebrates, especially polychaetes, in environments with relatively high concentrations of dissolved organic matter, such as estuarine sediments, may take up significant amounts of amino acids which constitute an important contribution to the metabolism and the need for nitrogen to the animals (Jørgensen, 1976; Stephens, 1975, and papers cited therein).

Hobbie and Wright (1965) and Wright and Hobbie (1966) found that planktonic algae cannot compete with bacteria with respect to the uptake of glucose and acetate at natural substrate levels. However, some microalgae and flagellates are known to depend on dissolved organic matter under special conditions where concentrations are very high, such as in decaying seaweeds (e.g., Lewin and Lewin, 1960). Williams (1970) found that in sea-water two-thirds of the uptake of labeled organic compounds was due to organisms smaller than 3 $\mu$m. Allen (1971), on the other hand, found that in lake water most of the uptake was due to microflagellates.

The quantitative role of fungi in aquatic environments is not understood, although chytrids and other phycomycetes as well as yeasts are known and sometimes abundant in aquatic environments; the general impression is, however, that their role is modest compared to that of bacteria.

## 3.3. Bacterial Colonization and Decomposition of Particulate Detritus

The study of microbial colonization and decomposition of particulate detritus is rendered difficult by the heterogeneous nature of such material. The combined use of epifluorescence microscopy for the observation and quantification of the associated microorganisms and various chemical and radiotracer experiments has given much new information (e.g., Barsdate et al., 1974; Fenchel, 1970, 1972, 1973, 1977; Fenchel and Harrison, 1976; Hargrave, 1972; Harrison and Mann, 1975b; Odum and de la Cruz, 1967; Oláh, 1972).

Natural, particulate detritus consists mainly of mechanically broken down tissue of dead leaves, roots, stems, or thallus of macrophytes (in the sea, mainly seagrasses, macroalgae, and mangroves; in freshwater, remains of terrestrial plants play a large role in addition to aquatic macrophytes) mixed with smaller amounts of organic debris of other origin, e.g., crustacean exuviae. Such material may form a considerable part of the surface sediment or appear as suspended material; in shallow waters, the material is often accumulated and partly sorted according to particle sizes due to local hydrographical conditions. Dead tissue of macrophytes may also be transported to deep sea sediments and become an important base of food chains there (Menzies et al., 1967).

Microscopic observations have revealed that the associated microbial communities are complex but with a relatively constant composition, qualitatively as well as quantitatively (Fig. 3). Most microbial activity is associated with the surfaces of the individual particles or, as is the case with

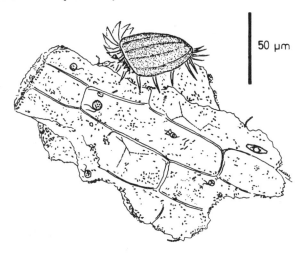

50 μm

**Figure 3.** A detrital particle (approx. 200 μm long) derived from eelgrass with bacteria, zooflagellates, a diatom, and a ciliate (*Euplotes*).

detrital particles derived from vascular plants, with the one or two outermost cell layers. Typically between 2 and 15 bacterial cells are found per 100 μm² surface area, somewhat depending on the nature of the surface and other factors. Since a bacterial cell only covers about 1 μm² of surface, only a small part of the surface area is usually covered by bacteria. The bacterial flora is, as judged by morphological criteria, diverse and contains rods, cocci, filamentous and mycelial forms, as well as myxobacteria.

The above-mentioned bacterial density per surface area corresponds to a quantity of $5 \times 10^8$ to $10^{10}$ bacteria/g dry wt. of detritus according to the detrital particle size, which determines the total surface area. The observation that the bacterial number is mainly a function of surface area is consistent with the finding that the decomposition rate and oxygen uptake is nearly inversely proportional to the particle size of detritus (Fenchel, 1970; Hargrave, 1972; Odum and de la Cruz, 1967).

In addition to bacteria, detritus harbors other microorganisms. Fungi are often observed; they have not been quantified but seem to play a relatively small role in seawater, where they are mainly represented by chytrids. However, Fell and Master (1973) and Newell (1973) have shown that a variety of fungi are important in the breakdown of mangrove tissue, and Meyers and Hopper (1973) have demonstrated the importance of fungi on decomposing *Thalassia* and cellulose in seawater. Among the protozoa, small zooflagellates (genera such as *Bodo, Monas, Oikomonas, Rhynchomonas*, choanoflagellates, and colorless euglenoids) are dominant. They seem to feed exclusively on bacteria and occur typically in numbers between $5 \times 10^7$ and $5 \times 10^8$/g dry wt. of detritus. Ciliates (mainly bacterivorous forms such as *Cyclidium*, *Euplotes*, *Holosticha*, and *Uronema*, but also some predators on other protozoa) are also abundant, whereas small amoebas and heliozoans are less numerous. Diatoms and unicellular blue-green algae may be frequent. Among metazoans the whole size range is found from gastrotrichs and rotifers with a size range and biology comparable to that of large protozoa, over the usually abundant nematodes, harpacticoids and oligochaetes to large animals such as snails and amphipods, which ingest and mechanically rework the whole complex of detritus and microorganisms. The role of animals in relation to the microbial community will be discussed in Sections 5.1 and 5.2.

If sterilized natural detritus, previously dried, fragmented, and leached plant material, or particles of pure cellulose or xylan are placed in sea- or freshwater and inoculated with a small amount of natural detritus, a characteristic succession of organisms can be followed. This succession eventually leads to a microbial community which closely resembles that of natural detritus. Bacteria occur in small numbers on the particles after 6–8 hr; with doubling times of 10–20 hr, they reach their maximum number after a period of 50–150 hr after the inoculation and then decrease in numbers again to a relatively stable level after about 200 hr. The small zooflagellates appear about 20 hr after the inoculation and show a maximum population size after 100–200 hr. Ciliates appear after about 100 hr and reach their maximum numbers between 200 and 300 hr after the inoculation. The initial behavior of these laboratory systems can be described as a damped prey–predator system with bacteria as prey and the protozoans (in particular the zooflagellates) as predators as discussed in Section 5.1. The other groups of detritus microorganisms (rhizopods, diatoms, and small metazoans) usually appear late in the succession. The exact time sequence of the succession, of course, depends on temperature (the above mentioned times apply to room temperature, i.e., 20–25 °C) and on inoculum size, and the absolute population sizes attained depend on a number of factors, in particular on the concentration of mineral nutrients in the water. However, the close resemblance of the microbial communities of such laboratory model systems with those of natural detritus may justify that general properties of the latter may be inferred from the former.

The rate of mineralization and thus the rate of microbial activity of such systems may be measured as the oxygen consumption, as disappearance of organic material with time, or as the release of $^{14}CO_2$ from organic substrates labeled homogeneously with $^{14}C$. Using such substrates, Fenchel (1973, 1977) showed that the rate of mineralization, after an initial lag period (which corresponds to the time in which the particles are colonized by bacteria, i.e., about 50 hr), may be described as a first-order process, and it may proceed rapidly depending on particle size distribution and the availability of mineral nutrients (Figs. 7 and 8). The substrates were derived from barley hay or were pure cellulose or xylan. Some other plant materials,

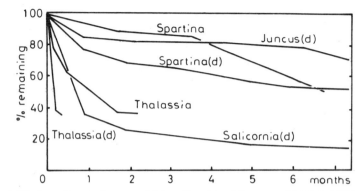

**Figure 4.** The loss of particulate material from litter bags. Plant material labeled (d) had been dried prior to the experiment. Redrawn from Wood *et al.* (1969).

e.g., seagrasses, seem to decompose at a slower rate in such laboratory systems (Harrison and Mann, 1975b; see also Fig. 5). This may be due to a lower content of nutrients and a higher content of structural plant compounds in this material, but other compounds, e.g., tannins, which inhibit bacterial activity may also play a role.

The turnover of particulate detritus in nature is very difficult to estimate. Whereas more or less crude estimates of macrophyte productivity in shallow water areas can be made, the standing crop of detrital material is very difficult to measure because of its very patchy distribution. Much of this material is decomposed under anaerobic conditions in the sediments (see Section 4). Some information may be derived from litter bag experiments, although many sources of error are inherent in this type of experiment. The material is in part broken down mechanically, and some particulate material is lost during the experiment. On the other hand, the mechanical protection of the bags may slow down the decomposition rate relative to what occurs in nature. Figure 4 shows a compilation of such data for various vascular plants under aerobic conditions; the figure does indicate that there is a great variation in turnover according to plant species and other factors. It also illustrates an important property of decomposer food chains with bacteria as the base: The availability of food for consumers is not restricted to the growing season in seasonal climates. The energy tied up in the dead plant material is slowly released, depending on the rate of the primary microbial degradation, to become available to higher trophic levels throughout the year.

### 3.4. The Significance of Mineral Nutrients

Microorganisms which decompose nutrient-poor organic substrates meet their needs for certain elements by assimilating dissolved mineral nutrients; the importance of this principle has long been established in soil microbiology and in agricultural practice (Alexander, 1971). More recently, this aspect of microbial activity has also been studied in aquatic environments.

The uptake of dissolved mineral nutrients by bacteria may be studied from different viewpoints. By assimilating nutrients, bacteria render them available to bacterivorous or detritivorous animals and enrich initially nutrient-poor detrital material. It may be asked how the uptake of mineral nutrients relates to the classical concept of mineralization, i.e., that decomposers release nutrients bound in the organic substrates to become available to primary producers. It could also be discussed to what extent the availability of mineral nutrients limits the rate of decomposition, and finally to what extent decomposers compete with primary producers for these nutrients in nature.

There are several examples to show that the content of protein in detrital material increases during the process of decomposition and may become considerably higher than in the original dead plant material (Fig. 5; other examples are given by Fenchel and Harrison, 1976; Harrison and Mann, 1975b; Hynes and Kaushik, 1969; Mann, 1972; and Odum and de la Cruz, 1967). In some cases, the analysis of total nitrogen of decomposing plant material has given less easily interpretable results (e.g.,

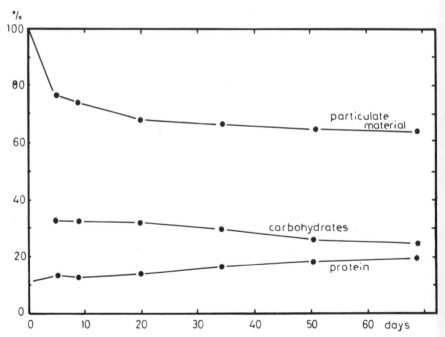

**Figure 5.** The loss of particulate material and the percentage of total carbohydrate and protein in fragmented eelgrass leaves decomposing in seawater. Redrawn from Fenchel (1973).

Kaushik and Hynes (1968) found that the leaves of some tree species lost nitrogen during decomposition). Obviously, the initial stages of decomposition may be complex, involving leaching of dissolved material and a rapid decomposition of amino acids. On the surfaces, however, the growth of microbial communities will result in a concentration of nutrients and, for aged detritus, also a net increase of nitrogen.

The quantitative relation between the assimilation of mineral nutrients from the water and the simultaneous mineralization of detritus has been studied in the case of phosphorus. These studies are based on laboratory microcosms consisting of detrital material, natural sea- or freshwater, and different microbial inoculants and the application of radiotracer techniques (Barsdate *et al.*, 1974; Fenchel, 1977; Fenchel and Harrison, 1976; Fenchel and Rahn, in preparation).

A typical experimental system consists of 200 mg of water-extracted plant material and 0.5 liter of natural seawater in flasks. In such systems, the largest pool of phosphorus is initially bound in the organic substrate (typically 70–80 $\mu$g phosphorus), whereas the water prior to the microbial inoculation may contain of the order 15 $\mu$g of phosphate-P. After the inoculation, the concentration of $PO_4^{3-}$ in the water rapidly decreases simultaneously with an increase in microbial population size; it can be shown that the loss in the water closely corresponds to the phosphorus content of the total microbial biomass. As the substrate is decomposed, the pool of dissolved inorganic phosphorus slowly increases again (Fig. 6).

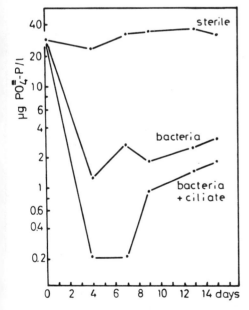

**Figure 6.** The concentration of dissolved $PO_4^{3-}$ in model laboratory systems based on seawater and fragmented, water-extracted barley hay and with different inoculants. Redrawn from Fenchel and Harrison (1976).

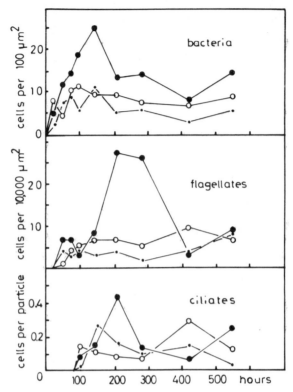

**Figure 7.** The numbers of bacteria, zooflagellates, and ciliates in laboratory model systems based on fragmented, water-extracted barley hay and seawater and inoculated with natural detritus. Points, enriched with $PO_4^{3-}$; open circles, enriched with $NO_3^-$; and filled circles, enriched with $PO_4^{3-}$ and $NO_3^-$. Note that the mineral nutrient enrichment has the greatest effect in terms of biomass on the bacterial grazers. Redrawn from Fenchel and Harrison (1976).

If the initial $PO_4^{3-}$ content of the water is low and if the system has initially been enriched with an inorganic nitrogen source, then the concentration of $PO_4^{3-}$ may decrease to very low levels ($<1$ μg/liter). This shows that the decomposing bacteria are extremely efficient in taking up this nutrient and that its availability may limit the size of the microbial biomass present (Fig. 7).

Using pulse injections of $^{32}PO_4^{3-}$ to such systems, Barsdate et al. (1974) and Fenchel and Rahn (in preparation) have shown that there is a very rapid turnover of the pool of dissolved $PO_4^{3-}$ (turnover times of down to 2 min at low concentrations and a turnover of the bacterial pool of phosphorus in less than 1 hr). Thus, a rapid uptake and release of phosphorus takes place even in phorphorus-limited systems with bacteria as the only living constituent, and the gross uptake of phosphorus exceeds the net uptake needed for bacterial growth. By using a combination of radiotracer and chemical techniques, Barsdate *et al.* (1974) could measure the rate of mineralization of phosphorus from the substrate independently of the bacterial phosphorus cycling. In a system such as that described above, it was found that about 0.84 μg phosphorus/liter/hr) was mineralized from the substrate. Simultaneously, however, the bacteria in the system took up and excreted about 147 μg phosphorus/liter/hr. Thus, there is a rapid mineral cycling involved in decomposition which far exceeds the actual mineralization rate.

Some authors (Johannes, 1965, 1968; Pomeroy, 1970) have suggested that in the absence of bacterial grazers, nutrients (in particular phosphorus) would become tied up in the bacterial populations, and further decomposition would then be slowed down by the low availability of nutrients. In this way, these authors explained the stimulation of bacterial grazing on the decomposition rate and bacterial activity, a phenomenon which is discussed in Section 5.2. This interpretation is in part based on an experiment by Johannes (1965) which shows that the liberation of $PO_4^{3-}$ from bacteria took place only when grazed by a ciliate (*Euplotes*) but not in the absence of the grazer. In this experiment, however, the bacteria had previously been grown in a batch culture based on easily utilized substrates (glucose and amino acids), and when the experiments took place the bacteria were deprived of a carbon source. The experiment, therefore did not represent a steady-state situation. In the grazed system, the bacteria were simply mineralized but could not show any compensatory growth; in the ungrazed system, the bacteria might well have excreted and assimilated $PO_4^{3-}$ while keeping the external concentration of this nutrient low or have become totally inactive. Barsdate *et al.* (1974) showed that $PO_4^{3-}$ is indeed cycled more rapidly in grazed than in ungrazed systems, but this seems to be the result of rather than the cause for a higher bacterial activity. Even in ungrazed systems, there is a rapid turnover of the pool of dissolved $PO_4^{3-}$, and the concentration is not higher in grazed systems (Fig. 6) in accordance with the fact that although the bacterial populations are considerably smaller in grazed systems, the total living biomass (when the protozoa are included) is of the same magnitude or somewhat higher in the grazed systems. In a particular system consisting of dead *Carex* leaves, bacteria, and the protozoan *Tetrahymena*, it was shown that while the turnover time of bacterial phosphorus was about 1 hr, the bacterial population was turned over only once every 24 hr due to ciliate grazing; the phosphorus excretion due to the ciliates could therefore at most account for approximately 5% of the phosphorus cycling of the system.

Figure 8 shows how the enrichment of laboratory microcosms with mineral nutrients above the level of natural seawater may increase the rate of decomposition of dead plant material. Fenchel (1977) found that initial

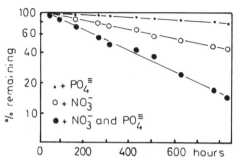

**Figure 8.** The decomposition rates of the systems shown in Fig. 7 and given as percent of remaining organic carbon. The rate was measured as the release of $^{14}CO_2$ from the homogeneously labeled barley hay. Redrawn from Fenchel and Harrison (1976).

concentrations of up to 10 mg $PO_4^{3-}$-P/liter or up to 16 mg $NO_3^-$-N/liter will stimulate the mineralization rate of water-extracted barley hay.

The most interesting question in this context, however, remains essentially unresolved: To what extent does nutrient limitation control the rate of microbial decomposition and thus the utilization of detritus by animals in natural ecosystems? There is only little evidence to throw light on this question. It has been shown that the enrichment of natural stream water with $PO_4^{3-}$ may stimulate the mineralization rate of dead leaves (Kaushik and Hynes, 1971), and Egglishaw (1972) found a clear correlation between the rate of cellulose decomposition and the $NO_3^-$ concentration in a number of freshwater localities. Similarly, while the competition between primary producers and bacteria for a mineral nutrient has been studied in experimental systems (Rhee, 1972), nearly nothing is known about the importance of this in natural ecosystems.

## 4. Anaerobic Decomposition

### 4.1. Pathways of Carbon during Anaerobic Decomposition

Parallel to the succession of decomposing organisms, detrital particles also undergo a succession of physical–chemical environments in which they are decomposed. Due to sedimentation and water movement, the particulate detritus is transported from the photic zone where the organic material was originally produced down to the aphotic water masses and sediments. Already at this stage, mineralization is in rapid progress due to autolysis and bacterial attack of soluble cell components. Macrophytes generally reach the sediment while still in a relatively undegraded state, whereas residues from planktonic organisms leave the photic zone mainly as a rain of empty algal cell walls, zooplankton exuviae, and fecal pellets (Krause, 1964). Thus, the detritus which is introduced into the sediments and deep waters is relatively enriched in structural carbohydrates and scleroproteins such as cellulose, chitin, etc.

With the exception of parts of the deep sea, aquatic sediments are anoxic below a thin, oxidized surface layer (Fenchel and Riedl, 1970; Hutchinson, 1957). In addition, many lakes and stagnant marine basins have anoxic bottom waters, This is due to the activity of heterotrophic organisms which consume oxygen. Because of the burrowing activity of the benthic macrofauna and the gradual accumulation of fresh sediment, particulate detritus is readily transported into the anoxic sediment zone. Thus in many aquatic environments a main part of the mineralization seems to take place only after the detritus has reached this anoxic zone (see Section 6.2).

The flow diagram in Fig. 9 illustrates the main pathways of carbon during the anaerobic decomposition of detritus. It also stresses how the interaction between anaerobic and aerobic processes leads to a complete mineralization of detritus. Due to this interaction, the redox discontinuity layer of sediments becomes a zone of high metabolic activity influencing the whole aquatic ecosystem. Although in Fig. 9 this zone is situated within the sediment, it could as well have been shown situated in the water column, illustrating the chemocline of a stratified lake.

The thin, oxidized surface layer of sediments is inhabited by a variety of heterotrophic organisms which constitute a complete, aerobic detritus food chain. These organisms are all physiologically similar in having a respiratory metabolism which can catalyze, within each individual, the complete oxidation of organic substrates. At least in this respect, aerobic food chains are relatively simple.

In the anaerobic environment, mineralization has been divided into a sequence of metabolic steps. Each step carries through only an incomplete part of the oxidation process, and each has its particular, physiological type of organisms. Thus, the anaerobic food chain becomes an energy flow between different bacteria via extracellular intermediates in the form of small, reduced molecules. Often these molecules are of inorganic nature and only carry the chemical energy of the detritus but not its carbon. This whole complex of metabolic processes includes various types of fermentations, anaerobic respiration, chemosynthetic oxidation, and even bacterial photosynthesis. Through many of these processes, the anaerobic food chain becomes closely connected to the cyclic transformations of inorganic nitrogen and sulfur compounds which serve as substrates in the energy metabolism (see Section 4.2).

The anaerobic degradation of detritus starts just as in the aerobic environment with a hydrolytic cleavage of the particulate material into small molecules which can then be assimilated by the bacteria. Several extracellular enzymes, including protease, cellulase, and maltase, have been demonstrated in aquatic habitats (Litchfield, 1973; Reichardt and Simon, 1972). Some of the exoenzymes have been shown not to appear in free solution but are

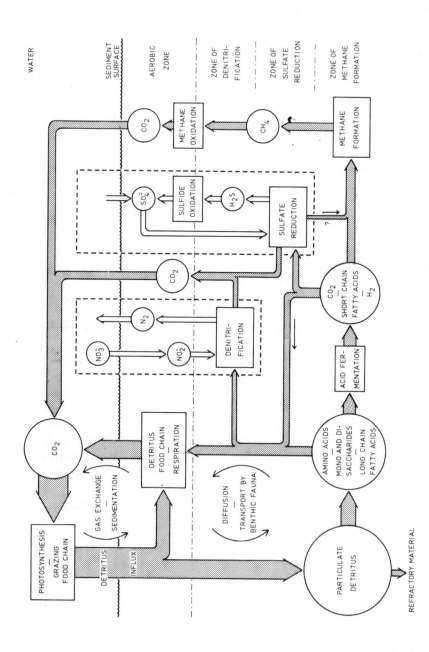

**Figure 9.** Pathways of carbon during anaerobic decomposition in an aquatic sediment. Pathways leading to the nitrogen or sulfur cycles are also indicated.

bound to the outer surface of bacteria which are sitting directly on the surface of detrital particles. This is the case with, for example, the protease of *Cytophaga* (Golterman, 1972).

The end-products of extracellular hydrolysis are mostly amino acids, mono- and disaccharides, and long-chain fatty acids. These appear in the pore water of sediments in concentrations which are 100- to 1000-fold higher than in the free water above. Although they are readily assimilated by anaerobic organisms, a considerable fraction may be transported to the aerobic zone by diffusion along steep surface gradients.

Within the reducing zone, the hydrolytic products are mainly taken up by bacteria which catalyze various types of fermentation. Little is known about the microbiology or biochemistry of these processes in natural sediments. Most of our knowledge still derives from pure culture studies and from studies of anaerobic digestion in sewage (Smith and Mah, 1966; Chynoweth and Mah, 1971) and in the rumen (Hungate, 1966). Only quite recently have kinetic studies of fermentation in the field been initiated (Cappenberg, 1974a,b; Cappenberg and Prins, 1974).

The main fermentation products from complex substrates like natural detritus are various short-chain fatty acids including acetic, propionic, butyric, formic, and lactic acids (Toerien and Hattingh, 1969). In radiotracer experiments, both lactic and acetic acid have been shown to be important intermediates in the anaerobic metabolism of lake muds. Both compounds were found in the pore water in concentrations of 2–20 mg/liter (Cappenberg, 1974b; Cappenberg and Prins, 1974).

Large amounts of $H_2$ and $CO_2$ are also produced during acid fermentation. The $H_2$ formation mainly serves to remove excess electrons produced during the fermentation of formate and pyruvate (Gray and Gest, 1965). Although it is expectedly an important end-product in natural sediments, $H_2$ seldom accumulates in significant quantities, at least not in the marine environment (Kaplan, 1974). This is partly because the presence of $H_2$ itself inhibits further $H_2$ formation and partly because $H_2$ is utilized by other bacteria for methane formation, sulfate reduction, or aerobic oxidation (Mechalas, 1974).

Since the energy yield in fermentation processes is relatively small, a large amount of substrate must be processed relative to the production of bacterial biomass. McCarty (1971) showed from thermodynamic calculations that the expected cell yield in anaerobic digestion varied within a range of 2–10% depending upon the substrate. Thus, most of the chemical energy of the organic matter is still left in the products of acid fermentation.

A fraction of the organic acids will diffuse to the surface layer of the sediment and serve as substrates for aerobic heterotrophs. Hall *et al.* (1972) have demonstrated a rapid turnover of acetate in the upper few centimeters of muds from Marion Lake, Canada.

Within the anaerobic sediment, the carbon pathway may proceed to the stage of methane formation. Methane-producing bacteria represent a very specialized group of strict anaerobes. They only grow under very reducing conditions $(E_h \leq -300 \text{ mV})$, and their substrate requirements are extraordinarily simple. Thus, only formate, methanol, and acetate can be fermented to methane, but in addition methane may also be formed by oxidation of $H_2$ with $CO_2$. These few substrates are utilized as both carbon and energy sources (Bryant et al., 1971; Zeikus et al., 1975).

The relative importance of the different substrates for methanogenesis in natural environments is not clear. Cappenberg (1974a) counted the number of bacteria in lake muds that would produce methane from each of the four types. Although all substrates yielded numbers up to $1-3 \times 10^4$ cells/ml, there was a slight differentiation in the vertical distribution of the different types. The $CO_2/H_2$ and formate-fermenting bacteria dominated at 2–3 cm depth, and the acetate- and methanol-fermenting ones dominated at 4–5 cm depth. Experiments with $^{14}$C-labeled substrates indicated that acetate alone was the precursor of 70 % of the methanogenesis in the mud (Cappenberg and Prins, 1974). Similar results have been obtained from tracer experiments in anaerobic sewage digesters (Jeris and McCarty, 1965; Smith and Mah, 1966). In the marine environment, the $CO_2/H_2$ metabolism was suggested as the most important mechanism of methane production (Claypool and Kaplan, 1974).

As shown in Fig. 9, the pathway of carbon via fermentation and methane formation is not the only possibility within the anoxic sediment. Two alternative routes are presented, leading to denitrification and sulfate reduction, respectively. The bacteria which catalyze these two processes have an anaerobic type of respiratory metabolism, utilizing nitrate or sulfate as the terminal electron acceptors. In contrast to the fermentation pathway, these processes cause a net oxidation of the organic matter.

An interesting relationship exists between sulfate and methane metabolism in anoxic sediments. The methane-producing bacteria seem to be inhibited in the presence of sulfate. Thus, in natural environments, the zone of active methane production is generally found below the zone of sulfate reduction. This is indicated by the vertical distribution of sulfate and methane in sediments and water (Claypool and Kaplan, 1974; Atkinson and Richards, 1967) and by the bacterial distribution (Cappenberg, 1974a). Martens and Berner (1974) showed that the methane concentration in an isolated mud sample only started to increase after sulfate had been depleted by bacterial reduction. The reason for this mutual exclusion of sulfate reduction and methanogenesis is not quite clear. Mechanisms such as competition for $H_2$, inhibition of methane producers by $H_2S$, and utilization of methane by sulfate reducers, together with thermodynamic calculations of the respective energy yields, have been suggested as possibilities (Cappenberg, 1974a;

Martens and Berner, 1974; Richards, 1965). The latest results indicate that anaerobic methane oxidation actually takes place in the sulfate zone (Reeburgh and Heggie, unpublished manuscript).

Apart from the antagonistic relation between sulfate reduction and methane formation, a syntrophic relation has also been found. Due to an incomplete enzyme system of the tricarboxylic acid cycle, the sulfate-reducing bacteria are not able to catalyze a complete substrate oxidation. They excrete large amounts of acetate as the end-product of their energy metabolism. As acetate constitutes one of the main substrates of the methane-producing bacteria, these would benefit from living in close contact with the sulfate reducers. In freshwater sediments, the vertical separation of the two types is only a few centimeters (Cappenberg, 1974a) long, and acetate may well be transported over this distance by diffusion. However, in marine sediments, the spacing is usually a meter or more (Kaplan, 1974), which should exclude a direct syntrophism.

Methane seems to be a true end-product in the anaerobic pathway of decomposition since it cannot readily be metabolized by other anaerobic organisms. It often accumulates to high concentrations in the environment where it is produced. Up to 150 ml methane/liter or more have been measured in marine sediments (Claypool and Kaplan, 1974; Reeburgh, 1969), and in lake muds methane may sometimes reach over-saturation and be released in the form of bubbles (Ohle, 1958). Methane that ascends slowly into oxygenated waters is finally oxidized to $CO_2$ by special bacteria, of which some are even obligate methane oxidizers. High rates of methane oxidation have been measured in the chemocline of stratified lakes (Rudd et al., 1974).

A small but significant fraction of the detritus in aquatic ecosystems never reaches a final mineralization but accumulates, mainly within the anoxic environment. It is gradually transformed into organic complexes which are refractory to microbial attack. The study of these complexes is a traditional field of organic geochemistry, and the subject has been reviewed by Bordovskiy (1965) and others.

## 4.2. Element Cycles Associated with Anaerobic Decomposition

Much of the chemical energy of detritus is conserved in the products of fermentation, and these become important substrates for the respiratory metabolism of the benthic community. The respiring organisms use various electron acceptors to oxidize the organic compounds. The highest energy yield is obtained when $O_2$ is the oxidizing agent. The thermodynamic drive for this aerobic respiration is expressed by the $E_h$ gradient: An electric potential difference of 500 mV or more exists between the surface and the deeper sediment layers.

As schematically shown in Fig. 9, there is a vertical sequence of the dominant electron acceptors in the sediment. Thus $O_2$, $NO_3^-/NO_2^-$, $SO_4^{2-}$, and $CO_2$ successively become the oxidizing agents with increasing depth. However, this whole sequence may not be sufficiently spaced vertically in natural sediments to allow the observation of all zones and should therefore be considered as an idealized model.

The same succession of electron acceptors as in Fig. 9 can be observed as a time sequence in the decomposition of detritus which is produced in the photic zone and gradually sinks into deep waters and anoxic sediments. Initially oxygen is available, and aerobic mineralization takes place. When oxygen is almost depleted, nitrate takes over the role as the oxidizing agent. Sulfate reduction starts later in the sequence than nitrate because it requires more reducing conditions. Even more so does methane formation from $CO_2$ oxidation of $H_2$. This thermodynamic differentiation of the processes may be the main reason for their succession with time and depth (Richards, 1965). Redox potentials typical of the zones where the four electron acceptors dominate are $+400$, $-100$, $-200$, and $-300$ mV, respectively.

An important functional aspect of the anaerobic electron acceptors is that they are diffusible substances that carry reducing energy up to the oxidized surface. Here a part of the energy becomes available for the biosynthesis of organisms, which thus continue the detrital electron flux towards a completion of the mineralization process. In this way the nitrogen and sulfur cycles become integrated parts of the detritus food chain. Interesting in this context is the fact that the processes of the two cycles are almost exclusively mediated by bacteria.

### 4.2.1. The Nitrogen Cycle

Bacterial denitrification has in recent years been shown to be an important, integrated part of the nitrogen cycle in aquatic environments. The main requirement of the process is an almost complete depletion of oxygen and the presence of suitable substrates. These conditions are mostly met within the upper reducing zone of sediments and in deep or stagnant water bodies of lakes and marine basins and fjords (Richards, 1965). For example, denitrification has been demonstrated in the hypolimnion and sediment of stratified lakes (Brezonik and Lee, 1968; Goering and Dugdale, 1966b), in some bottom waters and sediments of coastal marine areas (Goering and Dugdale, 1966a; Goering and Pamatmat, 1971), and even in the oxygen minimum layer of the tropical eastern Pacific Ocean (Goering, 1968).

The denitrifying bacteria are all facultative anaerobes which have a normal, aerobic respiration in the presence of oxygen. They include many bacilli and pseudomonads which are common in aquatic habitats. When the oxygen concentration in the water is reduced below about 5 $\mu$mol $O_2$/liter, they will switch to nitrate respiration. The organic substrates for the denitrifiers include sugars, organic acids, and alcohols (Payne, 1973). One species, *Thiobacillus denitrificans*, utilizes reduced, inorganic sulfur compounds as the energy substrate (Baalsrud and Baalsrud, 1954). It is unique in being an anaerobic chemautotroph.

As shown in Fig. 9 the organic substrates for the denitrifying bacteria are derived from the hydrolysis and fermentation of detritus within the anaerobic zone. The organic carbon is either built into the bacterial cells or it may be completely oxidized to $CO_2$.

As a consequence of the bacterial respiration, nitrate is reduced to $N_2$ (Fig. 10). Nitrite appears as an intermediate step in the reduction and is commonly found in the upper layers of oxygen-deficient waters and sediments. The formation of a gaseous end-product causes a loss of nitrogen from the ecosystem. In Lake Mendota, Brezonik and Lee (1968) estimated that denitrification provided a loss of 11% of the combined nitrogen entering the lake yearly. Unlike the end-product of sulfate respiration, denitrification does not provide a reduced compound which may serve as an

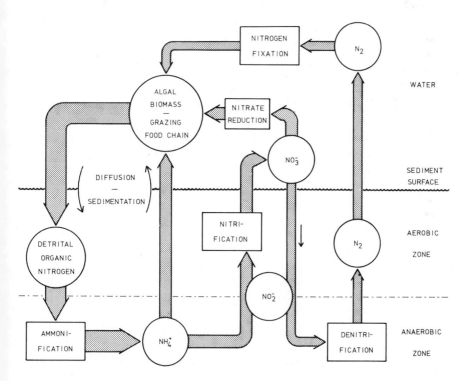

**Figure 10.** The nitrogen cycle of aquatic sediments.

energy substrate for chemosynthesis. Although nitrate is the thermo-
dynamically stable oxidation state in oxic waters, no bacteria are found
which will catalyze the oxidation of $N_2$.

However, ammonia, which is liberated during the hydrolytic decomposition of proteinaceous material, readily serves as a substrate for chemosynthetic bacteria. The oxidation is carried out in two steps by two different
bacterial groups. The first group (e.g., *Nitrosomonas*) oxidizes ammonia to
nitrite, while the second (e.g., *Nitrobacter*) oxidizes nitrite to nitrate. Apart
from these autotrophs, also many heterotrophs are able to oxidize nitrogen
compounds (Keeney, 1972). Autotrophic nitrification has been demonstrated
in surface sediments of lakes by Chen *et al.* (quoted by Keeney, 1972). It is
a key process in the nitrogen budget of aquatic ecosystems, but little is
known about its quantitative importance for the chemosynthetic productivity
of bacteria. Since nitrification and denitrification occur close together in
surface sediments, they may in combination lead to the conversion of
ammonia via nitrate to $N_2$.

As a counterbalance to the activity of denitrifying bacteria, $N_2$ may
also be assimilated by various procaryotes. Blue-green algae, e.g., *Anabaena*,
are important nitrogen fixers in lake water, but nitrogen fixation has also
been demonstrated in sediments, presumably due to anaerobic bacteria.

The aquatic nitrogen cycle is completed by the assimilation of combined
nitrogen. Algae and bacteria preferentially take up ammonia as the inorganic
nitrogen source. Assimilation of nitrate involves an initial reduction to
ammonia, which is then incorporated into proteins. Like the dissimilatory
process, the assimilatory nitrate reduction also occurs via nitrite but with
hydroxylamine as an additional intermediate. The process occurs both in
oxic and in anoxic environments but is inhibited by the presence of ammonia.

A recent review of the nitrogen cycle in aquatic ecosystems is given by
Keeney (1972).

### 4.2.2. The Sulfur Cycle

The sulfate-reducing bacteria are strictly anaerobic heterotrophs which
reduce sulfate to sulfide in their respiratory metabolism. They seem to be
present wherever anoxic conditions develop. Most sulfate reducers belong
to the genus *Desulfovibrio*; in fact, this seems to be the only genus living
in marine environments. Only few organic compounds serve as energy
substrates for sulfate reduction. These are mainly products of acid fermentation such as lactate, succinate, malate, and pyruvate (LeGall and Postgate,
1973).

As discussed previously, the organic substrates are only oxidized to the
acetate level. This would seem to set an upper limit to the fraction of
organic matter which could theoretically be oxidized anaerobically by

*Desulfovibrio*. Even if all detritus was converted to lactate, the sulfate reducers could only oxidize this to about one-third of completion. However, quantitative measurements indicate that up to 60 % of the organic detritus in marine sediments may be oxidized by sulfate-reducing bacteria (Jørgensen and Fenchel, 1974; Jørgensen, in preparation). This can only be explained by a bacterial syntrophism in the sediment in which acetate is recycled into suitable substrates for the sulfate reducers. Indications that acetate itself may be utilized, although rather inefficiently, were given by Selwyn and Postgate (1959). The actual substrates for sulfate reduction in natural environments have not yet been determined.

Quantitatively the sulfur cycle is more dominant in the sea than in freshwater. This is due to the high sulfate concentration in seawater of 20–30 mmol/liter, corresponding to an oxidation capacity 100-fold higher than that of oxygen in fully aerated seawater. In freshwater, sulfate occurs in concentrations similar to that of oxygen. This difference is perhaps the most important chemical factor distinguishing the anaerobic food chains in lakes and in the sea. In marine sediments, sulfate is usually found down to 1 m depth or more (Goldhaber and Kaplan, 1974). In lakes, it is often not detectable right below the sediment surface (Cappenberg, 1974a), and the diffusion of sulfate into the sediment thus becomes limiting for the rate of reduction.

In addition to dissimilatory sulfate reduction, sulfide is produced by the hydrolytic breakdown of detritus (Fig. 11). All organisms contain about 1 % of their dry weight as sulfur, mostly as a constituent of their proteins. The main sulfur source of autotrophic organisms is sulfate, which after assimilation is reduced to sulfide and incorporated into the amino acids cysteine, cystine, and methionine. During anaerobic decomposition of the organisms, the organic sulfur is released as free or methyl sulfides.

The energy yield of sulfate respiration is relatively small compared with that of oxygen and nitrate respiration. However, the end-product, $H_2S$, contains a large part of the chemical energy which can be exploited by other bacteria as the sulfide diffuses upwards and becomes oxidized. The mechanism of oxidation depends on environmental factors such as light and oxygen (Fig. 11). In some lakes, mainly of the meromictic type, and in the shallow sediments of sheltered, coastal areas, the anoxic environment with free $H_2S$ may reach into the photic zone. In these areas, photosynthetic sulfur bacteria may develop (Fenchel, 1969; Kondratjeva, 1965; Pfennig, 1975). They utilize the $H_2S$ for the photoreduction of $CO_2$. The green sulfur bacteria excrete the sulfur in the elementary state ($S^0$), whereas the purple ones oxidize it completely to sulfate (Pfennig, 1967). Also one species of blue-green algae (*Oscillatoria limnetica*) has recently been shown to catalyze a photosynthetic oxidation of $H_2S$ to $S^0$ (Cohen *et al.*, 1975). Although bacterial photosynthesis is restricted to rather special environ-

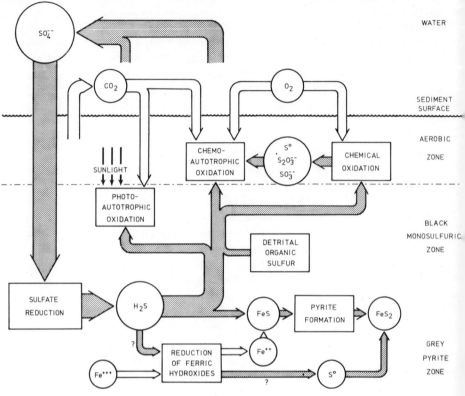

**Figure 11.** The sulfur cycle of aquatic sediments.

ments, it may become a dominating process for the whole carbon cycle where it occurs (Culver and Brunskill, 1969; Jørgensen and Cohen, 1977; Takahashi and Ichimura, 1968).

In most aquatic systems, the oxidation of $H_2S$ is a dark process with $O_2$ as the oxidizing agent. The process may be purely chemical since $H_2S$ reacts spontaneously with oxygen to form $S^0$, $S_2O_3^{2-}$, or $SO_3^{2-}$ depending on factors like pH, rate of oxidation, etc. (Chen and Morris, 1972; Cline and Richards, 1969). The half-life of sulfide in natural, oxygenated water is usually about 1 hr. Because of the chemical oxidation, the free energy of the reaction is lost for the living organisms. However, the process may also be biologically catalyzed by colorless sulfur bacteria. Since these are always competing with the chemical process, they can only live in the narrow gradient where $H_2S$ and $O_2$ meet. Here they will often proliferate into dense "bacterial plates." Thus the filamentous sulfur bacterium *Beggiatoa* has been shown to occur with a biomass of 5–15 $g/m^2$ in some marine sediments (Jørgensen, unpublished). Other types, like *Thiobacillus*, mainly oxidize intermediates like $S_2O_3^{2-}$ and $SO_3^{2-}$. Experiments by Sorokin (1972) indicate that in the Black Sea it is only this last step in the oxidation which is biologically mediated. A recent review of the colorless sulfur bacteria is given by Kuenen (1975).

Not all of the sulfide which is produced in the sediment will be reoxidized. A part of it is precipitated within the reducing zone by metal ions, especially iron. Iron is introduced into the sediment, for example, as ferric hydroxide coatings on mineral grains. In the anoxic environment it is converted to ferrous iron which rapidly precipitates with $H_2S$ to form various "monosulfuric" iron sulfides with the approximate composition, FeS. This reduction may be coupled with an oxidation of $H_2S$ to elemental sulfur, $S^0$. The ferrous sulfide is then slowly converted to pyrite, $FeS_2$, probably by reaction with $S^0$ (Berner, 1970). The course of this process can be seen from the color change of the sediment with depth. The zone of intensive FeS production is usually black, whereas the deeper zone is gray due to an almost complete conversion to pyrite.

In a coastal marine sediment, Jørgensen (in preparation) found that iron precipitated 10 % of all the sulfide formed. Although this is a rather small percentage, the continuous recycling of sulfur may ultimately cause a large enrichment. In the sediments mentioned, pyrite constituted up to 4 % of the dry weight.

## 5. The Utilization of Bacteria by Animals

### 5.1. Animal Adaptation for Bacterivory

The importance of bacteria as food for animals is a large topic which cannot be reviewed in detail. Here we will discuss some physiological and morphological adaptations for feeding on bacteria and the quantitative importance of bacterivory. Following sections will treat the effect of bacterial grazing on bacterial growth and symbiotic relationships between aquatic animals and bacteria.

Bacteria can be concentrated by bacterivores in different ways; in particular for larger animals, special adaptations are required to concentrate a resource which is dispersed in the form of small particles. Thus, although bacteria do represent a concentration of organic material relative to dissolved organic matter, the problem of feeding on bacteria is related to that of utilizing dissolved organic matter, i.e., how to obtain energy from the resource without spending more than gained in the process of concentrating it. Many aquatic organisms have evolved complicated feeding mechanisms for filtering large amounts of water and concentrate suspended food particles such as bacteria. Other animals, viz., the detritus feeders, have specialized in ingesting larger mineral or detrital particles and assimilate the associated microflora; these animals utilize the natural concentration of microorganisms on particulate material. There is, however, no sharp distinction between the two ways of feeding since many filter feeders may collect suspended detrital aggregates with attached bacteria (cf. the example in Section 3.2). In the following discussion, we have grouped the animal kingdom into four classes according to a mixture of taxonomic and functional criteria, viz., protozoa, meiofauna (metazoans smaller than 1–2 mm), filter feeders, and detritus feeders.

### 5.1.1. Protozoans

Bacteria constitute an important food source for free-living protozoa. They may be the most important consumers of bacteria in nature and constitute a significant link in food chains between bacteria and metazoans. Within all major groups of protozoa (flagellates, rhizopods, and ciliates), single species and larger taxons are found which are totally specialized to a bacterial diet.

The small colorless flagellates belonging to different taxonomic groups (e.g., *Bodo, Monas, Oikomonas, Rhynchomonas*, choanoflagellates, and some of the colorless euglenoids) all seem to be bacterivorous. With the exception of a fixed cytostome (e.g., in the euglenoids), these forms do not usually possess any special adaptations for concentrating bacteria (Fig. 12, A and D).

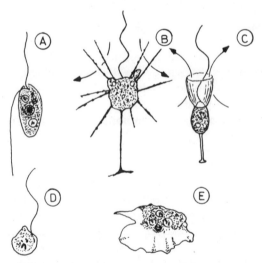

**Figure 12.** Some bacterivorous protozoa. A, *Bodo*; B, *Actinomonas*; C, *Monosiga*; D, *Oikomonas* (all flagellates); and E, the amoeba *Vahlkampfia*. Feeding water currents created by the flagellum is indicated for B and C.

The reason is, of course, that to a small flagellate ($<$5–15 $\mu$m) one bacterium represents a large food particle. Some of these flagellates, in particular the sessile ones, do have special feeding mechanisms, however. In *Actinomonas*, the flagellum creates water currents passing pseudopodia. Choanoflagellates have the flagellum surrounded by a collar of fine pseudopodia with a spacing of about 0.2 $\mu$m (Fjerdingstad, 1961). Water currents are drawn through the collar to which bacteria adhere (Fig. 12, B and C). Other, motile zooflagellates (e.g., *Mastigamoeba*) also use pseudopodia for feeding.

Among the rhizopods, many of the small amoebas (Fig. 12, E) feed on bacteria; while such forms are frequently seen in aquatic environments, nothing is known about their quantitative importance. Bacterivory is also known among the foraminifera (Lee *et al.*, 1966).

The most complex adaptations for feeding are found among the ciliates, and this biologically diverse group is the best known among the protozoa. Ciliates are relatively large protozoans, and the bacterivorous forms have therefore developed elaborate mechanisms for concentrating their food resource. The more primitive ciliate groups (e.g., the gymnostomatids) do not, with a few exceptions, feed on bacteria. Since these ciliates have not developed filtering mechanisms, they depend on larger food particles (flagellates, microalgae, other ciliates, and carrion). Within the trichostomatids and in particular within the hymenostomatids, more complex ciliary organelles are found in conjunction with the mouth. This apparatus is used for filtering and often also for sorting suspended food particles with respect to size. Within the hymenostomatids, the most primitive arrangement is found in forms such as *Tetrahymena* (Fig. 13, C), in which three ciliary membranelles on the left side of the mouth beat water against a ciliary undulating membrane. This membrane acts as a filter and transports the trapped bacteria to the cytostome. This arrangement has developed in the pleuronematid hymenostomatids, in which the undulating membrane has developed into a huge filtering organelle. These ciliates only feed when attached to a substrate; the undulating membrane is then unfolded, and water is passed through it by the activity of the membranelles and sometimes by some of the anterior somatic cilia (Fig. 13, B). This system is even further developed in the sessile peritrich ciliates. Here the undulating membrane and the membranelles form three circular membranes, of which the outermost (corresponding to the undulating membrane) acts as a filter. The ciliary organelles continue into the funnel-shaped vestibulum where they sort particles according to size and eject some of them (Fig. 13, D). The importance of filtering large amounts of water can be seen by the fact that peritrich ciliates (and many other filter feeders as well) are often found where there are strong water currents, for example, as epibionts on animals producing respiratory water currents. In the spirotrich ciliates, the hymenostome ciliary pattern of the mouth has been transformed into a long row of

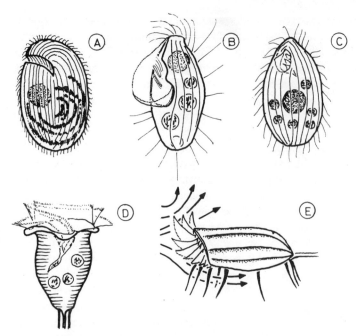

**Figure 13.** Some bacterivorous ciliates. A, *Sonderia schizostoma* which feeds nearly exclusively on filaments of *Beggiatoa*. B–E, filtrating ciliates: B, *Cyclidium glaucoma*; C, *Tetrahymena pyriformis*; D, *Vorticella campanularia*; E, *Euplotes* sp.

powerful membranelles which create water currents from which food particles can be collected (Fig. 13, E). In some of the groups mentioned (e.g., the peritrichs), all the species are bacterivorous; in other groups some species, in particular the larger ones, use the described feeding mechanisms for collecting larger food particles (algae, flagellates, etc.).

The rate of bacterial consumption has been studied in some cases (e.g., Sleigh, 1973, and references therein). Barsdate *et al.* (1974) and Fenchel (unpublished observations) studied the bacterial consumption of *Tetrahymena pyriformis* in model systems consisting of decomposing plant material and bacteria and using the technique of Fenchel (1975) for quantifying bacterial consumption of ciliates. During the initial exponential growth of the ciliates they consumed 500–600 bacteria/individual/hr. Some days later, when population sizes of bacteria and ciliates were relatively stable, one *Tetrahymena* consumed 100–200 bacteria/hr. With a bacterial density of $5 \times 10^6$/ml and a bacterial consumption of 150 bacteria/ciliate/hr, it can be calculated that one ciliate filters and clears about $3 \times 10^{-2}$ µl water/hr or about 20,000 times its own volume. This ratio is some 20 times higher than that found in filter feeding lamellibranchs (Jørgensen, 1966). Larger or more efficient ciliates (peritrichs, *Spirostomum*) may consume several thousand bacteria per hour.

Although some ciliates are quite omnivorous, most show food selectivity with respect to different kinds of bacteria. Ciliates are often very selective with respect to particle size and may discriminate between different sizes of bacteria (Fenchel, 1968). Several groups of ciliates (Plagiopylidae within the trichostomatids, several genera within the Heterotrichida, and other forms) which are mainly found within or adjacent to anaerobic, sulfide-containing environments, feed dominantly or exclusively on sulfur bacteria, and many are even specialized to particular kinds of sulfur bacteria (Fenchel, 1968; see also Fig. 13, A). Other ciliates seem to have specialized on proteolytic bacteria (e.g., *Uronema marina* and related forms); these species are found browsing on bacteria of decomposing animal tissue.

It is also known that in cultures many ciliates will grow at different rates or not at all depending on which bacterial strain they feed on (e.g., Burbanck, 1942). An interesting example of the significance of the qualitative composition of bacterial food was given by Burbanck and Eisen (1960); *Paramecium aurelia* will thrive on a number of different bacterial strains, but its ciliate predator *Didinium nasutum* can only grow on paramecia which have been fed certain of these strains.

In various, more or less artificial laboratory model ecosystems (such as the ones described in Section 3.3.), it has been shown that bacterial grazing by protozoa is very important in controlling bacterial density (Barsdate et al., 1974; Bick, 1964; Fenchel, 1969, 1976; Fenchel and Harrison, 1976). In systems based on decomposing plant material, for example, bacterial densities are typically two to five times higher in ungrazed systems compared with systems in which flagellates and/or ciliates are present. The grazers are typically responsible for a bacterial turnover every 10 to 30 hr at room temperature. It is reasonable to assume that bacterial grazing in natural accumulations of detrital material is of the same magnitude, considering the close qualitative and quantitative resemblence of the microbial populations with those of the artificial systems. The significance in sewage treatment plants of protozoan grazing for reducing the bacterial density of the effluent has long been known (McKinney, 1971, and references therein).

With respect to nature, our knowledge on the quantitative significance of protozoan grazing is very incomplete. In sediments, the protozoan fauna is very dependent on the mechanical properties of the sediment (Fenchel, 1969, 1975). In well-sorted sands, ciliates often dominate the microfauna with respect to numbers and biomass. They occur in up to 2000 individuals/ $cm^2$ which may comprise up to 50 species, and they will contribute significantly to the "community respiration." The "food spectra" of such different ciliate faunas of three different marine sand bottoms are shown on Fig. 14. In these examples, the bacteria constitute from 17 to 62 % of the food of the total ciliate faunas.

In less-sorted, very fine-grained, or detrital sediments, ciliates play a more modest quantitative role, whereas smaller protozoans, in particular

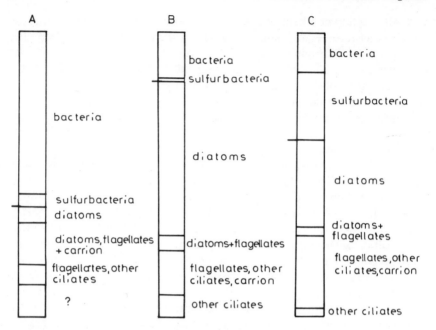

**Figure 14.** The "food spectra," i.e., the percentage of individuals of the total ciliate fauna, utilizing different food items for three different sandy sediments in Danish waters. A, an offshore locality at 10 m depth; B, a shallow sandy subtidal locality; C, an estuarine locality with a high organic content. All three food spectra are from summer; during winter algae play a relatively smaller role compared with bacteria. Data after Fenchel (1969).

zooflagellates, are important. The only attempt to quantify the different components of the microflora and -fauna and the consumption rate of bacteria is, to our knowledge, the study of Fenchel (1975) on the detrital sediments of an arctic tundra pond. Among the protozoa, the zooflagellates dominate quantitatively, constituting a biomass about 20 times larger than that of the ciliates. Other protozoan groups (amoebas and heliozoans) are even less numerous. It was estimated that about 8.5% of the standing stock of bacteria is consumed by zooflagellates every day (temperature 12°C), whereas ciliates play a modest role as consumers of bacteria. Altogether, protozoans are responsible for about 75% of the total grazing of bacteria, the rest being due to rotifers, chironomid larvae, copepods, etc. Apparently grazing can only account for about 50% of the bacterial production; it was conjectured that the remaining part is lost through sediment suspension and drainage from the pond and by burial and lysis in the sediment.

Although such estimates are extremely crude, they do indicate that in sediments protozoans are the most important group of bacterial grazers and significantly influence the standing stock of bacteria. In planktonic systems, very little is known about the importance of protozoan grazing of bacteria; some planktonic protozoans are known to be bacterivorous, but their quantitative importance is totally unknown.

### 5.1.2. Meiofauna

By meiofauna is meant benthic metazoa measuring less than 1 or 2 mm, and it includes groups such as rotifers, turbellarians, nematodes, gastrotrichs, tardigrades, harpacticoid copepods, and (small) oligochaetes. This heterogeneous assemblage of organisms undoubtedly plays a large role as a consumer of bacteria in nature. However, nearly all evidence is based on isolated observations on the feeding of different species or on laboratory experiments under somewhat unrealistic conditions. Some rotifers are bacterivorous, although most species specialize on larger food particles (Thane-Fenchel, 1968). Turbellarians are nearly all carnivorous or sometimes herbivorous (Straarup, 1970), while harpacticoids are often referred to as bacterivorous or detritivorous (see Fenchel, 1969, 1972, for references). Among the nematodes, many species have been shown to feed and grow on a pure bacterial diet (Duncan *et al.*, 1974; Tietjen *et al.*, 1970) or to feed on bacteria (and fungi) in the field (Meyers and Hopper, 1973; Perkins, 1958). However, there is a long way to go before a complete picture of the feeding biology of this species-rich and quantitatively important animal group is achieved. The oligochaetes are probably the best studied group; they seem to be important bacterial consumers in freshwater and marine sediments (Brinkhurst *et al.*, 1972; Giere, 1975; Wavre and Brinkhurst, 1971).

### 5.1.3. Filter Feeders

The biology of metazoan filter feeding (including considerations of the importance of bacteria as food) has been reviewed by Jørgensen (1966) and will therefore not be treated in detail here. Within all major groups of invertebrates, forms have developed which filter large amounts of water (typically from 0.1 to 3 liter/g body wt./hr) and utilize the suspended material which is caught by the filtering organs. The mechanism of the filtration varies for different groups and includes the filtration through mucus (some polychaetes and larvaceans), ciliary filtering and sorting mechanisms (molluscs, ascidians, and polychaetes), choanocytes (cells morphologically and functionally identical to choanoflagellates; found in Porifera), and filters based on bristles and hairs (arthropods).

It has long been suggested that bacteria (in addition to or instead of phytoplankton cells) may play an important role for certain filter feeders (e.g., ZoBell and Feltham, 1942; and ZoBell and Landon, 1937, who showed that *Mytilus* may grow on a bacterial diet). Studies on the retention size of filters may yield some information; thus, many forms can efficiently filter particles down to or smaller than 1 μm and thereby also retain suspended bacteria. These filter feeders include sponges, some polychaetes, cladocerans, cirripedians, black-fly larvae, some lamellibranchs, and tunicates. Other forms (e.g., copepods, euphausians, and chironomids) only retain larger particles, usually 5–10 μm. The former group will certainly retain bacteria, but it is necessary to know the density of bacteria in natural water in order to evaluate their nutritive importance. Also, the species with a coarser filter may well utilize suspended bacterial–detrital aggregates. An example of this was discussed in the case of the amphipod *Corophium* in Section 3.2. This problem has also been studied by Seki (1972), who argues that the formation of aggregates will under all circumstances enhance the efficiency of bacterial consumption by filter feeders.

There are several examples of filter feeders which will grow or reproduce on a diet of bacteria in natural concentrations. Thus, Manuvilova (1958) found that cladocerans could grow on the bacterial flora of natural freshwater. Freeden (1960) showed that the larvae of *Simulium* (black-fly) could grow on a suspension of *Bacillus subtilis* at concentrations of $1.6 \times 10^5$/ml. That bacteria are the natural food of black-fly larvae could be inferred from their dense occurrence in streams strongly polluted by organic matter.

Bacteria and bacterial–detrital aggregates are probably to a large extent consumed by filter-feeding animals. How large a fraction of the bacterial production which is utilized in this way and to what extent bacteria make up the diet of different filter feeders in nature, however, cannot yet be answered in general.

### 5.1.4. Detritus Feeders

A large number of species belonging to all major invertebrate groups more or less indiscriminately ingest bottom sediment or deposited organic detritus. This habit is even found among a few species of fish such as the mullet (Mann, 1972; Odum, 1970). It is now generally recognized that detritus feeders only to a small extent digest the bulk of ingested organic material but mainly utilize the living (microbial) components (for discussion and references see Fenchel, 1970, 1972; Hargrave, 1970a; Hylleberg Kristensen, 1972; Kofoed, 1975a, b; Newell, 1965; and see Section 2.2).

The question of which and how many of the different living components are in fact utilized in different species is often difficult to answer experimentally. Only a small fraction of the total organic carbon ingested is assimilated, and in many cases fecal material has proved to contain more organic matter than the substrate. This, of course, is due to a certain degree of selectivity with respect to the particles actually ingested. Such selectivity, especially with respect to particle size, has been demonstrated (e.g., Fenchel *et al.*, 1975; Hylleberg and Gallucci, 1975). Also some detritus feeders browse on larger particles and thus mainly ingest the surface layers, which are particularly rich in microorganisms.

Kofoed (1975b) recently demonstrated the significance of of microorganisms for detritus feeders. Feeding sterile, $^{14}$C-labeled hay particles, similar particles on which bacteria were growing, and pure $^{14}$C-labeled bacteria to the prosobranch snail *Hydrobia ventrosa*, he could show that the assimilation efficiencies were 34, 56, and 70%, respectively, for the three types of food. (The hay was derived from young, growing barley plants and thus contained a large amount of compounds, for example, starch, which the snail can assimilate. Sterile, natural detritus would be assimilated to a much lower extent.) More interestingly, however, it could be shown that while only 15% of the assimilated material from the sterile hay was used for growth (the remaining being respired immediately or excreted), the similar figures for hay with bacteria and for pure bacteria were 34 and 33%, respectively. Thus, the presence of bacteria in the diet has a significant qualitative aspect in addition to rendering the food digestible for the snail.

Microscopic observations (Fenchel, 1970, 1972) have indicated a high efficiency of assimilation of different types of microorganisms by some detritus feeders. More careful experiments (Hargrave, 1970a; Kofoed, 1975a) have shown high assimilation efficiencies for a wide variety of bacteria and microalgae (with the exception of blue-green algae) for an amphipod and a snail, respectively. Other studies, however, indicate that in some species there may be a differential digestion of different types of microorganisms and that some may pass the digestive apparatus in a living state (Hylleberg, 1975; Hylleberg and Gallucci, 1975; Wavre and Brinkhurst, 1971).

The particle size selection of detritus feeders may indirectly lead to some degree of qualitative selection of the ingested microorganisms. Thus, Fenchel *et al.* (1975) found that the amphipod *Corophium*, which selects relatively small, in part suspended, detrital particles, ingests a relatively larger amount of bacteria than the snail *Hydrobia*, which tends to enrich its diet with diatoms by selecting larger particles.

If we consider a hypothetical (but in no way unrealistic) detritus feeder which ingests detrital particles with sizes of up to say 200 $\mu$m, it will ingest about 4 mg of bacteria, 5 mg of flagellates, 0.5 mg of ciliates, and some other protozoans (as based on the figures given in Section 3.3, see also Fig. 3) per gram of consumed detritus. If the detritus feeder can ingest even larger particles, it will also include small metazoans, representing a considerable biomass, in its diet. (It may finally also ingest a considerable number of diatoms and other microalgae.) Thus, detritus feeders feed not only on bacteria but on a whole food web. The base of this food web is composed of bacteria (and perhaps fungi and microalgae), but the actual ingested material may to a large extent consist of other organisms.

## 5.2. The Effect of Bacterial Grazing on Bacterial Activity

The activity of bacterivorous (and other) animals may directly or indirectly affect the growth rate and metabolic activity of bacterial populations. Several mechanisms, some of which are not completely understood, may be responsible. In terrestrial environments, it has been shown that while microorganisms are responsible for the chemical degradation of plant litter, this process is accelerated by mechanical degradation due to soil animals (Edwards and Heath, 1963). Similar mechanisms are also important in aquatic environments. Thus, leaves of trees are processed by a number of invertebrates in streams and lakes (Kaushik and Hynes, 1968; Petersen and Cummins, 1974). Fenchel (1970) showed that the processing of detrital particles by a marine amphipod results in a significant decrease in particle size. A natural density of these amphipods relative to the detritus may result in a doubling in microbial population sizes and thus microbial breakdown rates in about 4 days. This is due to the increase in the surface area of detritus. Since the amphipod feeds on the microbial flora, it also stimulates the growth of its own food.

The many burrowing invertebrates, in particular polychaetes and bivalves, are often found in anaerobic, sulfide-containing sediments. However, since such animals are continuously pumping oxygenated water through their burrows, they are surrounded by a more or less oxidized layer of sediment. This effect must increase the rate of oxidation of reduced low molecular weight compounds of reduced sediments. The effect has not been quantified, but since, for example, worm burrows may be very dense, the total boundary surface between the reduced and oxidized zones of the sediment may be greatly enlarged. It has been shown that this boundary zone has a high bacterial productivity, which again results in large populations of protozoa (Fenchel, 1969; see also Section 4). Hylleberg (1975) suggested that this principle is utilized by lugworms for enriching the sediment which they ingest with microorganisms. These polychaetes create an oxidized zone in front of the mouth by pumping oxygenated water from the free water above the sediment. In this zone, increased populations of bacteria and of protozoans, nematodes, etc., are found.

There is a large body of evidence showing that bacterial grazing by protozoa and other animals somehow increases bacterial growth rate or metabolic activity (as measured by $O_2$ uptake, decomposition rate, or nutrient uptake). This has been shown for a detritivore amphipod (Hargrave, 1970b) and has especially been studied in laboratory ecosystem

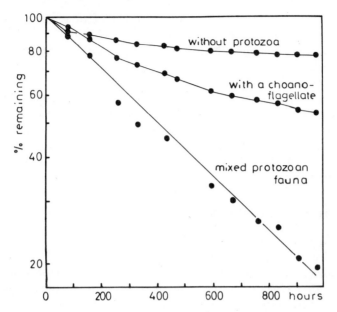

**Figure 15.** The decomposition rate (given as percent organic material remaining and measured as the release of $^{14}CO_2$ from homogeneously labeled plant material) in model laboratory systems based on fragmented, water-extracted barley hay and seawater inoculated with a mixed population of bacteria, bacteria and a choanoflagellate, and bacteria and a mixed population of protozoa. Redrawn from Fenchel and Harrison (1976).

models and sewage treatment plants (e.g., Javornický and Prokesová, 1963; McKinney, 1971). Fenchel (1977) demonstrated the stimulating effect on bacterial decomposition rate by the presence of bacterivorous protozoa. In the experiment shown in Fig. 15, the bacterial population sizes were about 50 and 70% in the systems grazed by a mixed protozoan fauna and by a single protozoan species, respectively, as compared with the ungrazed system.

That grazing (i.e., a constant rate of removal of individuals) of a population increases its production is not strange; this applies to any density-regulated population (cf. the well-known fishery-yield curves). The problem is to understand the precise nature of the mechanism, i.e., what limits the bacterial populations. As already discussed in Section 3.4, experimental evidence has shown that at least in the studied laboratory model systems the mechanism does not seem to be nutrient regeneration by the grazers. The fact that bacterial density on plant-derived detritus is closely related to the free surface could suggest that available surface area may be limiting (rarely more than 20% of the surface of detrital particles with a mixed bacterial flora, but in the absence of grazers this seems to be covered by bacteria as based on light microscopic observations). This limiting density could be set by the sorption of dissolved nutrients or concentration of hydrolyzed organic matter derived from the substrate. An improved understanding of the relationship between the single bacterial cell and its substrate is necessary; the systematic observation by scanning electron microscopy of detrital surfaces under different conditions might accomplish this in the near future. Another possibility is that the micro-turbulence created by protozoa stimulates bacteria. Whatever the exact mechanism is (and there may be different ones for different systems), it is clear that animal activity has an important influence on bacterial growth in nature, and this effect will often exceed the metabolic activity of the animals themselves.

### 5.3. Symbiotic Relationships between Bacteria and Herbivores

The digestive apparatus of ruminant animals may be considered as the most perfect example of the use of microorganisms for the exploitation of the tissue of higher plants (Hungate, 1966). In several respects this system exemplifies principles discussed in the previous sections: the significance of bacteria for the primary decomposition of structural plant constituents and the synthesis of protein after a mechanical degradation of the substrate by the animal itself and also that protozoa in part play a role as a link between bacteria and the metazoan in this food chain. It may be asked whether similar systems have evolved in aquatic animals. In a way, the examples given in the previous section (e.g., the amphipod enhancing the growth of its own food organism by mechanical activity) represent a very primitive version of such a system.

Unfortunately, not much is known about symbiotic bacteria in digestive systems of aquatic animals. There is a large body of literature (e.g., Bernard, 1970; Colwell and Liston, 1962; Hylleberg and Gallucci, 1975) which reports the presence of bacteria in intestinal tracts of various aquatic creatures. Sometimes various physiological characteristics of bacterial strains are reported, but there is usually no evidence with respect to their relation to the host; or pure conjectures suggesting mutualistic relationships are offered.

Regular sea urchins, however, may represent an interesting example. Echinoids are among the very few marine invertebrates which feed on macrophytes. Lasker and Giese (1954) showed that *Strongylocentrotus purpuratus* does possess amylases and proteinases but not enzymes attacking agar or tissue of the red alga *Iridophycus*, which is eaten by the urchins. However, bacteria which hydrolyze agar and digest whole pieces of algae could be isolated from the gut. By direct counts, the authors found $2 \times 10^{11}$ cells/ml of intestinal fluid. While these observations do not prove a mutualistic relationship, they are certainly suggestive. More recently,

Prim and Lawrence (1975) have confirmed the presence of a bacterial flora of echinoids which attacks a number of algal polysaccharides. Since sea urchins are more handy than cows as experimental animals, they seem to be promising objects for further studies.

The general absence of evidence like the one described for sea urchins for other, smaller herbivorous invertebrates may, as already indicated, be due to the lack of adequate studies, but it may also be real. In ruminants, the rumen system can roughly be compared to a chemostat with a turnover of about 24 hr. In small herbivorous or detritivorous invertebrates, however, the intestinal contents may be turned over every 20 min or so (Fenchel *et al.*, 1975) and cannot, therefore, sustain an intestinal microflora in a similar way. The presence of mutualistic bacteria in aquatic vertebrates feeding on seagrasses (sea turtles, some fish, and manatees) could be predicted, but to our knowledge this has not yet been investigated.

## 6. The Quantitative Importance of Bacteria

### 6.1. Measurements of Heterotrophic Bacterial Activity

The increase in our quantitative knowledge of detritus food chains during the last years has been closely connected to the development of new experimental techniques. The aim in the development of many of these techniques has been to obtain integrated measures of the heterotrophic activity in the ecosystem. Often it has been possible from such measurement to distinguish between the microbial or even the bacterial activity and that of the larger organisms.

The use of radioactive substrates to measure the metabolism of specific organic compounds has already been discussed and so has the inherent difficulty in not being able to work at natural substrate concentrations or even to measure these. Thus, the few reported rates of *in situ* metabolism are still subject to some uncertainty, and little can be said about the quantitative relationship between these results and the total carbon flow. Measurements of the turnover of glucose have shown rates ranging from about 0.001 $\mu$g carbon/liter/hr in the Atlantic Ocean to 0.1 $\mu$g carbon/liter/hr in coastal waters and lakes (Hobbie, 1971; Vaccaro *et al.*, 1968; Wood, 1973). Crawford *et al.* (1974) found a turnover of about 1 $\mu$g carbon/liter/hr for the sum of 12 amino acids in an estuary. In sediments, the metabolic rate is about 1000-fold higher. Thus, Wood (1973) has found a turnover of 1 mg carbon/liter/hr in coastal sediments and in the muds of a eutrophic lake. Cappenberg and Prins (1974) found rates for lactate, acetate, and methane of about 10, 1, and 0.5 mg carbon/liter/hr, respectively. Although the sediment seems much more active than the water, its relative importance must be compared on an areal basis. In sediments most of the biological activity is found in a narrow surface layer, whereas in the water it is distributed over much greater depths. Simultaneous tracer measurements of the vertical distribution of organic solute metabolism in both sediment and water are still lacking.

The bacterial decomposition of detritus occurs via a large number of intermediate compounds including sugars, amino acids, fatty acids, etc. Obviously it is an impossible task to measure the turnover of all these compounds, and one would therefore search for a few key compounds through which most of the decomposition processes are chanelled. A great variety of organic molecules is produced from the extracellular hydrolysis of detritus. In aerobic environments, these are taken up directly by heterotrophic bacteria, and the further metabolism is intracellular. Turnover measurements will therefore have to be carried out for many hydrolytic products in order to obtain a complete picture of the carbon flow through bacterial food chains. In anaerobic waters and sediments, however, the hydrolytic products are substrates for a series of fermentation and anaerobic respiration processes which have a very narrow spectrum of extracellular compounds as dominating intermediates or end-products. (This fact is one of the basic working principles in anaerobic treatment plants, where sewage is efficiently transformed into methane.) Thus, the use of a few tracer substrates should be sufficient to quantify the major part of the carbon flow in the detritus food chain of sediments. Lactate, acetate, and methane are probably among such substrates. Turnover rates, which for glucose and acetate were in the order of 1 mg/liter/hr, are already extremely high. Even if this activity is restricted to the uppermost 4 cm of the sediment, it would still correspond to 1 g carbon/m$^2$/day, or the total productivity of a eutrophic body of water. The lactate turnover measured by Cappenberg and Prins (1974) was tenfold higher.

A way to escape the problem of organic substrate diversity is the use of $^{14}CO_2$ dark assimilation as a measure of heterotrophic activity. Many bacteria have been observed to take up $CO_2$ in proportion to their organic carbon metabolism. Mainly Russian workers have used this uptake to estimate the bacterial productivity of water bodies assuming a $CO_2$ fixation ratio of 3–5% (Sorokin and Kadota, 1972). In the chemocline of lakes and sediments where methane oxidation and chemosynthesis take place, a much higher percentage of dark $CO_2$ fixation is found (Sorokin, 1964, 1965, 1972).

The method is very appealing as a relatively simple and fast measure of heterotrophic production parallel to the $^{14}C$ method developed by Steemann Nielsen (1952) for measuring photosynthetic production. However, the ratio between $CO_2$ fixation and total, heterotrophic carbon assimilation has recently been found to be quite variable (0.5–12%) both in natural lake samples and in cultures (Overbeck, 1972; Overbeck and Daley, 1973). The method should therefore be used with some precaution until more is known about normal $CO_2$ fixation ratios.

The bacterial production estimated with this technique is high. In the Black Sea, Sorokin (1972) found a bacterial production of about 100 mg carbon/$m^2$/day for the whole water column. This is one-third of the daily algal photosynthesis.

A rather new method which has been applied for estimating the microbial respiration in aquatic ecosystems involves the measurement of respiratory electron transport. The reduction of a tetrazolium salt added to the sample is measured after a short incubation, and this should give an indication of the potential dehydrogenase activity of the system (Hobbie et al., 1972; Packard et al., 1971). Since the method does not directly show in situ activity, it is still of limited use in quantitative studies of mineralization processes. Pamatmat and Bhagwat (1973) calibrated the method by microcalorimetry for use in anaerobic lake sediments. However, a direct comparison of the sulfate respiration and dehydrogenase activity in an estuarine sediment did not show proportionality between the two (Jørgensen and Perry, unpublished results).

## 6.2. The Importance of Anaerobic Mineralization

Since the complete mineralization of detritus always involves some kind of respiratory metabolism, it is important to obtain quantitative measures of natural respiration rates. Most of the available data only show rates of aerobic respiration. In the following, a brief review is given also on measurements of anaerobic respiration.

Natural rates of nitrate respiration have been measured by using the stable isotope, $^{15}N$, as a tracer. The activity ranges from less than 0.1 $\mu mol$ $NO_3^-$/liter/day in oxygen minimum layers of the Pacific Ocean to about 1 $\mu mol$ $NO_3^-$/liter/day in an island bay and in lakes (Brezonik and Lee, 1968; Goering, 1968; Goering and Dugdale, 1966a,b). More data are needed to show whether nitrate respiration contributes significantly to the mineralization of detritus. Especially the role of denitrification in sediments needs investigation.

Partly due to the availability of a suitable radioisotope of sulfur, somewhat more is known about the rate of sulfate respiration. Thus, $^{35}SO_4^{2-}$ has been used as a tracer, mainly in marine sediments but also in lakes. Estimates of natural rates of sulfate reduction have also been obtained more indirectly by means of a mathematical model describing the sulfate gradient, the rate of diffusion and sedimentation, and the rate of reduction (Berner, 1964). The activity in marine surface sediments varies widely, from less than 1 $\mu mol$ $SO_4^{2-}$/liter/day in oceanic sediments to about

100 $\mu$mol $SO_4^{2-}$/liter/day in coastal areas (Berner, 1964; Ivanov, 1968; Sorokin, 1962). Even rates up to 5000 $\mu$mol $SO_4^{2-}$/liter/day have been registered in a saline pond in the Sinai Desert (Jørgensen and Cohen, 1977). Although freshwater has a much lower sulfate concentration than the sea, the reduction rate in some lake sediments may be comparable to coastal sediments (Stuiver, 1967).

Sufficient data are available on sulfate reduction to compare this process with the carbon budget of various aquatic ecosystems. Trudinger et al. (1972) quote references indicating that 20% of the primary production in the Black Sea is mineralized by sulfate-reducing bacteria, while the percentage in Linsley Pond is 35%. Jørgensen (in preparation) found in coastal sediments that sulfate reduction was equivalent to the total annual plankton production, but in this area an additional high productivity of eelgrass was probably sustaining the intensive sulfur cycle. In this investigation the oxygen uptake of the sediment was also measured. The mean daily oxygen uptake was 34 mmol $O_2$/m²/day, while the sulfate respiration in the whole sediment column was 10 mmol $SO_4^{2-}$/m²/day. As 1 mol of sulfate is linked with the oxidation of twice as much organic substrate as 1 mol of oxygen, this means that anaerobic respiration was equivalent to 60% of the aerobic respiration.

This type of comparison is important since the oxygen uptake of sediments is frequently used as a measure of "total community metabolism" of the benthic organisms (Teal and Kanwisher, 1961). It should include also the anaerobic processes since these result in reduced end-products that are ultimately consuming oxygen when they diffuse to the oxidized zone. However, a fraction of the sulfide never reaches the surface but is precipitated by metal ions within the reduced zone. The oxygen uptake thus underestimates the total community metabolism corresponding to the accumulation of metal sulfides. In the investigation of Jørgensen (in preparation), this amounted to 6% of the oxygen uptake. Furthermore, denitrification produces $N_2$, which is not oxidized back to nitrate at the sediment surface and therefore totally escapes measurements of community metabolism. Little is known about the quantitative importance of denitrification in sediments, so the error introduced cannot be estimated.

### 6.3. Trophic–Dynamic Aspects of Detritus Food Chains

The trophic–dynamic concept of Lindeman has succesfully been used for grazing food chains to illustrate the transfer of energy and organic matter from one trophic level to the next. The detritus food chain is seldom illustrated this way but appears in dynamic flow diagrams mainly as a black box with detrital input and respiratory output. This is partly because the detritus food chain is difficult to fit into the Lindeman model since it receives material from all trophic levels in the system and partly because so little is known about bacterial metabolism in nature. However, in order to understand how the large amounts of detritus are recycled into new living biomass, it is necessary to view also this food chain in a trophic–dynamic way.

The synthesis of bacterial biomass available for higher trophic levels in the detritus food chain is a function of both the production of detritus and of the growth yield of decomposing bacteria.

In the classical work of Odum (1957), it was shown that approximately half of the net primary production in Silver Springs ended up in the detritus food chain. A similar estimate was made by Saunders (1972) for lake phytoplankton. The extensive work by Wetzel et al. (1972) on the carbon budget of Lawrence Lake, Mich. showed that out of an average yearly photosynthesis of 171 g carbon/m$^2$, 20 g carbon/m$^2$ was respired by bacteria in the pelagic zone and 118 g carbon/m$^2$ was respired in the sediment. This indicates that 80% of the primary production was consumed via the detritus food chain. As previously discussed, less than 10% of macrophytes such as kelps and seagrasses are grazed directly. In tropical oceans, where the zooplankton grazing has reached its maximal degree of efficiency, Jørgensen (1966) estimated that 40% of the net production would still ultimately end up as detritus. Thus, 40% of the total ecosystem production seems to be the minimum input to the detritus food chain. The maximum is close to 100%, and the average is probably not far from 50%.

The efficiency of transformation of organic matter into bacterial biomass has been measured in many of the uptake experiments with labeled substrates. The bacterial growth yield has been calculated from the total assimilated carbon minus that respired to $CO_2$. The growth yield has been found to be in the range of 40–80% for amino acids (Burnison and Morita, 1974; Crawford et al., 1974; Wright, 1973) and 65–90% for sugars and acetate (Hall et al., 1972; Wood, 1973). On whole detrital particles, growth yields ranging from 24 to 66% were found by Gosselink and Kirby (1974). Since the growth yield will depend upon the growth rate of the bacteria, the type of substrate, and a number of other factors, it is difficult directly to use the quoted figures as typical values. But as a rough estimate for aerobic environments, the bacteria convert about 50% of their detritus food into bacterial biomass. With 50% of the net primary production ending up as detritus, this means that as an average for aquatic ecosystems the production of bacterial biomass is 25% of the photosynthetic production (including imported material).

It is important to note that this primary step in the detritus food chain operates with a much higher efficiency of transfer between trophic levels than Lindeman's model normally does. This is because losses in other levels are recycled at this step, and only the burial and gradual

## Table I.  Detritus Food Chain[a]

|                              | Bacterium            | Protozoan            | Invertebrate         | Fish                   |
|------------------------------|----------------------|----------------------|----------------------|------------------------|
| Metabolic rate of trophic level ($kcal/m^2/day$) | 5 | 0.5 | 0.05 | 0.005 |
| Specific metabolic rate ($kcal/g/day$) | 2 | 0.1 | 0.1 | 0.01 |
| Individual biomass (g) | $5 \times 10^{-13}$ | $10^{-7}$ | $10^{-3}$ | 10 |
| Total biomass ($g/m^2$) | 2.5 | 5 | 0.5 | 0.5 |
| Total number individuals per square meter | $5 \times 10^{12}$ | $5 \times 10^7$ | $5 \times 10^2$ | $5 \times 10^{-2}$ |

[a] The distribution of biomass, number of individuals, and metabolic rate of four trophic levels in a hypothetical detritus food chain from a coastal, marine ecosystem. For calculations, consult text.

fossilization of detritus together with other exports from the ecosystem represent a true loss. In spite of their usually low biomass and inconspicuous activity, the bacteria must therefore be among the most important consumers in all ecosystems.

This may be visualized from the following calculations of biomass, number, and metabolic rate of four trophic levels in a hypothetical detritus food chain (Table I). The actual figures used in the table are calculated in energy units and are comparable to a temperate, coastal system where the detritus decomposition mainly takes place in the sediment. A typical net primary production in such an area may be up to 1.5 g carbon/$m^2$/day or 20 kcal/$m^2$/day, of which 10 kcal/$m^2$/day is assumed to enter the detritus food chain. With an average growth yield of 50%, the respiratory metabolism and the production of bacterial biomass are both 5 kcal/$m^2$/day. This includes recycling from higher trophic levels but neglects export or accumulation. In the following calculations, a gross ecological efficiency of 10% has been assumed in the energy transfer from one trophic level to the next. Average metabolic rates of the weight classes at each level were estimated from the regression equations given by Hemmingsen (1960).

An average specific metabolic rate of bacteria is 2 kcal/g/day. Thus, the total respiring biomass of bacteria must be 2.5 g/$m^2$. This is considerably less than what is usually found by direct counting methods in coastal sediments (Dale, 1974, and references cited therein). The discrepancy may be due to an overestimation in Table I of the bacterial respiration in natural environments, possibly because a large part of the population is in an inactive state. With an individual weight of $5 \times 10^{-13}$ g, the number of bacteria is $5 \times 10^{12}$ (active) cells/m$^2$.

As the bacterial consumer we assume a protozoan, e.g., a ciliate. The energy metabolism of the second trophic level is reduced to 0.5 kcal/m$^2$/day. With an individual biomass of $10^{-7}$ g and a specific metabolic rate of 0.1 kcal/g/day, the total ciliate population will be $5 \times 10^7$ individuals/m$^2$ with a biomass of 5 g. In estuarine sands, where Fenchel (1969) found the ciliates to be among the primary consumers of bacteria, individual numbers were $5 \times 10^6$–$2 \times 10^7$/m$^2$ with a biomass of 0.2–2.3 g/m$^2$. When micro-metazoa ($< 10^{-5}$ g/individual) were included, the biomass was about 3 g/m$^2$.

As the next trophic level, we will assume a small invertebrate with a body weight of $10^{-3}$ g, e.g., a nematode, a harpacticoid, or an oligo-chaete. As these have the same specific rate of metabolism as the ciliate but only one-tenth of the energy available, the biomass is only 0.5 g/m$^2$ with a population density of 500 individuals/m$^2$. This is somewhat below natural population sizes, which is reasonable since these organisms also feed directly on bacteria (as well as on microalgae).

To carry our hypothetical detritus food chain one step further, we could include a fish which fed on the small worms and crustaceans. With a body weight of 10 g and a specific metabolism of 0.01 kcal/g/day, the fish stock would be one/20 m$^2$ with a biomass of 0.5 g/m$^2$.

The detritus food chain presented here is of course a simplification of the natural systems, which are better described as food webs without sharp divisions into specific trophic levels. Still, the example illustrates how bacteria due to their high metabolic rate can become the dominating organisms of detritus food chains, even though their biomass is seldom very impressive.

# References

Alexander, M., 1971, *Microbial Ecology*, John Wiley and Sons, New York.

Allen, H. L., 1971, Dissolved organic carbon utilization in size fractionated algal and bacterial communities, *Int. Rev. Gesamten Hydrobiol.* **56**:731.

Atkinson, L. P., and Richards, F. A., 1967, The occurrence and distribution of methane in the marine environment, *Deep-Sea Res.* **14**:673.

Baalsrud, K., and Baalsrud, K. S., 1954, Studies on *Thiobacillus denitrificans*, *Arch. Mikrobiol.* **20**:34.

Barsdate, R. S., Fenchel, T., and Prentki, R. T., 1974, Phosphorus cycle of model ecosystems: Significance for decomposer food chains and effect of bacterial grazers, *Oikos* **25**:239.

Bernard, F. R., 1970, Occurrence of the spirochaete genus *Cristispira* in western Canadian marine bivalves, *Veliger* **13**:33.

Berner, R. A., 1964, An idealized model of dissolved sulfate distribution in recent sediments, *Geochim. Cosmochim. Acta* **28**:1497.

Berner, R. A., 1970, Sedimentary pyrite formation, *Amer. J. Sci.* **268**:1.

Bertram, G. C. L., and Bertram, C. K. R., 1968, Bionomics of dugongs and manatees, *Nature (London)* **218**:423.

Bick, H., 1964, Die Sukzession der Organismen bei der Selbstreinigung von organisch verunreinigtem Wasser unter verschiedenen Milieubedingungen, Min. ELF des Landes Nordrhein/Westfalen, Düsseldorf, p. 139.

Bordovskiy, O. K., 1965, Accumulation and transformation of organic substances in marine sediments, *Mar. Geol.* **3**:3.

Boysen-Jensen, P., 1914, Studies concerning the organic matter of the sea bottom, *Rep. Danish Biol. Sta.* **22**:1.

Brezonik, P. L., and Lee, G. F., 1968, Denitrification as a nitrogen sink in Lake Mendota, Wisconsin, *Environ. Sci. Technol.* **2**:120.

Brinkhurst, R. O., Chua, K. E., and Kaushik, N. K., 1972, Interspecific interactions and selective feeding by tubificid oligochaetes, *Limnol. Oceanogr.* **17**:122.

Bryant, M. P., Tzeng, S. F., Robinson, I. M., and Joyner, Jr., A. E., 1971, Nutrient requirements of methanogenic bacteria, in: *Anaerobic Biological Treatment Processes* (F. G. Pohland, ed.), pp. 23–40, American Chemical Society, Washington.

Burbanck, W. D., 1942, Physiology of the ciliate *Colpidium colpoda*. I. The effect of various bacteria as food on the division rate of *Colpidium colpoda*, *Physiol. Zool.* **15**:342.

Burbanck, W. D., and Eisen, J. D., 1960, The inadequacy of monobacterially fed *Paramecium aurelia* as food for *Didinium nasutum*, *J. Protozool.* **7**:201.

Burnison, B. K., and Morita, R. Y., 1974, Heterotrophic potential for amino acid uptake in a naturally eutrophic lake, *Appl. Microbiol.* **27**:488.

Cairns, J. (ed.), 1971, The structure and function of fresh-water microbial communities, Research Division Monograph 3, Virginia Polytechnic Institute and State University, Blacksbury, Virginia.

Cappenberg, T. E., 1974a, Interrelations between sulfate-reducing and methane-producing bacteria in bottom deposits of a fresh-water lake. I. Field observations, *Antonie van Leeuwenhoek J. Microbiol. Serol.* **40**:285.

Cappenberg, T. E., 1974b, Interrelations between sulfate-reducing and methane-producing bacteria in bottom deposits of a fresh-water lake. II. Inhibition experiments, *Antonie van Leeuwenhoek J. Microbiol. Serol.* **40**:297.

Cappenberg, T. E., and Prins, R. A., 1974, Interrelations between sulfate-reducing and methane-producing bacteria in bottom deposits of a fresh-water lake. III. Experiments with $^{14}C$-labelled substrates, *Antonie van Leeuwenhoek J. Microbiol. Serol.* **40**:457.

Chen, K. Y., and Morris, J. C., 1972, Oxidation of sulfide by $O_2$-catalysis and inhibition, *J. Sanit. Eng.* **98**:215.

Chynoweth, D. P., and Mah, R. A., 1971, Volatile acid formation in sludge digestion, in: *Anaerobic Biological Treatment Processes* (F. G. Pohland, ed.), pp. 41–54, American Chemical Society, Washington.

Claypool, G. E., and Kaplan, I. R., 1974, The origin and distribution of methane in marine sediments, in: *Natural Gases in Marine Sediments* (I. R. Kaplan, ed.), pp. 99–140, Plenum Press, New York.

Cline, J. D., and Richards, F. A., 1969, Oxygenation of hydrogen sulfide in sea water at constant salinity, temperature, and pH, *Environ. Sci. Technol.* **3**:838.

Cohen, Y., Jørgensen, B. B., Padan, E., and Shilo, M., 1975, Sulfide-dependent anoxygenic photosynthesis in the cyanobacterium *Oscillatoria limnetica*, *Nature (London)* **257**:489.

Colwell, R. R., and Liston, J., 1962, The natural bacterial flora of certain marine invertebrates, *J. Insect Pathol.* **4**:23.

Colwell, R. R., and Morita, R. Y., 1974, *Effect of the Ocean Environment on Microbial Activities*, University Park Press, Baltimore.

Crawford, C. C., Hobbie, J. E., and Webb, K. L., 1974, The utilization of dissolved free amino acids by estuarine microorganisms, *Ecology* **55**:551.

Culver, D. A., and Brunskill, G. J., 1969, Fayetteville, Green Lake, New York. V. Studies of primary production and zooplankton in a meromictic marl lake, *Limnol. Oceanogr.* **14**:862.

Dale, N. G., 1974, Bacteria in intertidal sediments: Factors related to their distribution, *Limnol. Oceanogr.* **19**:509.

Duncan, A., Schiemer, F., and Klekowski, R. Z., 1974, A preliminary study of feeding rates on bacterial food by adult females of a benthic nematode, *Plectus palustris* de Man 1880, *Pol. Arch. Hydrobiol.* **21**:249.

Edwards, A. C., and Heath, W. G., 1963, The role of soil animals in breakdown of leaf material in: *Soil Organisms* (J. Docksen and J. van der Drift, eds.), pp. 27–34, North Holland, Amsterdam.

Egglishaw, H. J., 1972, An experimental study of the breakdown of cellulose in fast-flowing streams, *Mem. Ist. Ital. Idrobiol., Suppl.* **29**:405.

Fell, J. W., and Master, I. M., 1973, Fungi associated with the degradation of mangrove (*Rhizophora mangle* L.) leaves in South Florida, in: *Estuarine Microbial Ecology* (L. H. Stevenson and R. R. Colwell, eds.), pp. 455–465, University of South Carolina Press, Columbia, S.C.

Fenchel, T., 1968, The ecology of marine microbenthos. II. The food of marine benthic ciliates, *Ophelia* **5**:73.

Fenchel, T., 1969, The ecology of marine microbenthos. IV. Structure and function of the benthic ecosystem, its chemical and physical factors and the microfauna communities with special reference to the ciliated protozoa, *Ophelia* **6**:1.

Fenchel, T., 1970, Studies on the decomposition of organic detritus derived from the turtle grass *Thalassia testudinum*, *Limnol. Oceanogr.* **15**:14.

Fenchel, T., 1972, Aspects of decomposer food chains in marine benthos, *Verh. Dtsch. Zool. Ges. 65 Jahresversamml.* **14**:14.

Fenchel, T., 1973, Aspects of the decomposition of sea-grasses, International Seagrass Workshop, Leiden, Netherlands.

Fenchel, T., 1975, The quantitative importance of the benthic microflora of an arctic tundra pond, *Hydrobiologia* **46**:445.

Fenchel, T., 1977, The significance of bacterivorous protozoa in the microbial community of detrital particles, in: *Freshwater Microbial Communities*, 2nd Ed. (J. Cairns, ed.), Garland Publishing, Inc., New York.

Fenchel, T., and Harrison, P., 1976, The significance of bacterial grazing and mineral cycling for the decomposition of particulate detritus, in *The Role of Terrestrial and Aquatic Organisms in Decomposition Processes* (J. M. Anderson, ed.), pp. 285–299, Blackwell Scientific, Oxford.

Fenchel, T. M., and Riedl, R. J., 1970, The sulfide system: A new biotic community underneath the oxidized layer of marine sand bottoms, *Mar. Biol.* **7**:255.

Fenchel, T., Kofoed, L. H., and Lappalainen, A., 1975, Particle size-selection of two deposit feeders: The amphipod *Corophium volutator* and the prosobranch *Hydrobia ulvae*, *Mar. Biol.* **30**:119.

Field, J. G., 1972, Some observations on the release of dissolved organic carbon by the sea urchin *Strongylocentrotus droebachiensis*, *Limnol. Oceanogr.* **17**:759.

Fjerdingstad, E. J., 1961, Ultrastructure of the collar of the choanoflagellate *Codonosiga botrytis* (Ehrenb.), *Z. Zellforsch. Mikrosk. Anat.* **54**:499.

Frankenberg, D., and Smith Jr., K. L., 1967, Coprophagy in marine animals, *Limnol. Oceanogr.* **12**:443.

Freeden, F. J. H., 1960, Bacteria as a source of food for black-fly larvae, *Nature (London)* **187**:963.

Giere, O., 1975, Population structure, food relations and ecological role of marine oligochaetes, with special references to meiobenthic species, *Mar. Biol.* **31**:139.

Goering, J. J., 1968, Denitrification in the oxygen minimum layer of the eastern tropical Pacific Ocean, *Deep-Sea Res.* **15**:157.

Goering, J. J., and Dugdale, R. C., 1966a, Denitrification rates in an island bay in the equatorial Pacific Ocean, *Science* **154**:505.

Goering, J. J., and Dugdale, V. A., 1966b, Estimates of the rates of denitrification in a subarctic lake, *Limnol. Oceanogr.* **11**:113.

Goering, J. J., and Pamatmat, M. M., 1971, Denitrification in sediments of the sea off Peru, *Invest. Pesq.* **35**:233.

Goldhaber, M. B., and Kaplan, I. R., 1974, The sulfur cycle, in: *The Sea* (E. D. Goldberg, ed.), pp. 569–655, John Wiley and Sons, New York.

Golterman, H. L., 1972, The role of phytoplankton in detritus formation, *Mem. Ist. Ital. Idrobiol., Suppl.* **29**:89.

Gordon, G. C., Robinson, G. G. C., Hendzel, L. L., and Gillespie, D. C., 1973, A relationship between heterotrophic utilization of organic acids and bacterial populations in West Blue Lake, Manitoba, *Limnol. Oceanogr.* **18**:264.

Gosselink, J. G., and Kirby, C. J., 1974, Decomposition of salt marsh grass, *Spartina alterniflora* Loisel, *Limnol. Oceanogr.* **19**:825.

Gray, C. T., and Gest, H., 1965, Biological formation of molecular hydrogen, *Science* **148**:186.

Hall, K. S., Kleiber, P. M., and Yesaki, I., 1972, Heterotrophic uptake of organic solutes by microorganisms in the sediment, *Mem. Ist. Ital. Idrobiol., Suppl.* **29**:441.

Hargrave, B. T., 1970a, The utilization of benthic microflora by *Hyalella azteca* (Amphipoda), *J. Anim. Ecol.* **9**:427.

Hargrave, B. T., 1970b, The effect of a deposit-feeding amphipod on the metabolism of benthic microflora, *Limnol. Oceanogr.* **15**:21.

Hargrave, B. T., 1971, An energy budget for a deposit-feeding amphipod, *Limnol. Oceanogr.* **16**:99.

Hargrave, B. T., 1972, Aerobic decomposition of sediment and detritus as a function of particle surface area and organic content, *Limnol. Oceanogr.* **7**:583.

Harrison, P. G., and Mann, K. H., 1975a, Chemical changes during the seasonal cycle of growth and decay in eelgrass (*Zostera marina*) on the Atlantic coast of Canada, *J. Fish. Res. Board Can.* **32**:615.

Harrison, P. G., and Mann, K. H., 1975b, Detritus formation from eelgrass (*Zostera marina* L.): The relative effects of fragmentation, leaching, and decay, *Limnol. Oceanogr.* **20**:924.

Hemmingsen, A. M., 1960, Energy metabolism as related to body size and respiratory surfaces, and its evolution, *Rep. Steno. Hosp., Copenhagen* **9**:1.

Hobbie, J. E., 1967, Glucose and acetate in freshwater: Concentrations and turnover rates, in: *Chemical Environment in the Aquatic Habitat* (H. L. Golterman and R. S. Clymo, eds.), pp. 245–251, N. V. Noord-Hollandsche Uitgevers Maatschappij, Amsterdam.

Hobbie, J. E., 1971, Heterotrophic bacteria in aquatic ecosystems; some results of studies with organic radioisotopes, in: *The Structure and Function of Freshwater Microbial Communities* (J. Cairns Jr., ed.), pp. 181–194, Research Division Monograph 3, Virgina Polytechnic Institute and State University, Blacksbury, Virginia.

Hobbie, J. E., and Crawford, C. C., 1969, Respiration corrections for bacterial uptake of dissolved organic compounds in natural water, *Limnol. Oceanogr.* **14**:528.

Hobbie, J. E., and Wright, R. T., 1965, Competition between planktonic bacteria and algae for organic solutes, *Mem. Ist. Ital. Idrobiol., Suppl.* **18**:175.

Hobbie, J. E., Holm-Hansen, O., Packard, T. T., Pomeroy, L. R., Sheldon, R. W., Thomas, J. P., and Wiebe, W. J., 1972, A study of the distribution and activity of microorganisms in ocean water, *Limnol. Oceanogr.* **17**:544.

Hungate, R. E., 1966, *The Rumen and its Microbes*, Academic Press, New York.

Hutchinson, G. E., 1957, *A Treatise on Limnology*, Vol. I, *Geography, Physics and Chemistry*, John Wiley and Sons, New York.

Hylleberg Kristensen, J., 1972, Carbohydrases of some marine invertebrates with notes on their food and on the natural occurrence of the carbohydrates studied, *Mar. Biol.* **14**:130.

Hylleberg, J., 1975, Selective feeding by *Abarenicola pacifica* with notes on *Abarenicola vagabunda* and a concept of gardening in lugworms, *Ophelia* **14**:113.

Hylleberg, J., and Gallucci, V. F., 1975, Selectivity in feeding by the deposit-feeding bivalve *Macoma nasuta*, *Mar. Biol.* **32**:167.

Hynes, H. B. N., and Kaushik, N. K., 1969, The relationship between dissolved nutrient salts and protein production in submerged autumnal leaves, *Verh. Int. Verein. Limnol.* **17**:95.

Ivanov, M. V., 1968, *Microbiological Processes in the Formation of Sulfur Deposits*, Israel Program for Scientific Translations, Jerusalem.

Jannasch, H. W., and Pritchard, P. H., 1972, The role of inert particulate matter in the activity of aquatic microorganisms, *Mem. Ist. Ital. Idrobiol., Suppl.* **29**:289.

Javornický, P., and Prokesová, V., 1963, The influence of protozoa and bacteria upon the oxidation of organic substances in water, *Int. Rev. Gesamter Hydrobiol.* **48**:335.

Jeris, J. S., and McCarty, P. L., 1965, The biochemistry of methane fermentation using [14]C tracers, *J. Wat. Pollut. Control Fed.* **37**:178.

Johannes, R. E., 1965, Influence of marine protozoa on nutrient regeneration, *Limnol. Oceanogr.* **10**:434.

Johannes, R. E., 1968, Nutrient regeneration in lakes and oceans, *Adv. Microbiol. Sea* **1**:203.

Johannes, R. E., and Satomi, M., 1967, Measuring organic matter retained by aquatic invertebrates, *J. Fish. Res. Board Can.* **24**:2467.

Jørgensen, B. B., and Cohen, Y., 1977, Solar Lake (Sinai). V. The sulfur cycle of the benthic mats, *Limnol. Oceanogr.* (in press).

Jørgensen, B. B., and Fenchel, T., 1974, The sulfur cycle of a marine sediment model system, *Mar. Biol.* **24**:189.

Jørgensen, C. B., 1966, *Biology of Suspension Feeding*, Pergamon Press, Oxford.

Jørgensen, C. B., 1976, August Pütter, August Krogh, and modern ideas on the use of dissolved organic matter in aquatic environments, *Biol. Rev.* **51**:291.

Kaplan, I. R. (ed.), 1974, *Natural Gases in Marine Sediments*, Plenum Press, New York.

Kaushik, N. K., and Hynes, H. B. N., 1968, Experimental study of the role of autumn-shed leaves in aquatic environments, *J. Ecol.* **56**:229.

Kaushik, N. K., and Hynes, H. B. N., 1971, The fate of the dead leaves that fall into streams. *Arch. Hydrobiol.* **68**:465.

Keeney, D. R., 1972, The fate of nitrogen in aquatic ecosystems, Eutrophication Information Program, University of Wisconsin, Literature Review No. 3.

Kofoed, L. H., 1975a, The feeding biology of *Hydrobia ventrosa* (Montagu). I. The assimilation of different components of the food, *J. Exp. Mar. Biol. Ecol.* **19**:233.

Kofoed, L. H., 1975b, The feeding biology of *Hydrobia ventrosa* (Montagu). II. Allocation of the components of the carbon-budget and the significance of the secretion of dissolved organic material, *J. Exp. Mar. Biol. Ecol.* **19**:243.

Kondratjeva, E. N., 1965, *Photosynthetic Bacteria*, Israel Program for Scientific Translations, Jerusalem.

Krause, H. R., 1964, Zur Chemie und Biochemie der Zersetzung von Süsswasserorganismen, unter besonderer Berücksichtigung des Abbaues der organischen Phosphorkomponenten, *Verh. Int. Verein. Limnol.* **15**:549.

Kuenen, J. G., 1975, Colourless sulfur bacteria and their role in the sulfur cycle, *Plant Soil* **43**:49.

Lasker, R., and Giese, A. C., 1954, Nutrition of the sea urchin, *Strongylocentrotus purpuratus*, *Biol. Bull.* **106**:328.

Lee, J. J., McEnery, M. Pierce, S., Freudenthal, H. D., and Muller, W. A., 1966, Tracer experiments in feeding littoral foraminifera, *J. Protozool.* **13**:659.

LeGall, J., and Postgate, J. R., 1973, The physiology of sulphate-reducing bacteria, *Adv. Microb. Physiol.* **10**:81.

Lewin, J. C., and Lewin, R. A., 1960, Culture and nutrition of some apochloritic diatoms of the genus *Pitzschia*, *J. Gen. Microbiol.* **6**:127.

Litchfield, C. D., 1973, Interactions of amino acids and marine bacteria, in: *Estuarine Microbial Ecology* (L. H. Stevenson and R. R. Colwell, eds.), pp. 145–168, University of South Carolina Press, Columbia, S.C.

Mann, K. H., 1972, Macrophyte production and detritus food chains in coastal waters, *Mem. Ist. Ital. Idrobiol., Suppl.* **29**:353.

Manuvilova, E. F., 1958, The question of the role of bacterial numbers in the development of Cladocera in natural conditions, *Dokl. Biol. Sci. Sect.* **120**:438.

Martens, C. S., and Berner, R. A., 1974, Methane production in the interstitial waters of sulfate depleted marine sediments, *Science* **185**:1167.

McCarty, P. L., 1971, Energetics and kinetics of anaerobic treatment, in: *Anaerobic Biological Treatment Processes* (F. G. Pohland, ed.), pp. 91–107, American Chemical Society, Washington.

McKinney, R. E., 1971, Microbial relationships in biological wastewater treatment systems, in: *The Structure and Function of Freshwater Microbial Communities* (J. Cairns Jr., ed.), pp. 165–179, Research Division Monograph 3, Virginia Polytechnic Institute and State University, Blacksburg, Virginia.

McRoy, C. P., 1970, Standing stocks and other features of eelgrass (*Zostera marina*) populations on the coast of Alaska, *J. Fish. Res. Board Can.* **27**:1811.

Mechalas, B. J., 1974, Biogenic gas production, in: *Natural Gases in Marine Sediments* (I. R. Kaplan, ed.), pp. 11–25, Plenum Press, New York.

Melchiorri- Santolini, U., and Hopton, J. W. (eds.), 1972, *Detritus and its Role in Aquatic Ecosystems*, Proceedings of an IBP-UNESCO symposium, Pallanza, 1972, *Mem. Ist. Ital. Idrobiol., Suppl.* **29**.

Menzies, R. J., Zaneveld, J. S., and Pratt, R. M., 1967, Transported turtle grass as a source of organic enrichment of abyssal sediments off North Carolina, *Deep-Sea Res.* **14**:111.

Meyers, S. P., and Hopper, B. E., 1973, Nematological–microbial interrelationships and estuarine biodegradative processes, in: *Estuarine Microbial Ecology* (L. H. Stevenson and R. R. Colwell, eds.), pp. 483–489, University of South Carolina Press, Columbia, S.C.

Newell, R., 1965, The role of detritus in the nutrition of two marine deposit feeders, the prosobranch *Hydrobia ulvae* and the bivalve *Macoma balthica*, *Proc. Zool. Soc. London* **144**:25.

Newell, S. Y., 1973, Succession and role of fungi in the degradation of red mangrove seedlings in: *Estuarine Microbial Ecology* (L. H. Stevenson and R. R. Colwell, eds.), pp. 467–480, University of South Carolina Press, Columbia, S.C.

Nielsen, B. O., 1966, Carbohydrases of some wrack invertebrates, *Natura Jutl.* **12**:141.

Odum, E. P., and de la Cruz, A. A., 1967, Particulate detritus in a Georgia salt marsh-estuarine ecosystem, in: *Estuaries* (G. H. Lauff, ed.), pp. 383–388, *Amer. Assoc. Adv. Sci. Publ.*, **83**.

Odum, H. T., 1957, Trophic structure and productivity of Silver Springs, Florida, *Ecol. Monogr.* **27**:55.

Odum, W. E., 1970, Utilization of the direct grazing and plant detritus food chains by the striped mullet *Mugil cephalus*, in: *Marine Food Chains* (J. H. Steele, ed.), pp. 222–240, Oliver and Boyd, Edinburgh.

Ogura, N., 1975, Further studies on decomposition of dissolved organic matter in coastal seawater, *Mar. Biol.* **31**:101.

Ohle, W., 1958, Die Stoffwechseldynamik der Seen in Abhängigkeit von der Gasausscheiding ihres Schlammes, *Vom Wasser* **25**:127.

Oláh, J., 1972, Leaching, colonization and stabilization during detritus formation, *Mem. Ist. Ital. Idrobiol., Suppl.* **29**:105.

Oppenheimer, C. H. (ed.), 1963, *Symposium on Marine Microbiology*, Charles C. Thomas, Springfield, Ill.

Otsuki, A., and Hanya, T., 1972, Production of dissolved organic matter from dead green algal cells. I. Aerobic microbial decomposition, *Limnol. Oceanogr.* **17**:248.

Otsuki, A., and Wetzel, R. G., 1974, Release of dissolved organic matter by autolysis of a submerged macrophyte, *Scirpus subterminalis*, *Limnol. Oceanogr.* **19**:842.

Overbeck, J., 1972, Eksperimentelle Untersuchungen zur Bestimmung der bakteriellen Produktion im See, *Verh. Int. Ver. Limnol.* **18**:176.

Overbeck, J., and Daley, R. J., 1973, Some precautionary comments on the Romanenko technique for estimating heterotrophic bacterial production, *Bull. Ecol. Res. Comm. (Stockholm)* **17**:342.

Packard, T. T., Healy, M. L., and Richards, F. A., 1971, Vertical distribution of the activity of the respiratory electron transport system in marine plankton, *Limnol. Oceanogr.* **16**:60.

Paerl, H. W., 1974, Bacterial uptake of dissolved organic matter in relation to detrital aggregation in marine and freshwater systems, *Limnol. Oceanogr.* **19**:966.

Pamatmat, M. M., and Bhagwat, A. M., 1973, Anaerobic metabolism in Lake Washington sediments, *Limnol. Oceanogr.* **18**:611.

Parsons, T. R., 1963, Suspended organic matter in sea water, in: *Progress in Oceanography* (M. Sears, ed.), pp. 205–239, Vol. 1, Pergamon Press, New York.

Payne, W. J., 1973, Gas chromatographic analysis of denitrification by marine organisms, in: *Estuarine Microbial Ecology* (L. H. Stevenson and R. R. Colwell, eds.), pp. 53–71, University of South Carolina Press, Columbia, S.C.

Perkins, E. J., 1958, The food relationships of the microbenthos, with particular reference to that found at Whitstable, Kent, *Ann. Mag. Nat. Hist., Ser.* **13**:64.

Petersen, R. C., and Cummins, K. W., 1974, Leaf processing in a woodland stream, *Freshwater Biol.* **4**:343.

Pfennig, N., 1967, Photosynthetic bacteria, *Annu. Rev. Microbiol.* **21**:285.

Pfennig, N., 1975, The phototrophic bacteria and their role in the sulfur cycle, *Plant Soil* **43**:1.

Pomeroy, L. R., 1970, The strategy of mineral cycling, *Annu. Rev. Ecol. Syst.* **1**:171.

Prim, P., and Lawrence, J. M., 1975, Utilization of marine plants and their constituents by bacteria isolated from the gut of echinoids (Echinodermata), *Mar. Biol.* **33**:167.

Randall, J. E., 1965, Grazing effect on sea grasses by herbivorous reef fishes in the West Indies, *Ecology* **46**:255.

Reeburgh, W. S., 1969, Observations of gases in Chesapeake Bay sediments, *Limnol. Oceanogr.* **14**:368.

Reichardt, W., and Simon, M., 1972, Die Mettma-ein Gebirgsbach als Brauereivorfluter. Mikrobiologische Untersuchungen entlang eines Abwasser-Substratgradienten, *Arch. Hydrobiol., Suppl.* **42**:125.

Reilly, S. M., 1964, Importance of adsorbents in the nutrition of *Paramecium caudatum*, *J. Protozool.* **11**:109.

Rhee, G. -Y., 1972, Competition between an alga and an aquatic bacterium for phosphate, *Limnol. Oceanogr.* **17**:505.

Richards, F. A., 1965, Anoxic basins and fjords, in: *Chemical Oceanography* (J. P. Riley and G. Skirrow, eds.), pp. 611–645, Academic Press, New York.

Riley, G. A., 1970, Particulate organic matter in sea water, *Adv. Mar. Biol.* **8**:1.

Rodina, A. G., 1972, *Methods in Aquatic Microbiology*, Butterworths, London.

Rosswall, T. (ed.), 1973, *Modern Methods in the Study of Microbial Ecology, Bull. Ecol. Res. Comm. (Stockholm)* **17**.

Rudd, J. W. M., Hamilton, R. D., and Campbell, N. E. R., 1974, Measurement of microbial oxidation of methane in lake water, *Limnol. Oceanogr.* **19**:519.

Russell-Hunter, W. D., 1970, *Aquatic Productivity: An Introduction to Some Basic Aspects of Biological Oceanography and Limnology*, Macmillan, London.

Saunders, G. W., 1972, The transformation of artificial detritus in lake water, *Mem. Ist. Ital. Idrobiol., Suppl.* **29**:261.

Seki, H., 1972, The role of microorganisms in the marine food chain with reference to organic aggregate, *Mem. Ist. Ital. Idrobiol., Suppl.* **29**:245.

Selwyn, S. C., and Postgate, J. R., 1959, A search for the *rubentschikii* group of *Desulphovibrio, Antonie van Leeuwenhoek J. Microbiol. Serol.* **25**:465.

Sleigh, M., 1973, *The Biology of Protozoa*, Arnold, London.

Smith, P. H., and Mah, R. A., 1966, Kinetics of acetate metabolism during sludge digestion, *Appl. Microbiol.* **14**:368.

Sorokin, Yu. I., 1962, Experimental investigation of bacterial sulfate reduction in the Black Sea using $^{35}$S, *Mikrobiologiya* **31**:329.

Sorokin, Yu. I., 1964, On the trophic role of chemosynthesis in water bodies, *Int. Rev. Gesamter Hydrobiol.* **49**:307.

Sorokin, Yu. I., 1965, On the trophic role of chemosynthesis and bacterial biosynthesis in water bodies, *Mem. Ist. Ital. Idrobiol.* **18**:187.

Sorokin, Yu. I., 1972, The bacterial population and the process of hydrogen sulphide oxidation in the Black Sea, *J. Cons. Cons. Int. Explor. Mer.* **34**:423.

Sorokin, Yu. I., and Kadota, H., 1972, *Microbial Production and Decomposition in Fresh Waters*, IBP Handbook No. 23, Blackwell, Oxford.

Spector, W. S. (ed.), 1956, *Handbook of Biological Data*, Saunders, Philadelphia.

Steele, J. H., and Baird, I. E., 1972, Sedimentation of organic matter in a Scottish sea loch, *Mem. Ist. Ital. Idrobiol., Suppl.* **29**:73.

Steemann Nielsen, E., 1952, The use of radioactive carbon ($^{14}$C) for measuring organic production in the sea, *J. Cons. Cons. Int. Explor. Mer.* **18**:117.

Stephens, G. O., 1975, Uptake of naturally occuring primary amines by marine annelids, *Biol. Bull.* **149**:397.

Straarup, B. J., 1970, On the ecology of turbellarians in a sheltered brackish shallow-water bay, *Ophelia* **7**:185.

Stuiver, M. 1967, The sulfur cycle in lake waters during thermal stratification, *Geochim. Cosmochim. Acta* **31**:2151.

Takahashi, M., and Ichimura, S., 1968, Vertical distribution and organic matter production of photosynthetic sulfur bacteria in Japanese lakes, *Limnol. Oceanogr.* **13**:644.

Teal, J. M., 1962, Energy flow in the salt marsh ecosystem of Georgia, *Ecology* **43**:614.

Teal, T. M., and Kanwisher, J., 1961, Gas exchange in a Georgia salt marsh, *Limnol. Oceanogr.* **6**:388.

Thane-Fenchel, A., 1968, Distribution and ecology of nonplanktonic brackish water rotifers from Scandinavian waters, *Ophelia* **5**:273.

Tietjen, J. H., Lee, J. J., Rullman, J., Greengart, A., and Trompeter, J., 1970, Gnotobiotic culture and physiological ecology of the marine nematode *Rhabditis marina* Bastian, *Limnol. Oceanogr.* **15**:535.

Toerien, D. F., and Hattingh, W. H. J., 1969, Anaerobic digestion. I. The microbiology of anaerobic digestion, *Water Res.* **3**:385.

Trudinger, P. A., Lambert, I. B., and Skyring, G. W., 1972, Biogenic sulfide ores: A feasibility study, *Econ. Geol.* **67**:1114.

Vaccaro, R. F., Hicks, S. E., Jannasch, H. W., and Carey, F. G., 1968, The occurrence and role of glucose in sea water, *Limnol. Oceanogr.* **13**:356.

Vinogradov, A. P., 1953, *The Elementary Composition of Marine Organisms*, Sears Foundation for Marine Research, Memoir two, New Haven.

Wavre, M., and Brinkhurst, R. O., 1971, Interaction between some tubificid oligochaetes and bacteria found in the sediments of Toronto harbour, Ontario, *J. Fish. Res. Board Can.* **28**:335.

Webb, K. L., and Johannes, R. E., 1967, Studies of the release of dissolved free amino acids by marine zooplankton, *Limnol. Oceanogr.* **12**:376.

Wetzel, R. G., Rich, P. H., Miller, M. C., and Allen, H. L., 1972, Metabolism of dissolved and particulate detrital carbon in a temperate hard-water lake, *Mem. Ist. Ital. Idrobiol., Suppl.* **29**:185.

Wiegert, R. G., and Owen, D. F., 1971, Trophic structure, available resources and population density in terrestrial vs. aquatic ecosystems, *J. Theor. Biol.* **30**:69.

Williams, P. J. le B., 1970, Heterotrophic utilization of dissolved organic compounds in the sea. I. Size distribution of population and relationship between respiration and incorporation of growth substrates, *J. Mar. Biol. Assoc. U.K.* **50**:859.

Williams, P. J. le B., 1973, The validity of the application of simple kinetic analysis to heterogeneous microbial populations, *Limnol. Oceanogr.* **18**:159.

Wood, E. J. F., 1965, *Marine Microbial Ecology*, Chapman and Hall, London.

Wood, E. J. F., 1967, *Microbiology of Oceans and Estuaries*, Elsevier, Amsterdam.

Wood, E. J. F., Odum, W. E., and Zierman, J. C., 1969, Influence of sea grasses on the productivity of coastal lagoons, in: *Lagunas Costeras, un Simposio. Mem. Simp. Intern. Lagunas Costeras*, UNAM-UNESCO, Mexico, D.F., pp. 495–502.

Wood, L. W., 1973, Monosaccharide and disaccharide interactions on uptake and catabolism of carbohydrates by mixed microbial communities, in: *Estuarine Microbial Ecology* (L. H. Stevenson and R. R. Colwell, eds.), pp. 181–197, University of South Carolina Press, Columbia, S.C.

Wright, R. T., 1973, Some difficulties in using $^{14}$C-organic solutes to measure heterotrophic bacterial activity, in: *Estuarine Microbial Ecology* (L. H. Stevenson and R. R. Colwell, eds.), pp. 199–217, University of South Carolina Press, Columbia, S.C.

Wright, R. T., and Hobbie, J. E., 1965, Uptake of organic solutes in lake water, *Limnol. Oceanogr.* **10**:22.

Wright, R. T., and Hobbie, J. E., 1966, Use of glucose and acetate by bacteria and algae in aquatic ecosystems, *Ecology* **47**:447.

Wright, R. T., and Shah, N. M., 1975, The trophic role of glycolic acid in coastal seawater. I. Heterotrophic metabolism in seawater and bacterial cultures, *Mar. Biol.* **33**:175.

Zeikus, J. G., Weimer, P. J., Nelson, D. R., and Daniels, L., 1975, Bacterial methanogenesis: Acetate as a methane precursor in pure culture, *Arch. Microbiol.* **104**:129.

ZoBell, C. E., 1943, The effect of solid surfaces upon bacterial activity, *J. Bacteriol.* **46**:39.

ZoBell, C. E., and Feltham, C. B., 1937–38, Bacteria as food for certain marine invertebrates, *J. Mar. Res.* **1**:312.

ZoBell, C. E., and Feltham, C. B., 1942, The bacterial flora of a marine mud flat as an ecological factor, *Ecology* **23**:69.

ZoBell, C. E., and Grant, C. W., 1943, Bacterial utilization of low concentrations of organic matter, *J. Bacteriol.* **45**:555.

ZoBell, C. E., and Landon, W. A., 1937, The bacterial nutrition of the California mussel, *Proc. Soc. Exp. Biol. Med. N.Y.* **36**:607.

# 2

# Ecological Studies with the Chemostat

## H. VELDKAMP

## 1. Introduction

Most natural and seminatural environments are inhabited by a large variety of microorganisms. The functional status of a particular organism is difficult to assess in the highly complex field situation. Therefore, methods were developed to isolate microbes in pure culture and to study their characteristics in the laboratory. Many pure cultures of bacteria were isolated from batch enrichments and subsequently studied in a closed culture system. This approach was highly successful. In this way the various types of energy metabolism occurring in prokaryotic organisms were discovered so that deductions could be made about the role specific microorganisms play in the cycling of the elements. In addition, many details could be elucidated about metabolic pathways and their regulation.

From an ecological point of view, however, this approach had some serious drawbacks, as was pointed out by Tempest (1970a). One disadvantage of batch enrichments is that organisms have to be grown at nutrient concentrations that are much higher than those occurring in environments such as soil or natural waters. Therefore, the question is to what extent organisms isolated from such cultures are identical with those playing a major role in mineralization processes in habitats in which the concentration of dissolved organic and inorganic nutrients is much lower.

In a batch culture, microorganisms grow at maximal rate in the presence of an excess of all nutrients. This is a situation which they rarely encounter

H. VELDKAMP ● Department of Microbiology, University of Groningen, Biological Centre, Haren (Gr.), The Netherlands.

in nature, where nutrient limitation is the rule rather than the exception. Only at the end of the phase of exponential growth may a component of the growth medium become growth-rate-limiting, but this limitation will be only for a brief period and under conditions of continuous change of the chemical composition of the culture medium.

Therefore, it is impossible to study microbial growth under nutrient limitation in batch culture under properly controlled conditions. Studies of this kind can, however, be carried out with a flow-controlled continuous culture system, as first described by Monod (1950) and Novick and Szilard (1950). The latter authors called it "chemostat."

A wealth of information on continuous culture studies is to be found in the proceedings of the international symposia devoted to this topic (Málek, 1958; Anonymous, 1961; Málek *et al.*, 1964; Powell *et al.*, 1967; Málek *et al.*, 1969; Dean *et al.*, 1972, 1976). A review of application of the chemostat in studies of microbial ecology and physiology was published recently by Veldkamp (1976a). Abstracts of continuous culture literature were given annually in *Folia Microbiologica* up to 1975.

As the chemostat can be used for a variety of purposes and much has been written about results obtained in recent years, this review is not meant to be exhaustive. In Section 2 a brief description is given of the principle of nutrient-limited growth in the chemostat. In Sections 3–5 some recent developments are described in studies with microbial populations grown in homogeneous open systems. Symbols used in mathematical equations are those used by the authors referred to, unless different symbols were used for the same parameters by different authors. When this was the case, and it often was, the author has arbitrarily chosen one of the symbols to indicate a particular parameter for all equations given here.

## 2. The Chemostat

A chemostat with commonly used ancillary equipment is shown in Fig. 1. Technical details about chemostats were given by Řičica (1966), Baker (1968), Evans *et al.* (1970), Veldkamp and Kuenen (1973), and Harder *et al.* (1974).

Sterile culture medium is pumped from a reservoir at rate $F$ (ml/hr) into the culture vessel. As culture liquid is removed at the same rate, the culture volume $V$ (ml) is constant. The dilution rate is defined as $D = F/V$ $(hr^{-1})$. The composition of the sterile medium is such that all components needed for growth are present in excess except one, the growth-limiting substrate.

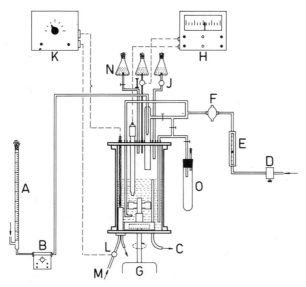

**Figure 1.** Schematic diagram of single-stage chemostat. Details of fermentor described by Harder *et al.* (1974). A, Burette (connected to medium reservoir) for measuring medium flow rate; B, medium pump; C, outlet for medium and air; D, gas-flow controller; E, gas-flow meter; F, filter; G, AC motor of magnetic drive; H, pH controller; I, peristaltic pump or magnetic valve connected to inlet tube for acid or alkali; J, peristaltic pump for antifoam addition; K, temperature controller; L, circulation pump; M, connection to temperature-controlled water bath; N, inoculation flask; O, sampling flask (Veldkamp, 1976a).

The specific growth rate ($\mu$) is defined as

$$\mu = \frac{1}{x} \cdot \frac{dx}{dt} = \frac{\ln 2}{t_d} \, (\mathrm{hr}^{-1}) \tag{1}$$

where $x$ is biomass (mg dry wt./ml) at time $t$, and the doubling time $t_d$ is the time required for the biomass to double.

The specific growth rate ($\mu$) of the culture in the chemostat depends on the concentration of the growth-limiting substrate ($s$). A typical example of this relation is given in Fig. 2.

The relation between $\mu$ and $s$ is described empirically by the equation (Monod, 1942, 1950)

$$\mu = \mu_{max}\left(\frac{s}{K_s + s}\right) \tag{2}$$

where $\mu_{max}$ is the maximum specific growth rate ($\mathrm{hr}^{-1}$) and $K_s$ is a saturation constant, numerically equal to the concentration of the growth-limiting nutrient at 0.5 $\mu_{max}$.

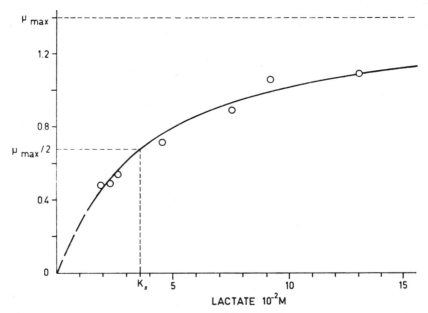

**Figure 2.** Relation between specific growth rate ($\mu$) and concentration of growth-rate-limiting substrate (lactate) in *Propionibacterium shermanii* (Jerusalimsky, 1967).

When a chemostat is put on stream and inoculated with a sample from a pure culture of bacteria, a steady state will be reached whatever the size of the inoculum. The chemostat is a self-regulating system.

The increase in biomass is given by

$$\frac{dx}{dt} = \mu x \tag{3}$$

and the washout of cells is

$$-\frac{dx}{dt} = Dx \tag{4}$$

Net growth therefore is

$$\frac{dx}{dt} = (\mu - D)x \tag{5}$$

Figure 3 shows that the system automatically adjusts to steady state in which $\mu = D$. To obtain a steady state, the dilution rate ($D$) should not exceed a certain critical value ($D_c$) which is determined by the concentration of the growth-limiting substrate in the reservoir ($S_R$):

$$D_c = \mu_m \left( \frac{S_R}{K_s + S_R} \right) \tag{6}$$

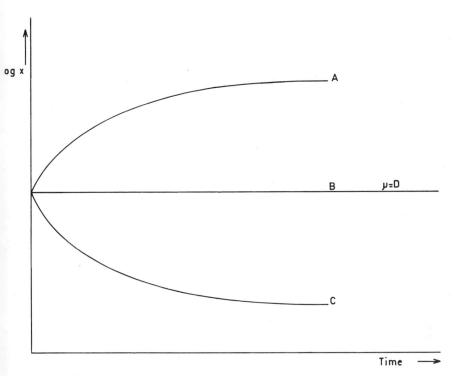

**Figure 3.** Change of bacterial concentration after inoculation of chemostat. A, initially $\mu > D$; $x$ increases and $s$ decreases until $\mu = D$. B, steady state is obtained immediately. C, initially $\mu < D$; $x$ decreases and $s$ increases until $\mu = D$. Adapted from Jannasch (1969).

Once a steady state is reached, the culture has become time independent, and the steady-state concentration of biomass $(\bar{x})$ and growth-limiting substrate $(\bar{s})$ then are (Herbert *et al.*, 1956)

$$\bar{x} = Y(S_R - \bar{s}) \qquad (7)$$

where $Y$ is the growth-yield coefficient (weight of biomass formed/weight of substrate used). If not stated otherwise, the value of $Y$ always refers to the growth-limiting substrate, and

$$\bar{s} = K_s\left(\frac{D}{\mu_{max} - D}\right) \qquad (8)$$

Once $\mu_{max}$, $K_s$, and $Y$ are known, the values of $\bar{x}$ and $\bar{s}$ can be predicted for any dilution rate (Fig. 4). Experimental deviations from theoretical predictions with aid of equations (7) and (8) are often due to the fact that $Y$ is not a constant (see Sections 3 and 4).

For further details of chemostat theory, the appropriate literature

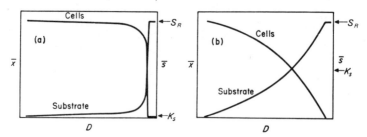

**Figure 4.** Influence of dilution rate ($D$) on the steady-state concentration of organisms ($\bar{x}$) and growth-limiting substrate ($\bar{s}$) in a chemostat culture: a, when $K_s$ (the bacterial saturation constant for growth-limiting substrate) is small relative to $S_R$ (the growth-limiting substrate concentration in the feed medium); and b, where $K_s$ is large relative to $S_R$ (Tempest 1970b).

should be consulted (Herbert *et al.*, 1956; Fencl, 1966; Kubitschek, 1970; Tempest, 1970b; Pirt, 1975).

From the above considerations it is clear that cells of prokaryotic and eukaryotic organisms which can be grown in homogeneous submerged culture can be exposed in the chemostat to a wide variety of conditions. They can be grown in steady state at growth rates ranging from nearly zero to nearly maximal, and any of the nutrients needed by the cells can be made growth limiting. Steady-state cultures can be exposed to any of a variety of controlled changes, supposed to occur in the natural environment. If the introduced physical or chemical change does not cause complete washout, a new steady state will be reached, again with a phenotypically unique type of cells, and integration of results of steady-state and transient-state studies with the chemostat will give a better insight into the functional status or niche of growing cells of a particular organism than can be obtained with batch cultures. Batch-culture conditions represent only a limited number of extreme conditions which are rare in nature. An important restriction of the chemostat is that it only allows studies of growing cells.

The chemostat offers many possibilities to study the interrelations between microorganisms and their abiotic environment. It also provides the microbial ecologist with a tool to study microbial interactions, as will be shown in Section 5.

## 3. Nutrient Limitation in Algae

In a study of a vitamin-$B_{12}$-limited culture of the chrysomonad *Monochrysis lutheri*, Droop (1968) observed that the vitamin concentration in the culture medium increased with decreasing dilution rates. According to equation (2) this would mean that $K_s$ increases with decreasing dilution

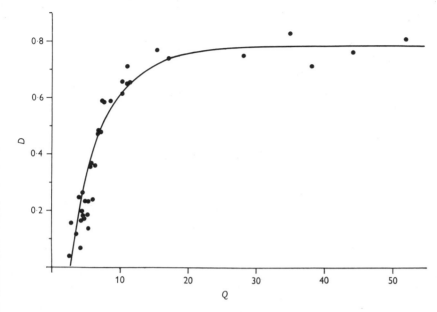

**Figure 5.** Relation between dilution rate ($D$; vol/day) and vitamin $B_{12}$ cell quota ($Q$; pg/$10^6$ cells). Steady-state values of vitamin-$B_{12}$-limited *Monochrysis lutheri* (Droop, 1968).

rate (cf. Droop, 1970). The apparent anomaly was explained as follows. The cells were shown to excrete some vitamin and to release a heat-labile, nondialyzable compound, probably a protein, which combined with the vitamin, rendering it unavailable. The excretion apparently occurred at a constant relative rate and therefore became particularly evident at high cell concentration and low dilution rate. As will be shown below, the cell concentration is highest at the lowest dilution rates. Excretion of and sensitivity to the binding factor were not species specific nor were they confined to vitamin-$B_{12}$-requiring algae. Possible ecological consequences of this phenomenon were discussed by Droop (1968).

The relation between the internal concentration of growth-limiting vitamin $B_{12}$, termed cell quota or demand coefficient ($Q$ = mass of internal nutrient per unit biomass; Droop, 1968, 1974), and dilution rate is given by a hyperbolic curve (Fig. 5). Note that $Q$ is the reciprocal of the yield coefficient ($Y$). The $Q$ intercept $K_Q$ can be regarded as the "subsistence quota," below which no growth will take place. The equation for the hyperbola (Droop, 1968, 1970) is

$$D = D_{max}\left(1 - \frac{K_Q}{Q}\right) \qquad (9)$$

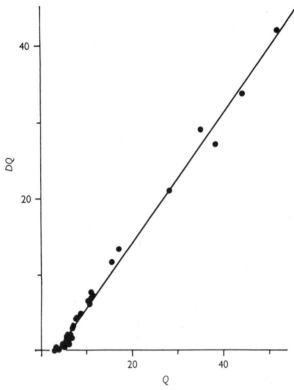

**Figure 6.** Regression of $DQ$ on $Q$. Steady-state values of vitamin-$B_{12}$-limited *Monochrysis lutheri* (Droop, 1968).

And the equation of regression is

$$DQ = D_{max}(Q - K_Q) \tag{10}$$

Regression of $DQ$ on $Q$ is shown in Fig. 6. In the chemostat (assuming $D = \mu$), we thus have

$$\mu = \mu'_{max}\left(1 - \frac{K_Q}{Q}\right) \tag{11}$$

where $\mu'_{max}$ is related to the internal concentration of the limiting nutrient; $\mu'_{max}$ is slightly higher than $\mu_{max}$ [equation (2)], as indicated by Droop (1973a). $K_Q$ is the saturation constant for the limiting nutrient and is characterized as follows: When $D = 0.5D_{max}$, $Q = 2K_Q$ (Droop, 1968).

The empirical relation [equation (11)] found by Droop (1968) in vitamin-$B_{12}$-limited *Monochrysis lutheri* appears to have a much wider scope. As indicated by Droop (1974), it also holds for the following nutrient limitations in algae: vitamin $B_{12}$ with *Skeletonema costatum*

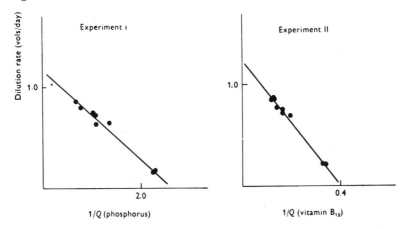

**Figure 7.** Experiment I, phosphorus-limited culture of *Monochrysis lutheri*; regression of *D* on $1/Q$ for phosphorus; *D* in vol/day, $1/Q$ in million cells/nmol. Experiment II, vitamin-$B_{12}$-limited culture of *Monochrysis lutheri*; regression of *D* on $1/Q$ for vitamin $B_{12}$; *D* in vol/day, $1/Q$ in million cells/fmol (Droop, 1974).

(Droop, 1970), nitrate with *Isochrysis galbana* (Caperon, 1968), phosphate with *Cyclotella nana* (Fuhs, 1969), nitrate with *M. lutheri* and *C. nana* (Caperon and Meyer, 1972a,b), phosphate with *Nitzschia actinasteroides* (Müller, 1972), iron with *M. lutheri* (Droop, 1973a), and silicon with *Thalassiosira pseudonana* (Paasche, 1973a,b). An obvious implication of equation (11) is that the growth yield with respect to the limiting substrate decreases linearly with increasing growth rate. Examples are given in Fig. 7. The fact that *Q* is variable does not invalidate equation (2) (Droop, 1973b).

Though the above data show the interdependence of specific growth rate and internal nutrient concentration, there seems to be little doubt that the specific growth rate is related to the free pool of nutrient rather than to the total amount per unit biomass (Paasche, 1973a; Droop, 1974). As pointed out by Droop (1974), we have to suppose that the active pool is proportional to the total in cells growing under steady-state conditions.

The specific substrate uptake rate ($q = -(1/x) \times (ds/dt)$) of growth-limiting substrates is related to the external substrate concentration according to a Monod equation [cf. equation (2)] (Droop, 1968, 1973a,b; Perry, 1976).

The relation between the rates of substrate uptake and growth (Droop, 1973b) is

$$\frac{dQ}{dt} = q - \mu Q \qquad (12)$$

so that in steady state we have

$$q = \mu Q \tag{13}$$

Thus, combining equations (11) and (13), we have the following equation, which relates cell quota ($Q$), specific growth rate ($\mu$), and specific rate of uptake ($q$):

$$q = \mu'_{max}(Q - K_Q) \tag{14}$$

As was shown by Droop (1973a,b), equation (11) not only holds for a limiting but also for a nonlimiting nutrient, as is shown in Figs. 8 and 9, though it is clear that limitation by one nutrient leads to "luxury consumption" (Droop, 1973b, 1974; Perry, 1976) of others. This means that in the chemostat during steady state, we have for the nutrients A and B, which may or may not be limiting (Droop, 1974),

$$\frac{D}{D'_M} = 1 - \frac{K'_{Q_A}}{Q_A} = 1 - \frac{K'_{Q_B}}{Q_B} \tag{15}$$

or

$$\frac{Q_A}{Q_B} = \frac{K'_{Q_A}}{K'_{Q_B}} \tag{16}$$

where $D'_{max}$ is the value of $D$ if $Q$ were infinite. The subsistence quota of the limiting nutrient, $K_Q$, is a constant, whereas the apparent subsistence quota, $K'_Q$, of a nonlimiting nutrient is larger and variable. $K_Q$ is the value

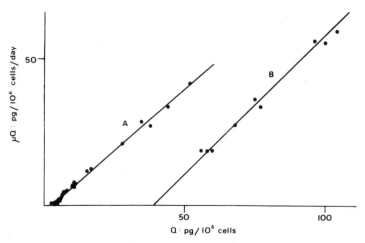

**Figure 8.** Specific rate of uptake ($\mu Q$) of vitamin $B_{12}$ by *Monochrysis* as a function of the vitamin $B_{12}$ cell quota ($Q$). A, in a chemostat with limiting vitamin $B_{12}$. B, in a chemostat with nitrate limiting (Droop, 1973b).

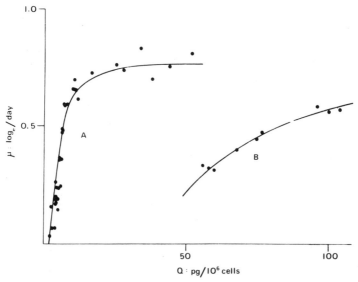

**Figure 9.** Specific growth rate ($\mu$) of *Monochrysis* as a function of the vitamin $B_{12}$ cell quota ($Q$). A, in a chemostat with limiting vitamin $B_{12}$. B, in a chemostat with limiting nitrate (Droop, 1973b).

$K'_Q$ takes when a nutrient is limiting growth rate. For a given nutrient, the ratio $K'_Q/K_Q$ is a measure of the degree of excess. Droop (1974) coined the term "luxury coefficient" ($R$) for this ratio.

Equations (15) and (16) were put to test in a series of experiments with *Monochrysis lutheri* grown in phosphorus- or vitamin-$B_{12}$-limited chemostats (Droop, 1974). Algal behavior was studied in conditions where the nonlimiting nutrient was present in both great and only slight excess. The experiments fully confirmed the above statements. This means that the ratio of the internal concentrations of nutrients is that of their apparent subsistence quotas. This is constant for any given input nutrient ratio and independent of the specific growth rate. In other words, the rates of uptake of two nutrients are in constant proportion, regardless of growth rate.

It was also shown that limitation for phosphate and vitamin $B_{12}$ followed a threshold pattern; that is, either one or the other effects control of growth rate. The controlling nutrient is the one which shows the smaller $Q/K_Q$ ratio, and for this nutrient $K_Q$ and $K'_Q$ are identical.

The experimental data and theoretical considerations discussed above form the basis of a model (Droop, 1974, 1976) describing phytoplankton growth in relation to nutrient supply. This model can in principle be used in multinutrient and multispecies situations and is capable of handling transient stages, as occur under natural conditions. One of the essential features of Droop's model, equation (11), was applied by Tett et al. (1975)

in a study of phosphorus quota and the chlorophyll/carbon ratio in marine phytoplankton.

As a result of continuous culture studies with phosphorus- and nitrogen-limited *Thalassiosira pseudonana*, Perry (1976) concluded that the chemical composition of phytoplankton, expressed as ratios, provides valuable information about the nutritional status of algal populations with respect to nitrogen and phosphorus, though a practical problem here is that analyses of oceanic particulate matter include material other than phytoplankton. An evaluation of possible means of identifying nutrient deficiencies in natural populations of algae was given by Healey (1973).

Until now bacteriologists have related the specific growth rate exclusively to the external concentration of limiting substrates. The observations of Droop with algae indicate that it may be of interest to pay more attention to internal concentrations. The few data that are available show, however, that in *Aerobacter aerogenes*, $DQ$ does not increase linearly with $Q$ as in Fig. 6 (Dicks and Tempest, 1966; Tempest *et al.*, 1966; Tempest and Dicks, 1967). The same seems to be true for the yeast *Candida utilis* (Aiking and Tempest, 1976).

## 4. Heterotrophic Production in Bacteria

A considerable part of organic carbon arising from plant photosynthesis becomes either directly or indirectly available for microbial breakdown. This holds both for terrestrial (Phillipson, 1973; Wagner, 1975) and aquatic (Tait, 1972; Steele, 1975) ecosystems. The question of how much bacterial cell material can be formed from organic substrates broken down, either aerobically or anaerobically, has been studied extensively and was reviewed by Stouthamer (1976), who considered the literature up to 1974. A short summary of past results will be given here, followed by a discussion of some recent developments.

Early studies on the amount of cell material formed during the breakdown of organic substrates indicated that in the relation

$$\frac{dx}{dt} = -Y\frac{ds}{dt} \tag{17}$$

the yield coefficient $Y$ is a constant (Monod, 1942).

Once the ATP yield of a variety of fermentative processes had been established, the amount of cell material formed per mole of ATP generated ($Y_{ATP}$) could be calculated (Bauchop and Elsden, 1960). For a number of organisms, a value of around 10 was found. This value was initially considered to be a universal constant, and this view was still held in recent reviews of the subject (Payne, 1970; Forrest and Walker, 1971). The

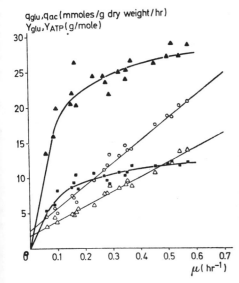

Figure 10. Effect of specific growth rate on the rates of glucose consumption and acetate production, on the molar growth yields, and on $Y_{ATP}$ by *Aerobacter aerogenes* during glucose-limited anaerobic growth. O – O, specific rate of glucose consumption; $\triangle - \triangle$, specific rate of acetate production; $\blacktriangle - \blacktriangle$, molar growth yield; $\blacksquare - \blacksquare$, $Y_{ATP}$ (Stouthamer and Bettenhaussen, 1975).

concept that there is a more or less fixed ratio between substrate carbon ending up in cell material and metabolic end-products found its origin in the way the experimental data were collected. This was done with batch cultures in which organisms were growing at maximal rate in the presence of an excess of nutrients needed for growth. However, continuous culture studies with the chemostat have shown that the growth yield, with respect to amount of substrate consumed or amount of ATP generated, depends on growth rate. A typical example is shown in Fig. 10 (Stouthamer and Bettenhaussen, 1975). Similar data were already obtained with aerobic cultures of glycerol-limited *Aerobacter aerogenes* (Herbert, 1959; Tempest *et al.*, 1967) and glucose-limited *Escherichia coli* (Schulze and Lipe, 1964). These data were consistent with the concept that a bacterial cell needs a certain constant amount of energy per unit time for purposes other than growth, collectively called maintenance purposes (Marr *et al.*, 1963). Schulze and Lipe (1964) expressed the specific rate of substrate uptake as follows:

$$q = \frac{\mu}{Y} = \frac{\mu}{Y_G} + q^m \qquad (18)$$

where $q$ is specific rate of substrate uptake (mmol of substrate used/mg dry wt./hr); the observed molar growth yield $Y = g$ dry wt./mol of substrate used; $Y_G$ is the molar growth yield corrected for maintenance requirement; $q^m$ is the maintenance coefficient (mmol substrate/mg dry wt./hr); and $\mu$ is specific growth rate (hr$^{-1}$). In equation (18) $\mu/Y_G$ and $q^m$ represent

the rate of substrate consumption for growth and for maintenance requirements, respectively. When $q$ is plotted against $\mu$ (Fig. 10), the slope of the
line equals the reciprocal of the true growth yield, and the intercept with
the ordinate gives the growth-rate-independent maintenance requirement.
With decreasing growth rates, this requirement gives in a carbon-limited
organism a drastic decrease in cell yield (Fig. 10).

Pirt (1965) obtained similar results when studying carbon-limited cells
and applied equation (18) as follows:

$$\frac{1}{Y} = \frac{1}{\mu} \times q^m + \frac{1}{Y_G} \tag{19}$$

The values found for $q^m$ in carbon-limited cultures are quite small (e.g.,
Pirt, 1965, 1975; Stouthamer, 1976), and thus assuming a growth-rate-
independent maintenance requirement, the percentage of the carbon and
energy source used for maintenance is rather small in cells growing at
maximal rate with an excess of nutrients. For example, Stouthamer and
Bettenhaussen (1975) found for $Aerobacter$ $aerogenes$ growing anaerobically
on glucose at maximal rate a $Y_{ATP}$ value of ca. 12 g/mol, whereas the
value for $Y'_{ATP}$ ($Y_{ATP}$ corrected for maintenance) was 14 g/mol.

The amount of ATP needed theoretically for the synthesis of a
bacterial cell was calculated by Gunsalus and Shuster (1961), Forrest and
Walker (1971), and Stouthamer (1973) with the aid of current knowledge
of biosynthetic pathways and energy needed for transport and turnover of
macromolecules. For a chemoorganotrophic organism utilizing glucose as
the carbon and energy source and provided with various supplements, the
theoretical $Y_{ATP}$ is around 30 g/mol according to Stouthamer (1973). When
comparing this figure with the $Y_{ATP}$ value of 10–13 g/mol as found for a
variety of glucose-fermenting organisms in batch culture (cf. Forrest and
Walker, 1971), one sees that there must be a considerable loss of ATP
which cannot be accounted for. To account for this gap, Harder and van
Dijken (1976) assumed a rather loose coupling between energy generation
and utilization and introduced a coupling factor $K$:

$$Y'_{ATP} = K \times Y^{max}_{ATP} \tag{20}$$

where $Y'_{ATP}$ is $Y_{ATP}$ corrected for maintenance and $Y^{max}_{ATP}$ is the theoretical
maximal amount of biomass formed per mole of ATP as calculated from
known metabolic pathways.

A variant of equation (19), introduced by Stouthamer and Bettenhaussen
(1973), would then according to Harder and van Dijken (1976) become

$$\frac{1}{Y_{ATP}} = \frac{1}{\mu} \times m_e + \frac{1}{K \times Y^{max}_{ATP}} \tag{21}$$

in which $m_e$ is a maintenance coefficient (gmole ATP/g dry wt./hr).

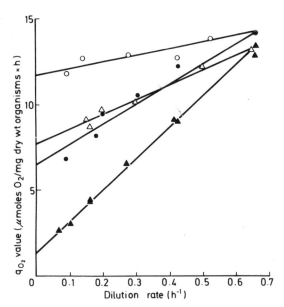

**Figure 11.** Relationship between the specific growth rate and the specific rate of oxygen consumption in variously limited chemostat cultures of *Aerobacter aerogenes*, growing in a glucose-containing medium. Cultures were, respectively, carbon limited (▲), ammonia limited (●), sulfate limited (△), and phosphate limited (○) (Neijssel and Tempest, 1976a).

Equations (18) and (19) were developed for growth of cultures limited by a carbon and energy source. When the parameters are applied to this substrate for cultures grown under other limitations, as was done by Stouthamer and Bettenhaussen (1973, 1975), the consequences are far reaching. This becomes clear when comparing the specific rates of substrate uptake (or of oxygen consumption in aerobic cultures) in cells growing under different nutrient limitations. An example is given in Fig. 11 (Neijssel and Tempest, 1976a), which shows that at low growth rates the oxygen consumption of carbon-sufficient cells is very much greater than the cell's minimum maintenance requirement found in carbon-limited cells. Assuming a growth-rate-independent maintenance requirement, the consequence would be that carbon-sufficient cells have a much larger maintenance requirement near $\mu_{max}$ than carbon-limited cells and at the same time use the remaining energy for growth with a much greater efficiency than carbon-limited cells. Neijssel and Tempest (1976a) considered this highly unlikely and assumed that the maintenance requirement near the maximal growth rate would be approximately the same for both types of cells. They pointed out that although it is generally assumed that because of the linearity of double reciprocal plots of $1/Y$ versus $1/\mu$ [according to equation (19)] the maintenance requirement is independent of growth rate (e.g.,

Hempfling and Mainzer, 1975), this need not necessarily be the case. The maintenance rate may well vary in a linear fashion with growth rate in such a way that cells limited by different nutrients show approximately the same maintenance energy requirement when growing at a rate close to maximal. Neijssel and Tempest (1976a) therefore proposed the following modification of equation (19):

$$\frac{1}{Y} = \frac{1}{\mu} \times q^m + \frac{1}{Y_G} + c \times q^m \qquad (22)$$

where $c$ is a constant defining the variation in maintenance rate with growth rate. The modified equation (18) thus becomes:

$$q = \mu \left( \frac{1}{Y_G} + q^m \times c \right) + q^m \qquad (23)$$

If the above concept of a variable maintenance requirement is correct, then an important consequence would be that the high values for true growth yields in cells limited by nutrients other than the carbon and energy source, as reported by Stouthamer and Bettenhaussen (1973, 1975), would be gross overestimates. Thus, the big difference between $Y'_{ATP}$ ($Y_{ATP}$ corrected for maintenance) and the calculated theoretical $Y_{ATP}^{max}$ would still remain.

When comparing the growth yield with respect to substrate and oxygen consumed, respectively, for *Aerobacter aerogenes* grown under mannitol-, glucose-, and gluconate-limitation, respectively, Neijssel and Tempest (1976b) observed that glucose- and mannitol-limited cells were carbon- rather than energy-limited. To test this finding, the cells were subsequently grown in a glucose-limited medium to which 2% NaCl was added. It appeared that at growth rates below 0.5/hr, the addition of NaCl did not affect the specific oxygen uptake rate. At higher dilution rates, however, addition of NaCl caused a rather steep increase in the specific oxygen consumption rate. At the higher growth rates, the excess of energy available for the increased maintenance requirement due to NaCl addition was apparently no longer high enough. This observation was interpreted by Neijssel and Tempest (1976b) as additional evidence that at low growth rates carbon and not energy was growth limiting, and that under carbon limitation the maintenance rate decreased with increasing growth rate. The fact that Watson (1970) observed an increased maintenance requirement at all growth rates after addition of 6% NaCl to a glucose-limited culture of *Saccharomyces cerevisiae* may have been due to the high salt concentration applied or, more likely, to the fact that a respiratory deficient mutant was used which at all growth rates was energy limited.

As pointed out by Neijssel and Tempest (1976b), the rate of uptake of carbon source and oxygen for maintenance purposes depends on a variety of factors, such as the identity of the carbon source (Hempfling and Mainzer, 1975; van Verseveld and Stouthamer, 1976), the nature of the growth limitation (Stouthamer and Bettenhaussen, 1975; Downs and Jones, 1975; Neijssel and Tempest, 1976a), the concentration of available oxygen (Rogers and Stewart, 1974), and the identity of the nitrogen source in carbon-limited organisms (Neijssel and Tempest, 1976b). The latter authors also observed that intermittent addition of a concentrated glucose solution to a glucose-limited culture of *Aerobacter aerogenes*, to which the other nutrients were added in the usual way, manifested itself as an increase in maintenance energy requirement. The loss of growth efficiency was higher at low growth rates than at high ones, which was in agreement with the observation that the difference between the actual specific oxygen uptake rate of glucose-limited cells in the chemostat and the maximal possible uptake rate immediately after release of glucose limitation was much greater at low than at high growth rates. Similar observations were obtained by Brooks and Meers (1973) with a methanol-limited *Pseudomonas* sp. culture. Luscombe (1974), however, did not find this effect with glucose-limited cultures of *Arthrobacter globiformis*. It would be of interest to extend such studies to other organisms, as in natural environments the irregular availability of growth-limiting substrates is the rule rather than the exception.

Modeling of aerobic bacterial growth is still handicapped by the fact that it is impossible to determine how much ATP is formed in the electron transport chain. This also holds for the stochastic model for heterotrophic growth in microorganisms developed by de Kwaadsteniet *et al.* (1976).

In summary, it can be stated that continuous culture studies on the effect of growth rate and kind of nutrient limitation on the growth yield of chemoorganotrophic bacteria have convincingly shown that both factors are of great importance. The yield with respect to the carbon and energy source decreases with decreasing growth rate, whatever the growth limitation, and in carbon-limited cells it is always higher than in carbon-sufficient cells. It is also clear that the maintenance energy requirement is markedly influenced by a variety of environmental parameters. Little is known about the fate of ATP not used for growth and of the function these ATP-consuming processes may have for the survival of a microbe. When a bacterial cell is able to generate ATP, it always does so, even when it cannot be applied for growth. And it seems that this may have survival value, for instance, by being able to keep carriers for the transport of essential molecules alert as much as possible for the uptake of nutrients present in extremely low concentrations in the environment. Experimental evidence for energy consumption by carriers in the cell membrane in the absence of the molecules they transport was given by Bisschop and Konings (1976).

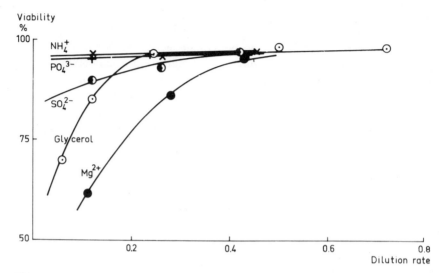

**Figure 12.** Effect of nutritional status on the steady-state viability of chemostat culture of *Aerobacter aerogenes*. Limiting nutrient indicated on each curve (Postgate, 1973).

Two final points may be made. The first is that in yield studies a practical problem may arise caused by death and lysis of cells at low growth rates. Postgate (1973) showed that this is highly dependent on the kind of growth limitation, as is shown in Fig. 12. Sinclair and Topiwala (1970) gave a mathematical description of viability as a function of specific growth rate that approximates the experimental data fairly well.

Another point is that, in bacteria, cell size is a function of growth rate (Herbert, 1959). Therefore, growth yield does not tell anything about cell numbers. The data presented by Herbert (1959) show that in glycerol-limited *A. aerogenes*, the growth yield (g cells formed/g glycerol consumed) decreased by a factor of 2 and the mean cell mass by a factor of 3 when the dilution rate was decreased from 0.9 to 0.05/hr. Thus, although at the low dilution rate only half the amount of cell material was formed per gram of glycerol consumed, the number of cells produced was larger. This may well have survival value as the cellular surface to volume ratio is larger at extremely low concentrations of the limiting nutrient (cf. Fig. 14). Secondly, it may favor a bacterial population in case the nutrient limitation becomes absolute. Assuming equal death rates, the length of the death phase of starving bacterial populations is a function of the initial number of cells rather than of the amount of biomass present at the moment when growth becomes impossible.

## 5. Microbial Interactions

The literature on mixed culture studies with the chemostat has been reviewed by Bungay and Bungay (1968), Veldkamp and Jannasch (1972), Meers (1973), Jannasch and Mateles (1974), Harrison (1976), and Veldkamp (1976b). Most of the mixed culture work to be mentioned here pertains to simple two- to three-membered cultures. As yet, phenomena occurring in cultures harboring more than a few species are very difficult to interpret.

It is rather difficult to classify microbial interactions, and no generally accepted classification scheme is available. As a matter of convenience, the nomenclature used by Meers (1973) is used here.

### 5.1. Competition

In the introduction some uncertainty was expressed as to what extent bacteria isolated from batch enrichments are involved in mineralization and other processes occurring in such environments as soil and natural waters; in other words, whether there is in these environments selection towards substrate concentration. When two or more organisms compete for the same growth-limiting substrate in the chemostat and no other interactions occur, the result of the competition can be predicted from the relation between their specific growth rates and the concentration of the growth-limiting substrate, as is shown in Fig. 13. With respect to chemo-organotrophic bacteria competing for a single limiting carbon and energy source, Jannasch (1967; cf. Jannasch and Mateles, 1974) was the first to show that in natural waters organisms occur showing crossing $\mu$–$s$ curves

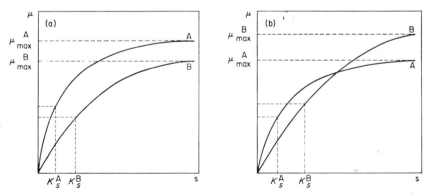

**Figure 13.** The $\mu$–$s$ relationship of two organisms $A$ and $B$. a, $K_s^A < K_s^B$ and $\mu_{max}^A > \mu_{max}^B$; b, $K_s^A < K_s^B$ and $\mu_{max}^A < \mu_{max}^B$ (Veldkamp, 1970).

for lactate and glycerol. Experiments of this type are carried out by connecting two chemostats to the same reservoir. One is run at high dilution rate (e.g., 0.3) and the other at low dilution rate (e.g., 0.03). Both are inoculated with a sample from the same source after filtering the sample to remove protozoa. After about five volume changes one particular organism, different for each vessel, generally has become dominant. When these are isolated in pure culture and the experiment is repeated with the mixture of pure cultures, the same result is obtained. Organisms growing faster at the lower concentration range are often spiral-shaped and show a relatively high surface-to-volume ratio (cf. Kuenen et al., 1976), as compared with organisms growing faster at the higher concentration range, which are often rod shaped. A high surface-to-volume ratio affects the maximum specific growth rate and not the $K_s$ for growth. This gives the organism with a high surface-to-volume ratio a selective advantage at the lower concentration extreme, as is shown in Fig. 14.

Competition for inorganic substrates has been little studied thus far. In an experiment of the type described above, Kuenen et al. (1976) studied competition for phosphate by a mixed population occurring in a sample of pond water. Again a *Spirillum* sp. came to the fore at the lower phosphate concentration range (i.e., at low dilution rate), whereas a rod-shaped bacterium became dominant at higher concentrations. The organisms were isolated in pure culture and exposed to competition for $K^+$, $Mg^+$, $NH_4^+$, succinate, lactate, aspartate, and glutamate. The *Spirillum* showed a selective advantage at very low concentrations of all these substrates.

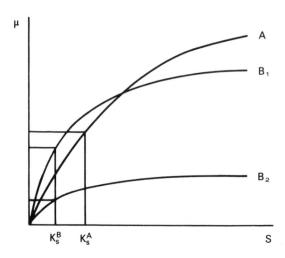

**Figure 14.** Specific growth rate $(\mu)$ as a function of substrate concentration(s) of organisms $A$, $B_1$, and $B_2$. Surface/volume, $B_1 > B_2$; $K_s^{B_1} = K_s^{B_2}$ (Kuenen et al., 1976).

In a comparative study of the obligately autotrophic bacteria *Thiomicrospira pelophila* and *Thiobacillus thioparus* (Kuenen *et al.*, 1976), it was found that the former organism showed a higher tolerance for the toxic energy source, sulfide. In an attempt to explain this behavior, competition experiments were carried out in an iron-limited chemostat. *Thiomicrospira pelophila* grew faster indeed at extremely low iron concentrations. Both organisms occur commonly in marine mud flats along the northern coastline of the Netherlands. Their distribution appeared to be in agreement with the predictions from laboratory studies, *Thiomicrospira* being more abundant in areas with particularly high sulfide concentrations, such as mussel banks. As sulfide tolerance may mean many things, it is as yet difficult to assess whether differences in specific growth rate of iron-limited cells are the only answer that explains the distribution of these organisms.

Only few studies on microbial selection have been made in which, in addition to substrate concentration, another abiotic parameter was introduced. Examples are temperature in competition for lactate by marine psychrophilic chemoorganotrophs (Harder and Veldkamp, 1971), presence of clay particles in competition for valeric acid in marine chemoorganotrophs (Jannasch and Pritchard, 1972), and a light–dark rhythm in competition for sulfide between two *Chromatium* species (van Gemerden, 1974). The latter study is of particular interest. Both organisms coexist in anaerobic aquatic environments into which light penetrates. Sulfide is the main electron donor for their photosynthesis. Nevertheless, in a continuously illuminated sulfide-limited chemostat one species, *Chromatium vinosum*, grows faster than the other, *Chromatium weissei*, at any sulfide concentration. Therefore, *C. weissei* is selectively eliminated at all dilution rates in continuous light. However, the organisms can coexist in the chemostat when this is illuminated intermittently (Fig. 15). During the dark period, the sulfide concentration increases as it does in their natural habitat, where *Desulfovibrio* continues to produce sulfide. *Chromatium weissei* shows its selective advantage during the first few hours of illumination after a dark period. Its specific sulfide uptake rate then is about twice that of *C. vinosum*, which means that during the first few hours after sunrise, it picks up two-thirds of the available sulfide, which is mainly oxidized to molecular sulfur. The elemental sulfur is stored intracellularly. The steady-state abundance of *C. vinosum* was found to be positively related to the length of the light period, and that of *C. weissei* to the length of the dark period.

Competition for sulfide among purple bacteria is probably not restricted to purple sulfur bacteria, as it was shown that at low sulfide concentrations, as occur in a sulfide-limited chemostat, purple nonsulfur bacteria can utilize sulfide as an electron donor in photosynthesis (Hansen and van Gemerden, 1972).

The competition studies of the last ten years can be summarized by

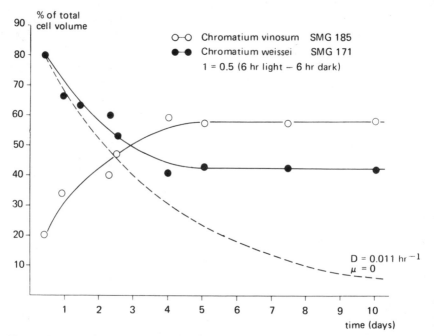

**Figure 15.** Time course of the relative abundance of *Chromatium vinosum* and *Chromatium weissei* in a mixed continuous culture ($D = 0.011$/hr) illuminated intermittently. Dotted line represents wash-out rate of *C. weissei* if $\mu$ were 0 (van Gemerden, 1974).

stating that at least part of the tremendous diversity in bacteria can be explained in terms of selection by substrate concentration. Organisms showing a selective advantage at extremely low concentrations of growth-limiting organic and inorganic substrates appear to be quite common. This was of course already recognized by Winogradsky (1949), who named this part of the microflora the "autochthonous" organisms. With the aid of the chemostat, many organisms of this type can now be easily enriched, and their properties can be studied under well-controlled and therefore reproducible conditions.

## 5.2. Predator–Prey Relations

As was shown above, the outcome of competition of two bacteria for a single growth-limiting substrate depends on the concentration of that substrate as well as on additional abiotic factors. Jost *et al.* (1973a) showed that a predator may also affect the outcome of bacterial competition. When *Azotobacter vinelandii* and *Escherichia coli* competed for glucose in a glucose-limited chemostat, the former organism was selectively eliminated at any dilution rate. However, introduction of the ciliate *Tetrahymena pyriformis* in the mixed culture had a stabilizing effect, and both bacteria could coexist in the three-membered culture.

Several studies have been made with a single-stage chemostat in which the growth rate of a bacterial culture was limited by a substrate present in the feed and that of a predator, e.g., a ciliate, by the bacterial concentration. Assuming that Monod's equation (2) (cf. Fig. 2) holds for both predator and prey, the equations of the model are (Bungay and Bungay, 1968; Jost *et al.*, 1973b; Bonomi and Fredrickson, 1976)

$$\frac{dp}{dt} = -\frac{1}{\theta}p + \frac{v_{max}\,xp}{L+x} \tag{24a}$$

$$\frac{dx}{dt} = -\frac{1}{\theta}x + \frac{\mu_{max}\,sx}{K_s+s} - \frac{1}{X}\frac{v_{max}\,xp}{L+x} \tag{24b}$$

$$\frac{ds}{dt} = \frac{1}{\theta}(S_f - s) - \frac{1}{Y}\frac{\mu_{max}\,sx}{K_s+s} \tag{24c}$$

where $p$ and $x$ are population densities of protozoa and bacteria, respectively; $s$ and $S_f[=S_R$; equation (6)] are concentrations of the rate-limiting substrate for bacterial growth in the chemostat and the feed to the chemostat, respectively; $\theta$ is the holding time (reciprocal dilution rate) of the chemostat; $v_{max}$ and $\mu_{max}$ are maximum specific reproduction rates of the protozoa and bacteria, respectively; $X$ and $Y$ are yield coefficients for reproduction of the protozoa and bacteria, respectively; and $L$ and $K_s$ are saturation constants of Monod's model for reproduction of the protozoa and bacteria, respectively. The results of a stability analysis of equation (24) were used by Tsuchiya *et al.* (1972) and Jost *et al.* (1973b) to construct an operating diagram, indicating the behavior of the system for different values of the two variables $D(=1/\theta)$ and $S_f$ for a given set of experimentally determined growth parameters. Death and maintenance were not taken into account. With some combinations of $D$ and $S_f$, a steady-state coexistence without oscillations is possible. However, in a wide range of operating conditions the system will show sustained oscillations. These were actually found experimentally by Tsuchiya *et al.* (1972) for the system *Dictyostelium discoideum–E. coli*, and by Jost *et al.* (1973a) for the systems *Tetrahymena pyriformis–E. coli*, and *T. pyriformis–A. vinelandii* (Fig. 16). However, the minimum bacterial densities predicted by the model were very much lower than those observed experimentally in the latter cases. To account for the discrepancy between experiment and theory, Jost *et al.* (1973a,b) replaced Monod's model by a "multiple saturation" model. Bonomi and Fredrickson (1976) subsequently pointed out that there is no need to reject the Monod model when it is assumed that wall growth

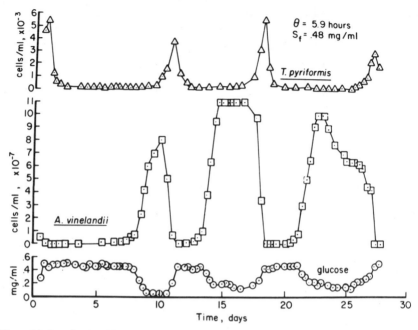

**Figure 16.** Sustained oscillations in the glucose–*Azotobacter*–*Tetrahymena* food chain; holding time 5.9 hr. *Azotobacter* grew under glucose limitation (Jost *et al.*, 1973a).

occurs with bacteria which escape the predator and which continually re-inoculate the culture liquid. Equation (24) was modified to take wall growth into account, and an operating diagram for this system was made. Bonomi and Fredrickson (1976) also gave a tentative explanation for the fact that in the *D. discoideum–E. coli* system (Tsuchiya *et al.*, 1972) Monod's model without wall growth fits the experimental data quite well. It was assumed that in that case the amoeboid cells of *Dictyostelium* keep the wall of the chemostat free of bacteria.

Wall growth also played an important role in a study of grazing of ciliates on blue-green algae (Bader *et al.*, 1976a). The unicellular blue-green alga *Anacystis nidulans* was grown under energy (light) limitation in a single-stage chemostat. Once a steady state was reached, cysts of the ciliate *Colpoda steinii* were introduced into the culture. Results of a typical experiment are shown in Fig. 17. A period of rapid increase of the ciliate population was followed by encystment and decline. The rate of decline was considerably greater than the washout rate, which was due to attachment of cysts to the vessel walls. Algal cells then began to stick to the attached cysts. After a subsequent increase of both population densities in the liquid phase, a steady state was eventually reached. Experiments of this type were carried out at two dilution rates and with three light intensities. In all cases, a steady state was eventually reached. At the lowest dilution rate applied (0.02/hr), the steady state initially reached appeared to have a transitional character. A week after the steady state was established, both population densities began to increase until a new steady-state level was reached. Further analysis of this complicated system is needed to assess the importance of various factors that bring about steady-state coexistence and of those responsible for its maintenance. It seems beyond doubt, however, that transfer organisms between liquid and solid phase does play a role in this particular predator–prey system.

A factor which exerts a destabilizing effect on the *Colpoda–Anacytis* relation was described by Bader *et al.* (1976b). It was found that the length of the period in which the algal culture was grown in steady state before the introduction of the ciliate affected the extent of decrease of the population density of *Anacystis* and also that of the subsequent decline of the *Colpoda* population. Because of shading, the algal cells were initially exposed to a relatively low light intensity. The rapid decrease of the algal population density due to grazing then caused a rapid increase in average light intensity. Bader *et al.* (1976b) observed that the algal population continued to decline after the bulk of the ciliate population had encysted

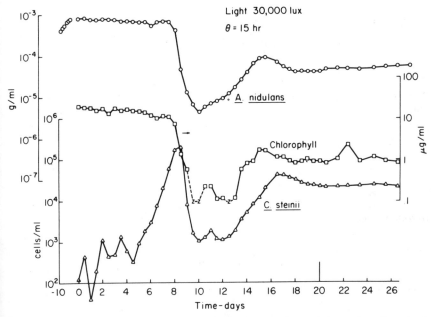

**Figure 17.** Result of inoculation of steady-state chemostat culture of *Anacystis* with cysts of *Colpoda*; incident light intensity was 30,000 lux. The symbol "z" indicates that the chlorophyll concentration was smaller than could be detected by the analytical technique used (Bader *et al.*, 1976).

in cases when the experiment was started with algal cells adapted to low light intensity. The longer the algal culture was grown in steady state before being inoculated with *Colpoda*, the lower was the level to which both populations decreased. It was shown that at the end of the grazing period, the algae did not grow anymore for some time, and it was assumed that this was due to the sudden exposure to higher light intensities. Experimental evidence which supported this assumption was obtained in dilution experiments with pure cultures of *Anacystis*. Light shock therefore seems to be responsible for the observed lag in growth response of *Anacystis* after the grazing pressure of the ciliate was released, as well as for a greater degree of encystment of the ciliate.

The experiments of Bader *et al.* (1976a,b) clearly illustrate how complicated even a simple two-membered system may be and how the chemostat can be applied in tracing factors involved in microbial interactions.

### 5.3. Other Interactions

Release of a metabolite by one organism which can be used by another is a phenomenon that frequently occurs in nature. A well-known example of a metabolic product that may be released in considerable amount is glycolate, which is produced during photorespiration by algae (Tolbert, 1974). Recent reports show that bacteria in which the Calvin cycle is operating may also release glycolate. Small amounts were produced by *Hydrogenomonas eutropha* (= *Alcaligenes eutropha*) when grown autotrophically (Codd *et al.*, 1975). Cohen and Kuenen (1976) grew the obligately chemolithotrophic *Thiobacillus neapolitanus* in a thiosulfate-limited chemostat and observed that organic compounds were excreted. One of the major components was glycolate, and its excretion was oxygen dependent. When the dilution rate was fixed at 0.07/hr, the total excretion at low $pO_2$ (5% of air saturation) was 4.9% of the total $CO_2$-C fixed. At high $pO_2$ (70% of air saturation), this value was 11.5%. At a fixed $pO_2$ (60% of air saturation), the excretion was growth-rate dependent and varied from 8.2–15% of total $CO_2$-C fixed. Wright (1975) showed that under natural conditions, chemoorganotrophic bacteria prevent glycolate from accumulating by mineralizing most of it to $CO_2$, and therefore commensalistic relations between glycolate-producing and -consuming organisms should be quite common.

The excretion of acetate under aerobic conditions by *Aerobacter aerogenes* was studied by Cooney and Mateles (1971) and Cooney *et al.* (1976). As is shown in Fig. 18, the excretion of acetate is a function of growth rate and nutrient limitation. Carbon-limited cells release acetate in relatively small amounts and only at high growth rates. Acetate excretion was highest in nitrogen-limited cells, whereas intermediate values were found for phosphorus limitation and various double nutrient limitations.

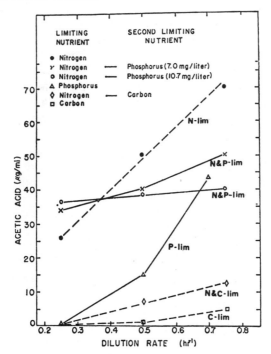

**Figure 18.** Excretion of acetic acid by aerobically grown *Aerobacter aerogenes* in variously limited chemostat cultures (Cooney and Mateles, 1971).

The latter conditions pertain to a situation in which one nutrient is growth limiting, while a second one is available to the cell in an amount less than it is capable of utilizing (cf. Cooney and Wang, 1976). Such conditions undoubtedly occur frequently in nature but have been very rarely applied in continuous culture studies. Generally speaking, only few analyses have been made of the effect of specific growth rate and kind of nutrient limitation on the release of metabolites of organisms grown aerobically, and very little is known about the various relationships with other microbes to which these excretions may give rise. A well-studied case of release of metabolic products in relation to growth rate and kind of nutrient limitation was reported by Neijssel and Tempest (1975a,b).

The excretion of metabolic products which may serve as a carbon and/or energy source for other organisms is a common feature of anaerobic metabolism. Cappenberg (1975) studied the utilization by *Methanobacterium* sp. of acetate produced by *Desulfovibrio desulfuricans*. The organisms were grown separately in a single-stage chemostat and in both cases the carbon and energy source (acetate for the methanogenic organism and lactate for *Desulfovibrio*) was growth limiting. The effluents of both cultures were fed into a third vessel in which additional growth of *Methanobacterium* on acetate produced by *Desulfovibrio* was observed. However, *Desulfovibrio* produces sulfide as well. By increasing the lactate level in the feed of the *Desulfovibrio* culture, the sulfide level at which methane production by *Methanobacterium* was completely inhibited in the mixed culture was found to be at $pS^{2-} = 10.5$. Field studies (Cappenberg, 1974a,b; Cappenberg and Prins, 1974) on the vertical distribution of *Desulfovibrio* and methanogenic bacteria in a lake sediment showed that maximum numbers of the latter occurred at depths of 3 to 6 cm. The redox potential at this level was $-250$ to $-300$ mV, and the $pS^{2-}$ value was 13.4 to 14.0. Maximum numbers of sulfate-reducing bacteria were found at depths of 0 to 2 cm, where redox potential values were $-100$ to $-150$ mV, and $pS^{2-}$ values of 11.4 to 12.2 were found. The mixed culture study with the chemostat indicates that the difference in vertical distribution may at least in part be explained in terms of sensitivity of methanogenic bacteria to sulfide.

Among the interactions between anaerobic organisms, those in which production and consumption of molecular hydrogen are involved are of particular interest. Ianotti *et al.* (1973) made a continuous culture study of interspecies hydrogen transfer with *Ruminococcus albus* and *Vibrio succinogenes*. The fermentation products of a pure culture of *R. albus* in a glucose-limited chemostat were ethanol, acetate and $H_2$. *Vibrio succinogenes* can obtain energy for growth by oxidizing $H_2$ coupled to reduction of fumarate to succinate, but it cannot grow with glucose, ethanol, or acetate, either in the presence or absence of fumarate. A mixed culture of the two organisms was grown in a glucose-limited chemostat, and fumarate was included in the feed. The fermentation products of the mixture were acetate and succinate. Thus, the presence of $H_2$-consuming *V. succinogenes* caused a shift in the fermentation pattern of *R. albus*. Ethanol, a major fermentation product in pure cultures of *R. albus*, was not produced in the mixture. Instead, an almost equivalent amount of extra acetate was formed, and from the amount of succinate produced it could be deduced that in the mixed culture *R. albus* showed an increase in $H_2$ production almost equivalent to the electron deficit caused by the disappearance of ethanol and the increase in acetate production. These results support the hypothesis of Hungate (1966), who tentatively explained the relatively low levels of reduced fermentation products, such as ethanol and lactate, in the rumen by assuming a shift towards $H_2$ production in the presence of $H_2$-consuming methanogenic bacteria.

Studies on the production of single-cell protein from methane showed that as a rule the cell yield is higher when mixed cultures are used (Harrison, 1976). A continuous culture study of a stable mixture of four bacterial species supplied with methane as the only carbon source was made by Wilkinson *et al.* (1974). The mixture, consisting of species of the genera

*Pseudomonas, Hyphomicrobium, Flavobacterium,* and *Acinetobacter,* was studied under methane and oxygen limitation. *Pseudomonas* was the only organism that could utilize methane as a carbon and energy source, and it always formed the dominant part (ca. 90%) of the mixed population. However, in liquid culture it required the presence of *Hyphomicrobium* for sustained growth on methane. The reason was that *Pseudomonas* released an inhibitory amount of methanol, which was removed by *Hyphomicrobium.* Even under $O_2$ limitation, *Hyphomicrobium* was methanol-limited as the growth medium contained nitrate, and this organism changes to denitrification at dissolved oxygen tensions below ca. 0.04 atm. In fact, anaerobic enrichment cultures with methanol as electron donor and nitrate as electron acceptor quite specifically yield *Hyphomicrobium* species (Attwood and Harder, 1972). *Flavobacterium* and *Acinetobacter,* which formed maximally 2.5% of the total cell numbers in the mixture, could not utilize either methane or methanol and were supposed to grow on complex organic compounds released by one or both of the other species. A simple mathematical model with which qualitative predictions could be made of the main interactions in the mixed culture under $O_2$-limiting conditions was developed by Wilkinson *et al.* (1974).

A theoretical study of the behavior in the chemostat of a mixed population showing complementary metabolism was made by Meyer *et al.* (1975). The situation considered is shown in Fig. 19. It was shown that a purely mutualistic relation in which the specific growth rate of organism $B_1$ is limited by product $s_1$ and that of $B_2$ by $s_2$ cannot give a stable steady state. A small perturbation will either lead to washout or to unbounded growth of both organisms. In the latter case, the organisms may encounter another growth-limiting substrate $(s_3)$ for which they compete and which is supplied in the culture feed. This situation was analyzed for the case in which the relation between the specific growth rates of $B_1$ and $B_2$ and the concentration of $s_3$ show crossing curves (cf. Fig. 13b). A

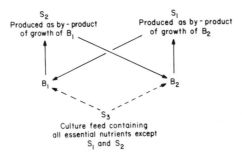

**Figure 19.** Scheme of complementary metabolism of organisms $B_1$ and $B_2$ grown in chemostat culture, analyzed by Meyer *et al.* (1975).

coexistence steady state which was stable with respect to a wide range of perturbations was possible when: (i) $s_3$ limits $B_1$ and $s_2$ limits $B_2$; (ii) $s_3$ limits $B_2$ and $s_1$ limits $B_1$; and (iii) $s_3$ limits both $B_1$ and $B_2$. The latter case is only stable when in addition either $B_1$ is also limited by $s_1$, or $B_2$ by $s_2$.

The interesting result of this analysis thus is that the combination of two intrinsically unstable interactions (competition and pure mutualism) may give stable steady states in the chemostat. As pointed out by Meyer et al. (1975), the reason for the stability is that an excess of either $s_1$ or $s_2$ always builds up when a steady state is approached. There were no cases in which $s_1, s_2$, and $s_3$ were present in the culture in very low concentrations simultaneously at steady state. It was also shown that the $\mu$–$s_3$ curves of organisms $B_1$ and $B_2$ need not necessarily cross to obtain a coexistence which is stable with respect to small perturbations.

Meyer et al. (1975) further showed that it is not only competition which may stabilize a purely mutualistic relation. If $s_1$ or $s_2$ or both substrates become toxic to their consumers above a certain concentration level, then coexistence may also occur, for instance, when the specific growth rate of $B_1$ is limited by scarcity of $s_1$ and that of $B_2$ by overabundance of $s_2$.

A point of practical importance is that several organisms which may potentially coexist under certain conditions may never reach steady-state coexistence in the chemostat if the initial cultural conditions are not properly chosen.

Megee et al. (1972) studied interrelations between *Saccharomyces cerevisiae* and *Lactobacillus casei*. In a glucose-limited chemostat with all growth factors in excess, the result of competition depended on the dilution rate applied, as the $\mu$–$s$ relations for glucose showed crossing curves (cf. Fig. 13b). However, if riboflavin was omitted from the feed, then *L. casei* depended on *Saccharomyces* for its riboflavin supply. Then steady-state coexistence could be obtained, showing thereby a combination of commensalism and competition. Under these conditions, both populations showed a linear response to the glucose concentration in the feed. When this concentration was decreased, *Lactobacillus* was washed out before *Saccharomyces*; this was directly related to the amount of riboflavin produced by the yeast population. In addition to the above factors, the pH also affected the relation, as *Saccharomyces* showed both a higher yield and specific growth rate at low pH. Thus, when the pH of the feed was relatively high (6.5) and no riboflavin was added to the culture, both organisms were competing for glucose and at the same time were of benefit to each other by producing riboflavin and lowering the pH. The results of Megee et al. (1972) clearly show how the type of interaction between two organisms may depend on their abiotic environment.

In conclusion, the results of mixed culture studies with the chemostat clearly show that progress depends on the harmonious interaction of experiment and theory. The better the biotic and abiotic components chosen mimic those actually occurring in natural situations, the greater the ecological importance of future studies will be.

ACKNOWLEDGMENTS

The author is indebted to A. Bisschop, W. N. Konings, O. M. Neijssel, and D. W. Tempest for providing manuscripts of papers that are in press, and thanks M. R. Droop, W. Harder, W. N. Konings, J. G. Kuenen, and D. W. Tempest for valuable comments and Mrs. J. W. Schröder-ter Avest for technical assistance in the preparation of the manuscript.

# References

Aiking, H., and Tempest, D. W., 1976, Growth and physiology of *Candida utilis* NCYC 321 in potassium-limited chemostat culture, *Arch. Microbiol.* **108**:117.

Anonymous, 1961, *Continuous Culture of Micro-organisms*, S.C.I. Monograph No. 12, Society of Chemical Industry, London.

Attwood, M. M., and Harder, W., 1972, A rapid and specific enrichment procedure for *Hyphomicrobium* spp., *Antonie van Leeuwenhoek J. Microbiol. Serol.* **38**:369.

Bader, F. G., Tsuchiya, H. M., and Fredrickson, A. G., 1976a, Grazing of ciliates on blue-green algae: Effects of ciliate encystment and related phenomena, *Biotechnol. Bioeng.* **18**:311.

Bader, F. G., Tsuchiya, H. M., and Fredrickson, A. G., 1976b, Grazing of ciliates on blue-green algae: Effects of light shock on the grazing relation and on the algal population, *Biotechnol. Bioeng.* **18**:333.

Baker, K., 1968, Low cost continuous culture apparatus, *Lab. Pract.* **17**:817.

Bauchop, T., and Elsden, S. R., 1960, The growth of micro-organisms in relation to their energy supply, *J. Gen. Microbiol.* **23**:457.

Bisschop, A., and Konings, W. N., 1976, Reconstitution of reduced nicotinamide adenine dinucleotide oxidase activity with menadione in membrane vesicles from the menaquinone-deficient *B. subtilis aro D*. Relation between electron transfer and active transport, *Eur. J. Biochem.* **67**:357.

Bonomi, A., and Fredrickson, A. G., 1976, Protozoan feeding and bacterial wall growth, *Biotechnol. Bioeng.* **18**:239.

Brooks, J. D., and Meers, J. L., 1973, The effect of discontinuous methanol addition on the growth of a carbon-limited culture of *Pseudomonas*, *J. Gen. Microbiol.* **77**:513.

Bungay, III, H. R., and Bungay, M. L., 1968, Microbial interactions in continuous culture, *Adv. Appl. Microbiol.* **10**:269.

Caperon, J., 1968, Population growth response of *Isochrysis galbana* to nitrate variation at limiting concentrations, *Ecology* **49**:866.

Caperon, J., and Meyer, J., 1972a, Nitrogen limited growth of marine phytoplankton. I. Changes in population characteristics with steady-state growth rate, *Deep-Sea Res.* **19**:601.

Caperon, J., and Meyer, J., 1972b, Nitrogen limited growth of marine phytoplankton. II. Uptake kinetics and their role in nutrient limited growth of phytoplankton, *Deep-Sea Res.* **19**:619.

Cappenberg, Th. E., 1974a, Interrelations between sulfate-reducing and methane-producing bacteria in bottom deposits of a fresh-water lake. I. Field observations, *Antonie van Leeuwenhoek J. Microbiol. Serol.* **40**:285.

Cappenberg, Th. E., 1974b, Interrelations between sulfate-reducing and methane-producing bacteria in bottom deposits of a fresh-water lake. II. Inhibition experiments, *Antonie van Leeuwenhoek J. Microbiol. Serol.* **40**:297.

Cappenberg, Th. E., 1975, A study of mixed continuous cultures of sulfate-reducing and methane-producing bacteria, *Microbial Ecol.* **2**:60.

Cappenberg, Th. E., and Prins, R. A., 1974, Interrelations between sulfate-reducing and methane-producing bacteria in bottom deposits of a fresh-water lake. III. Experiments with $^{14}$C-labeled substrates, *Antonie van Leeuwenhoek J. Microbiol. Serol.* **40**:457.

Codd, G. A., Bowien, B., and Schlegel, H. G., 1975, Glycollate formation and ribulose 1,5-disphosphate carboxylase/oxygenase in a facultative chemolitho-autotrophic bacterium, *Proc. Soc. Gen. Microbiol.* **3**:8.

Cohen, Y., and Kuenen, J. G., 1976, Growth yields and excretion products of *Thiobacillus neapolitanus* grown in the chemostat, in: *Abstracts of the Annual Meeting of the American Society for Microbiology*, 76th Annual Meeting, Atlantic City, N. J., p. 120.

Cooney, C. L., and Mateles, R. I., 1971, Fermentation kinetics, in: *Recent Advances in Microbiology* (A. Pérez-Miravete and Dionisio Peláez, eds.), pp. 441–449, Mexico.

Cooney, C. L., and Wang, D. I. C., 1976, Transient response of *Enterobacter aerogenes* under a dual nutrient limitation in a chemostat, *Biotechnol. Bioeng.* **18**:189.

Cooney, C. L., Wang, D. I. C., and Mateles, R. I., 1976, Growth of *Enterobacter aerogenes* in a chemostat with double nutrient limitations, *Appl. Environ. Microbiol.* **31**:91.

Dean, A. C. R., Pirt, S. J., and Tempest, D. W. (eds.), 1972, *Environmental Control of Cell Synthesis and Function, Proceedings of the 5th International Symposium on the Continuous Culture of Microorganisms*, Academic Press, London.

Dean, A. C. R., Ellwood, D. C., Evans, C. G. T., and Melling J. (eds.), 1976, *Application and New Fields, Proceedings of the 6th International Symposium on the Continuous Culture of Microorganisms*, Ellis-Horwood, Chichester, England.

Dicks, J. W., and Tempest, D. W., 1966, The influence of temperature and growth rate on the quantitative relationship between potassium, magnesium, phosphorus and ribonucleic acid contents of *Aerobacter aerogenes* growing in a chemostat. *J. Gen. Microbiol.* **45**:547.

Downs, A. J., and Jones, C. W., 1975, Energy conservation in *Bacillus megaterium, Arch. Microbiol.* **105**:159.

Droop, M. R., 1968, Vitamin B$_{12}$ and marine ecology. IV. The kinetics of uptake, growth and inhibition in *Monochrysis lutheri, J. Mar. Biol. Assoc. U.K.* **48**:689.

Droop, M. R., 1970, Vitamin B$_{12}$ and marine ecology. V. Continuous culture as an approach to nutritional kinetics, *Helgoländer Wiss. Meeresunters.* **20**:629.

Droop, M. R., 1973a, Some thoughts on nutrient limitation in algae, *J. Phycol.* **9**:264.

Droop, M. R., 1973b, Nutrient limitation in osmotrophic protista, *Amer. Zool.* **13**:209.

Droop, M. R., 1974, The nutrient status of algal cells in continuous culture, *J. Mar. Biol. Assoc. U.K.* **54**:825.

Droop, M. R., 1976, Towards a model for phytoplankton growth, *Proc. Soc. Gen. Microbiol.* **3**:105.

Evans, C. G. T., Herbert, D., and Tempest, D. W., 1970, The continuous cultivation of micro-organisms. 2. Construction of a chemostat, in: *Methods in Microbiology*, Vol. 2 (J. R. Norris and D. W. Ribbons, eds.), pp. 277–327, Academic Press, London/New York.

Fencl, Z., 1966, A theoretical analysis of continuous culture systems, in: *Theoretical and Methodological Basis of Continuous Culture of Microorganisms* (I. Málek and Z. Fencl, eds.), pp. 67–153, Publishing House of the Czechoslovak Academy of Sciences, Prague, and Academic Press, New York/London.

Forrest, W. W., and Walker, D. J., 1971, The generation and utilization of energy during growth, *Adv. Microb. Physiol.* **5**:213.

Fuhs, G. W., 1969, Phosphorus content and rate of growth in the diatoms *Cyclotella nana* and *Thalassiosira fluviatilis*, *J. Phycol.* **5**:312.

van Gemerden, H., 1974, Coexistence of organisms competing for the same substrate: an example among the purple sulfur bacteria, *Microb. Ecol.* **1**:104.

Gunsalus, I. C., and Shuster, C. W., 1961, Energy-yielding metabolism in bacteria, in: *The Bacteria*, Vol. 2 (I. C. Gunsalus and R. Y. Stanier, eds.), pp. 1–58, Academic Press, New York/London.

Hansen, T. A., and van Gemerden, H., 1972, Sulfide utilization by purple nonsulfur bacteria, *Arch. Mikrobiol.* **86**:49.

Harder, W., and van Dijken, J. P., 1976, Theoretical considerations on the relation between energy production and growth of methane-utilizing bacteria, in: *Proceedings of the Symposium of Microbial Production and Utilization of Gases*, E. Goltze KG, Göttingen, Germany.

Harder, W., and Veldkamp, H., 1971, Competition of marine psychrophilic bacteria at low temperatures, *Antonie van Leeuwenhoek J. Microbiol. Serol.* **37**:51.

Harder, W., Visser, K., and Kuenen, J. G., 1974, Laboratory fermenter with an improved magnetic drive, *Lab. Pract.* **23**:644.

Harrison, D. E. F., 1976, Production of protein from methane by the use of defined mixed bacterial culture, in: *Application and New Fields, Proceedings of the 6th International Symposium on the Continuous Culture of Microorganisms* (A. C. R. Dean, D. C. Ellwood, C. G. T. Evans, and J. Melling, eds.), Ellis-Horwood, Chichester, England.

Healey, F. P., 1973, The inorganic nutrition of algae, from an ecological viewpoint, *CRC Crit. Rev. Microbiol.* **3**:69.

Hempfling, W. P., and Mainzer, S. E., 1975, Effects of varying the carbon source limiting growth on yield and maintenance characteristics of *Escherichia coli* in continuous culture, *J. Bacteriol.* **123**:1076.

Herbert, D., 1959, Some principles of continuous culture, in: *Recent Progress in Microbiology* (G. Tunevall, ed.), pp. 381–396, Almqvist & Wiksell, Stockholm.

Herbert, D., Elsworth, R., and Telling, R. C., 1956, The continuous culture of bacteria; a theoretical and experimental study, *J. Gen. Microbiol.* **14**:601.

Hungate, R. E., 1966, *The Rumen and its Microbes*, Academic Press, New York/London.

Ianotti, E. L., Kafkewitz, D., Wolin, M. J., and Bryant, M. P., 1973, Glucose fermentation products of *Ruminococcus albus* grown in continous culture with *Vibrio succinogenes*: Changes caused by interspecies transfer of $H_2$, *J. Bacteriol.* **114**:1231.

Jannasch, H. W., 1967, Enrichments of aquatic bacteria in continuous culture, *Arch. Mikrobiol.* **59**:165.

Jannasch, H. W., 1969, Estimations of bacterial growth rates in natural waters, *J. Bacteriol.* **99**:156.

Jannasch, H. W., and Mateles, R. I., 1974, Experimental bacterial ecology studied in continuous culture, *Adv. Microb. Physiol.* **11**:165.

Jannasch, H. W., and Pritchard, P. H., 1972, The role of inert particulate matter in the activity of aquatic microorganisms, *Mem. Ist. Ital. Idrobiol., Suppl.* **29**:289.

Jerusalimsky, N. D., 1967, Bottle-necks in metabolism as growth rate controlling factors, in: *Microbial Physiology and Continuous Culture, Proceedings of the 3rd International Symposium* (E. O. Powell, C. G. T. Evans, R. E. Strange, and D. W. Tempest, eds.), pp. 23–33, Her Majesty's Stationery Office, London.

Jost, J. L., Drake, J. F., Fredrickson, A. G., and Tsuchiya, H. M., 1973a, Interactions of *Tetrahymena pyriformis, Escherichia coli, Azotobacter vinelandii*, and glucose in a minimal medium, *J. Bacteriol.* **113**:834.

Jost, J. L., Drake, J. F., Tsuchiya, H. M., and Fredrickson, A. G., 1973b, Microbial food chains and food webs, *J. Theor. Biol.* **41**:461.

Kubitschek, H. E., 1970, *Introduction to Research with Continuous Cultures*, Prentice-Hall, Englewood Cliffs, N.J.

Kuenen, J. G., Boonstra, J., Schröder, H. G. J., and Veldkamp, H., 1976, Competition for inorganic substrates among chemoorganotrophic and chemolithotrophic bacteria, *Microb. Ecol.* **3**: in press.

de Kwaadsteniet, J. W., Jager, J. C., and Stouthamer, A. H., 1976, A quantitative description of heterotrophic growth in micro-organisms, *J. Theor. Biol.* **57**:103.

Luscombe, B. M., 1974, The effect of dropwise addition of medium on the yield of carbon limited cultures of *Arthrobacter globiformis*, *J. Gen. Microbiol.* **83**:197.

Málek, I. (ed.), 1958, *Continuous Cultivation of Microorganisms, Proceedings of the 1st International Symposium*, Czechoslovak Academy of Sciences, Prague.

Málek, I., Beran, K., and Hospodka, J. (eds.), 1964, *Continuous Cultivation of Microorganisms, Proceedings of the 2nd International Symposium*, Academic Press, New York/London.

Málek, I., Beran, K., Fencl, Z., Munk, V., Řičica, J., and Smrčková, H. (eds.), 1969, *Continuous Cultivation of Microorganisms, Proceedings of the 4th International Symposium*, Academia, Prague.

Marr, A. G., Nilson, E. H., and Clark, D. J., 1963, The maintenance requirement of *Escherichia coli*, *Ann. N.Y. Acad. Sci.* **102**:536.

Meers, J. L., 1973, Growth of bacteria in mixed cultures, *CRC Crit. Rev. Microbiol.* **2**:139.

Megee, III, R. D., Drake, J. F., Fredrickson, A. G., and Tsuchiya, H. M., 1972, Studies in intermicrobial symbiosis. *Saccharomyces cerevisiae* and *Lactobacillus casei*, *Can. J. Microbiol.* **18**:1733.

Meyer, J. S., Tsuchiya, H. M., and Fredrickson, A. G., 1975, Dynamics of mixed populations having complementary metabolism, *Biotechnol. Bioeng.* **17**:1065.

Monod, J., 1942, *Recherches sur la Croissance des Cultures Bactériennes*, Hermann, Paris.

Monod, J., 1950, La technique de culture continue; théorie et applications, *Ann. Inst. Pasteur* **79**:390.

Müller, H., 1972, Wachstum und Phosphatbedarf von *Nitzschia actinasteroides* (Lemm.). V. Goor in statischer und homokontinuierlicher Kulture unter Phosphatlimitierung, *Arch. Hydrobiol.* **38**: Suppl. 399.

Neijssel, O. M., and Tempest, D. W., 1975a, Production of gluconic acid and 2-ketogluconic acid by *Klebsiella aerogenes* NCTC 418, *Arch. Microbiol.* **105**:183.

Neijssel, O. M., and Tempest, D. W., 1975b, The regulation of carbohydrate metabolism in *Klebsiella aerogenes* NCTC 418 organisms, growing in chemostat culture, *Arch. Microbiol.* **106**:251.

Neijssel, O. M., and Tempest, D. W., 1976a, Bioenergetic aspects of aerobic growth of *Klebsiella aerogenes* NCTC 418 in carbon-limited and carbon-sufficient chemostat culture, *Arch. Microbiol.* **107**:215.

Neijssel, O. M., and Tempest, D. W., 1976b, The role of energy-spilling reactions in the growth of *Klebsiella aerogenes* NCTC 418 in aerobic chemostat culture, *Arch. Microbiol.* **110**:305.

Novick, A., and Szilard, L., 1950, Experiments with the chemostat on spontaneous mutations of bacteria, *Proc. Nat. Acad. Sci., Wash.* **36**:708.

Paasche, E., 1973a, Silicon and the ecology of marine plankton diatoms. I. *Thalassiosira pseudomona* (*Cyclotella nana*) grown in a chemostat with silicate as limiting nutrient, *Mar. Biol.* **19**:117.

Paasche, E., 1973b, Silicon and the ecology of marine plankton diatoms. II. Silicate uptake kinetics in five diatom species, *Mar. Biol.* **19**:262.

Payne, W. J., 1970, Energy yields and growth of heterotrophs, *Annu. Rev. Microbiol.* **24**:17.

Perry, M. J., 1976, Phosphate utilization by an oceanic diatom in phosphorus-limited chemostat culture and in the oligotrophic waters of the central North Pacific, *Limnol. Oceanogr.* **21**:88.

Phillipson, J., 1973, The biological efficiency of protein production by grazing and other land-based systems, in: *The Biological Efficiency of Protein Production* (J. G. W. Jones, ed.), pp. 217–237, Cambridge University Press, London.

Pirt, S. J., 1965, The maintenance energy of bacteria in growing cultures, *Proc. R. Soc. Lond. B. Biol. Sci.* **163**:224.

Pirt, S. J., 1975, *Principles of Microbe and Cell Cultivation*, Blackwell Scientific, Oxford/London/Edinburgh/Melbourne.

Postgate, J. R., 1973, The viability of very slow-growing populations: A model for the natural ecosystem, in: *Modern Methods in the Study of Microbial Ecology* (T. Rosswall, ed.), *Bull. Ecol. Res. Comm. (Stockholm)* **17**:287.

Powell, E. O., Evans, C. G. T., Strange, R. E., and Tempest, D. W., (eds.), 1967, *Microbial Physiology and Continuous Culture, Proceedings of the 3rd International Symposium*, Her Majesty's Stationery Office, London.

Řičica, J., 1966, Technique of continuous laboratory cultivations, in: *Theoretical and Methodological Basis of Continuous Culture in Microorganisms* (I. Málek and Z. Fencl, eds.), pp. 155–313, Publishing House of the Czechoslovak Academy of Sciences, Prague, and Academic Press, New York/London.

Rogers, P. J., and Stewart, P. R., 1974, Energetic efficiency and maintenance energy characteristics of *Saccharomyces cerevisiae* (wild type and petite) and *Candida parapsilosis* grown aerobically and microaerobically in continuous culture, *Arch. Microbiol.* **99**:25.

Schulze, K. L., and Lipe, R. S., 1964, Relationship between substrate concentration, growth rate, and respiration rate of *Escherichia coli* in continuous culture, *Arch. Mikrobiol.* **48**:1.

Sinclair, C. G., and Topiwala, H. H., 1970, Model for continuous culture which considers the viability concept, *Biotechnol. Bioeng.* **12**:1069.

Steele, J. H., 1975, The structure of marine ecosystems, Harvard University Press, Cambridge, Mass./London.

Stouthamer, A. H., 1973, A theoretical study on the amount of ATP required for synthesis of microbial cell material, *Antonie van Leeuwenhoek J. Microbiol. Serol.* **39**:545.

Stouthamer, A. H., 1976, *Yield Studies in Microorganisms*, Meadowfield Press, Durham, England.

Stouthamer, A. H., and Bettenhaussen, C. W., 1973, Utilization of energy for growth and maintenance in continuous and batch cultures of microorganisms. A reevaluation of the method for the determination of ATP production by measuring molar growth yields, *Biochim. Biophys. Acta* **301**:53.

Stouthamer, A. H., and Bettenhaussen, C. W., 1975, Determination of the efficiency of oxidative phosphorylation in continuous cultures of *Aerobacter aerogenes*, *Arch. Microbiol.* **102**:187.

Tait, R. V., 1972, *Elements of Marine Ecology*, Butterworths, London.

Tempest, D. W., 1970a, The place of continuous culture in microbiological research, *Adv. Microb. Physiol.* **4**:223.

Tempest, D. W., 1970b, The continuous cultivation of micro-organisms. 1. Theory of the chemostat, in: *Methods in Microbiology*, Vol. 2 (J. R. Norris and D. W. Ribbons, eds.), pp. 259–276, Academic Press, London/New York.

Tempest, D. W., and Dicks, J. W., 1967, Inter-relationships between potassium, magnesium, phosphorus and ribonucleic acid in the growth of *Aerobacter aerogenes* in a chemostat, in: *Microbial Physiology and Continuous Culture, Proceedings of the 3rd International Symposium* (E. O. Powell, C. G. T. Evans, R. E. Strange, and D. W. Tempest, eds.), pp. 140–154, Her Majesty's Stationery Office, London.

Tempest, D. W., Dicks, J. W., and Hunter, J. R., 1966, The inter-relationship between potassium, magnesium and phosphorus in potassium-limited chemostat cultures of *Aerobacter aerogenes*. *J. Gen. Microbiol.* **45**:135.

Tempest, D. W., Herbert, D., and Phipps, P. J., 1967, Studies on the growth of *Aerobacter aerogenes* at low dilution rates in a chemostat, in: *Microbial Physiology and Continuous Culture, Proceedings of the 3rd International Symposium* (E. O. Powell, C. G. T. Evans, R. E. Strange, and D. W. Tempest, eds.), pp. 240–254, Her Majesty's Stationery Office, London.

Tett, P., Cottrell, J. C., Trew, D. O., and Wood, B. J. B., 1975, Phosphorus quota and the chlorophyll: Carbon ratio in marine phytoplankton, *Limnol. Oceanogr.* **20**:587.

Tolbert, N. E., 1974, Photorespiration, in: *Algal Physiology and Biochemistry* (W. D. P. Stewart, ed.), pp. 474–504, Blackwell Scientific, Oxford/London/Edinburgh/Melbourne.

Tsuchiya, H. M., Drake, J. F., Jost, J. L., and Fredrickson, A. G., 1972, Predator–prey interactions of *Dictyostelium discoideum* and *Escherichia coli* in continuous culture, *J. Bacteriol.* **110**:1147.

Veldkamp H., 1970, Enrichment cultures of prokaryotic organisms, in: *Methods in Microbiology*, Vol. 3A, (J. R. Norris and D. W. Ribbons, eds.), pp. 305–361, Academic Press, London/New York.

Veldkamp, H., 1976a, *Continuous Culture in Microbial Physiology and Ecology*, Meadowfield Press, Durham, England.

Veldkamp, H., 1976b, Mixed culture studies with the chemostat, in: *Application and New Fields, Proceedings of the 6th International Symposium of Continuous Culture of Microorganisms* (A. C. R. Dean, D. C. Ellwood, C. G. T. Evans, and J. Melling, eds.), Ellis-Horwood Chichester, England.

Veldkamp, H., and Jannasch, H. W., 1972, Mixed culture studies with the chemostat, in: *Environmental Control of Cell Synthesis and Function, Proceedings of the 5th International Symposium on the Continuous Culture of Microorganisms* (A. C. R. Dean, S. J. Pirt, and D. W. Tempest, eds.), pp. 105–123, Academic Press, London/New York.

Veldkamp, H., and Kuenen, J. G., 1973, The chemostat as a model system for ecological studies, in: *Modern Methods in the Study of Microbial Ecology* (T. Rosswall, ed.), *Bull. Ecol. Res. Comm. (Stockholm)* **17**:347.

van Verseveld, H. W., and Stouthamer, A. H., 1976, Oxidative phosphorylation in *Micrococcus denitrificans*. Calculation of the P/O ratio in growing cells, *Arch. Microbiol.* **107**:241.

Wagner, G. H., 1975, Microbial growth and carbon turnover, in: *Soil Biochemistry*, Vol. 3 (E. A. Paul, and A. D. McLaren, eds.), pp. 269–305, Marcel Dekker, New York.

Watson, T. G., 1970, Effects of sodium chloride on steady state growth and metabolism of *Saccharomyces cerevisiae*, *J. Gen. Microbiol.* **64**:91.

Wilkinson, T. G., Topiwala, H. H., and Hamer, G., 1974, Interactions in a mixed bacterial population growing on methane in continuous culture, *Biotechnol. Bioeng.* **16**:41.

Winogradsky, S., 1949, *Microbiologie du Sol*, Masson et Cie, Paris.

Wright, R. T., 1975, Studies on glycolic acid metabolism by freshwater bacteria, *Limnol. Oceanogr.* **20**:626.

# 3

# Microbial Transformations in the Phosphorus Cycle

## DENNIS J. COSGROVE

## 1. Introduction

As a result of the current spate of popular literature dealing with phosphorus in the aquatic environment, the expression "the phosphate problem" is firmly equated in most people's minds with the undesirable effects of the pollution of waterways and basins by phosphorus-containing effluents from urban areas and industrial sites. The importance of dealing with this problem cannot be decried, although some (Griffith, 1973) have deplored the rapid closing down of much research on industrial applications of phosphorus compounds which has unfortunately not been paralleled by a concomitant upsurge in the quantity and quality of such research on analytical methods, biochemistry, and microbiology as is still needed for a full understanding of the phosphorus cycle. Many of the pollution problems are in any case the result of the concentration of population in large conurbations and would undoubtedly have been less difficult to deal with if such socially and economically undesirable megalopolises had never evolved.

In the not too distant future, however, a "phosphate problem" of a much more serious nature is likely to become apparent, namely, a shortage of extractable mineral phosphate for the manufacture of fertilizers and other essential products. Forecasts of when serious shortages are likely to develop vary widely, being dependent on such factors as the level at which

**DENNIS J. COSGROVE** • CSIRO, Division of Plant Industry, Canberra, Australia.

the world's population is expected to stabilize, the expected increases in rate of usage, and estimates of the undiscovered reserves. The extremes given by Smith *et al.* (1972) range from 11,700 yr for a population of 1.8 billion using 1.7 million tons per year, to 17 yr for a population of 20 billion using 1170 million tons per year. If population (1.9%) and use of phosphate fertilizers (5.25%) continue to grow at constant rates, known reserves and supplies will be used up in about 100 yr—stranding a population of 20 billion. Thereafter, crop yields would be limited by the rate at which phosphate is released from insoluble forms in the soil and would suffice for a world population of 1–2 billion (Smith *et al.*, 1972).*

As is the case with the other important elements necessary for life, namely, carbon, hydrogen, oxygen, nitrogen, and sulfur, phosphorus circulates through air, land, water, and living systems in a complex interlocking biogeochemical cycle. In the case of phosphorus, movement in and out of the atmosphere is not important as the amounts involved are relatively small. The important pathways in the cycle are illustrated in Fig. 1, but inspection of this figure shows that phosphate eventually finds its way to the sea, where it is deposited as insoluble sediment and is thus removed from the cycle. The other "sink" of importance in the cycle is the accumulated "fixed" inorganic phosphorus in the soil. The total phosphorus content of many cultivated soils in industrialized countries has more than doubled since the introduction of chemical phosphatic fertilizers around the middle of the nineteenth century. The reason for this is that crops often remove as little as 10% of the applied phosphorus, the rest becoming rapidly "fixed" by chemical reactions with mineral oxides and other soil components (Larsen, 1967). Thus, of all the fertilizer applied over the past 120 yr a considerable proportion is still in the soil, in some soils not entirely unavailable to plants but in others apparently not available under ordinary conditions of cultivation. The economic exploitation of this reserve is not practicable at present, but stimulation of research in this field may well be an outcome of the inevitable increases in the cost of fertilizer phosphorus.

The processes involved in the transformations of phosphorus that are important in the cycling of this element can be summarized in the following equations.

| Organic P | $\rightarrow$ | Soluble inorganic P | (1) |
| Soluble inorganic P | $\rightarrow$ | Organic P | (2) |
| Insoluble inorganic P | $\rightarrow$ | Soluble inorganic P | (3) |
| Soluble inorganic P | $\rightarrow$ | Insoluble inorganic P | (4) |

---

* It is interesting to recall that long before much concern was expressed over possible phosphorus shortages, Aldous Huxley (1928) portrayed his scientifically minded Lord Edward Tantamount berating a political acquaintance for the latter's preoccupation with "votes and Bolshevism," when every year "a million tons of phosphorus pentoxide was running into the sea."

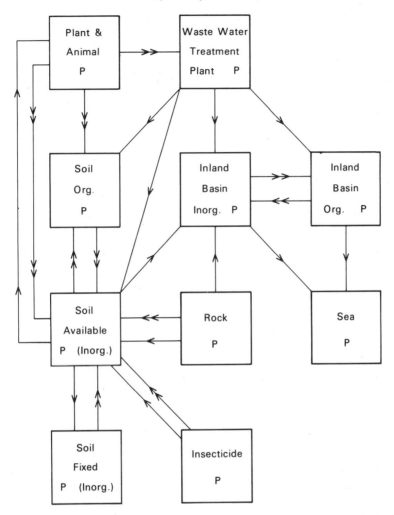

**Figure 1.** The phosphorus cycle. →→ Microbial processes. →— Chemical or physical processes.

The process designated in equation (1) is the hydrolysis of a phosphoric monoester (phosphomonoester) by a phosphoric monoester hydrolase (EC 3.1.3), an enzyme of the phosphatase group (Feder, 1973). Free inorganic orthophosphate is thus released (Fig. 2). Phosphoric diesters, such as nucleic acids, require preliminary hydrolysis by phosphoric diester hydrolases (EC 3.1.4) before their phosphorus can be finally released as inorganic orthophosphate by monoesterase action (Fig. 2). Equation (2) represents the formation of phosphate esters, such as nucleic acids, phospholipids, sugar phosphates etc., by the normal metabolic processes. Equation (3) represents the solubilization of inorganic phosphates, such as calcium phosphate and ferric phosphate, by organic acids or reducing

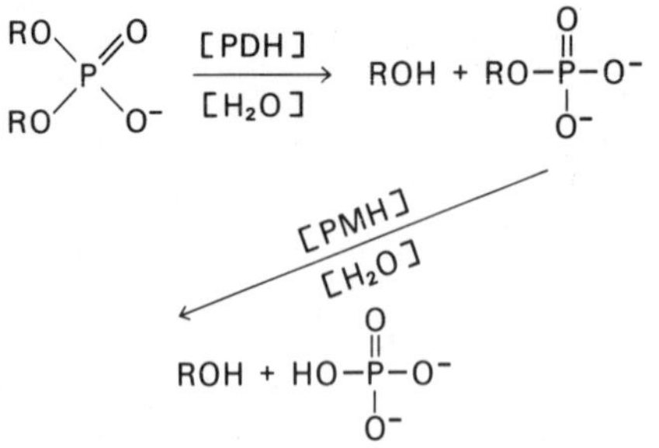

**Figure 2.** Enzymic hydrolysis of phosphate esters. [PDH], phosphoric diester hydrolase, [PMH], phosphoric monoester hydrolase.

conditions, produced as a result of microbial activity. Equation (4) represents a reaction that is normally a chemical precipitation.

As can be seen from Fig. 1, a number of important transformations in the cycling of phosphorus are microbially mediated. The results of recent investigation of these processes and of the nature of the phosphorus compounds involved together form the subject matter of this article.

## 2. Transformations of the Phosphorus Compounds in Plant and Animal Remains on Entering the Soil

The sources of soil organic matter are chiefly such materials as crop residues and roots, animal manures, green manures, composts and similar prepared manures, dead animals, organic fertilizers, and microorganisms. The rate and extent of transformation of the various plant and animal substances into gases and residual solids, i.e., humus, etc., are dependent on the conditions under which decay takes place.

The important factors in the processes are: (1) composition and particle size, (2) the microorganisms involved, (3) moisture level, (4) temperature, (5) pH, and (6) aeration (Allison, 1973). Simple carbohydrates and proteins are readily attacked, whereas lignin and waxes are resistant. Succulent plant material and finely ground material decompose more rapidly than large particles of dried material. A multitude of ecological factors will determine which kinds of microorganisms are most active in the decomposition of residues, but the two most important factors that determine the type of organism that will predominate in the presence of a mixed and adequate food supply are pH and oxygen. In a near-neutral environment bacteria usually predominate, whereas in an acid situation fungi are more important. If the oxygen supply is deficient anaerobic organisms will take over, provided the system is not too acidic. Moisture level, above a certain minimum, is not critical and in fact may reduce the oxygen supply in some soils if it is too high, thus restricting microbial growth. An increase in temperature usually leads to an increase in activity of all organisms present in soil.

Although important agronomic aspects of the transformation of the added organic matter in contact with or intimately incorporated in the soil, such as the overall effect of the C : N, C : P, and N : P ratios on the composition of the synthesized humic materials and the net mineralization or fixation of phosphorus, have been extensively investigated (Cosgrove, 1967), detailed studies of the fate of individual organic phosphorus compounds during transformations are few. The groups of microorganisms of importance in this stage of the phosphorus cycle are also largely unidentified, mainly because there has been a tendency for most workers in this field to use pure cultures of microorganisms and atypical substrates, e.g., chemically purified cellulose (Russell, 1961).

In all soils supporting plant growth a considerable proportion of the organic matter is derived from the decomposition of plant roots. Martin and Cunningham (1973) have recently studied the factors controlling the release of phosphorus from decomposing wheat roots. They concluded that root phosphorus can be released directly into soil solution, mainly as inorganic orthophosphate, as the result of autolysis and does not necessarily become incorporated into the organisms decomposing the roots. Air drying of the roots at 20–40°C prior to incubation with soil resulted in a rapid release of inorganic orthophosphate. This result resembled an observation made by Barr and Ulrich (1963) on the effect of air drying on plant tissue and is presumably due to rapid hydrolysis of phosphate esters by plant phosphatases. The ultimate fate of the inorganic phosphate released by roots depends on plant density, soil properties, etc. In most instances a proportion will become chemically fixed, but such fixed inorganic phosphate could subsequently become incorporated into microbial tissue by mechanisms to be discussed later (Section 5).

The effects of leaching, intermittent drying, and grazing on the transformations of plant and fertilizer phosphorus during the decomposition of hayed-off pasture plants have recently been studied (Jones and Bromfield, 1969; Bromfield and Jones, 1970, 1972). Pastures in most parts of southern Australia dry off during the late spring and early summer and at this stage may contain about 11 kg of phosphorus/hectare, an amount almost equal to the 130 kg/hectare of superphosphate commonly added to pastures each year. The rate and extent of the return of this plant phosphorus to the pool of available soil phosphorus are unknown, but it is known that organic phosphorus accumulates under these pastures in forms of low availability to plants (Donald and Williams, 1954). In laboratory studies, ground samples of hayed-off pasture plants were decomposed under continuously moist as well as intermittently moist and dry conditions. During the course of decomposition they were leached at different frequencies, and the resulting overall changes in inorganic and organic phosphorus were measured. About 60% of the phosphorus in the hayed-off plants was present as inorganic phosphorus, mainly in a water-soluble form; it was readily leached from the samples when microbial growth was suppressed. If not suppressed, microbial growth largely prevented the loss of inorganic phosphate by leaching over a period of 3 months. Intermittent drying increased the amount of phosphorus leached from decomposing plants, but the leaching frequencies examined had little effect. Further investigation of the effect of an initial prolonged leaching showed that more than 90% of the water-soluble phosphorus could be leached from intact plants, although in many cases the laboratory equivalent of prolonged heavy rain was required to do this (Bromfield and Jones, 1972). In the field, the intensity and duration of the first autumn rain will determine the proportion of inorganic phosphate leached directly into the soil and the amount left for conversion to organic forms during subsequent decomposition. The proportion of unaltered plant organic phosphorus to microbially synthesized organic phosphorus is not known, neither is the chemical nature of the compounds involved.

Floate (1970a,b,c) has reported the results of similar investigations conducted on materials derived from hill pastures in Sourhope, Roxburgh-shire, Scotland, and on dung from sheep grazed on these pastures. The soils under these pastures are characterized by a surface layer or mat ($A_0$ horizon) of partially decomposed organic matter derived largely from unconsumed plant material. The presence of this $A_0$ horizon suggests that the decomposition rate of the organic matter is slow. It also physically isolates freshly deposited organic materials from the mineral soil, at least in the early stages of decomposition. Floate (1970a,b,c) used a laboratory incubation technique to follow patterns of decomposition of pasture material and dung samples. Dried milled samples were wetted with $A_0$ soil inoculum to 100% moisture-holding capacity and were placed in stoppered incubation bottles together with tubes to absorb ammonia and carbon dioxide. Tubes of water to maintain high humidity were also placed in the bottles. The bottles and blanks were incubated at the required temperature and aerated twice weekly during the period of the experiment. Experiments were designed to study comparative rates of mineralization of carbon and nitrogen as well as phosphorus. Throughout most of the incubation period at 30°C, plant materials immobilized phosphorus, whereas between 3 and 34% of the total phosphorus in feces was mineralized. At lower temperatures phosphorus was increasingly immobilized in both plant and feces samples. These results bear a general similarity to those of earlier workers, who used similar techniques to follow phosphorus transformations in soil organic matter (Cosgrove, 1967).

## 3. Mineralization of Phosphorus-Containing Soil Organic Matter

Although it has never been fully resolved whether organic phosphorus must be mineralized prior to plant uptake, it is usually considered that such is the case. The transformations in the soil which are associated with the phosphorus nutrition of plants remain poorly defined, however, and in addition the chemical composition of soil organic phosphorus remains to a great extent unknown. Most of the reliable information on the quantities and composition of soil organic phosphorus is of comparatively recent origin; review articles on this topic have been published by Anderson (1967) and more recently by Halstead and McKercher (1975). The mineralization of organic phosphates in soil has also been reviewed (Cosgrove, 1967; Ramirez Martinez, 1968), and more recent work is dealt with in the article by Halstead and McKercher (1975).

It would appear from the information collected by Halstead and McKercher (1975) that approximately 1% of the organic phosphorus in soils is found in the phospholipid fraction, from 5 to 10% as nucleic acids or their degradation products (nucleotides, oligonucleotides) and up to 60% in the inositol polyphosphate fraction.

The inositol polyphosphates are present chiefly as the penta- and hexaphosphates of various inositol isomers (Halstead and McKercher, 1975), a mixture often referred to loosely as "soil phytate" or "soil phytin," although strictly speaking phytic acid is *myo*-inositol hexaphosphate and phytin its calcium/magnesium salt. Although mainly in the form of *myo*-inositol esters, soil phytate also contains a considerable proportion of the polyphosphates of *scyllo*-inositol and D-*chiro*-inositol, as well as about 1% of phosphate esters of *neo*-inositol (Fig. 3) (Cosgrove, 1966; Anderson, 1967; Halstead and McKercher, 1975). The presence of *chiro*-inositol was first reported in soil phytate as a phosphate ester of the racemic form, i.e., DL-*chiro*-inositol hexaphosphate (Cosgrove, 1963), but subsequent re-examination of material from two Australian soils showed that this constituent of soil phytate was in fact D-*chiro*-inositol hexaphosphate (Cosgrove, 1969). Small amounts of lower esters, e.g., monophosphates (Halstead and McKercher, 1975), have also been detected.

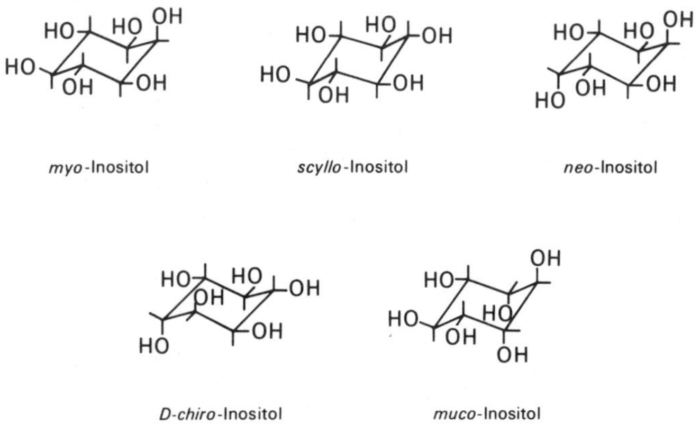

myo-Inositol          scyllo-Inositol          neo-Inositol

D-chiro-Inositol          muco-Inositol

Figure 3. Inositols.

For some time it was believed that inositols other than *myo*-inositol did not occur in phosphorylated forms in biological systems apart from in soil. Recently, Sherman *et al.* (1971) found that small amounts of *neo*-inositol could be detected in the brain and other tissues of the rat. They showed that an enzyme preparation from these tissues having D-glucose 6-phosphate-L-*myo*-inositol 1-phosphate activity was also capable of transforming D-mannose 6-phosphate to a *neo*-inositol 1-phosphate, presumably the L-enantiomer; the latter reaction proceded at about 1/200 the rate of the D-glucose 6-phosphate cyclization. The enzyme preparation was not capable of cyclizing D-galactose 6-phosphate to a D-*chiro*-inositol phosphate. It would be interesting to test the ability of the enzyme to cyclize D-idose 6-phosphate to *scyllo*-inositol monophosphate. The D-idose phosphates do not seem to have been prepared, although the synthesis of D-idose 6-phosphate should not be difficult. The range of inositol polyphosphates present in soil has recently been detected in organic phosphate preparations from activated sludge, where the polyphosphates presumably arise as a result of the activities of some organism, or group of organisms, whose identity has not been determined (Section 6.).

L'Annunziata and Fuller (1971) and L'Annunziata *et al.* (1972) have recently claimed to have shown the presence of *myo*- and D-*chiro*-inositol phosphates in the needles of *Pinus ponderosa* (pine) and the presence of *myo*-, D-*chiro*- and *muco*-inositol phosphates in the leaves of *Prosopsis juliflora* var. *velutine* (velvet mesquite). The existence of a phosphate of *muco*-inositol (Fig. 3) in nature has not previously been reported. In both cases the underlying soils contained only *myo*- and D-*chiro*-inositol phosphates; for this reason L'Annunziata and Fuller (1971) and L'Annunziata *et al.* (1972) speculated that *muco*-inositol phosphates were probably epimerized through the activities of microbial enzymes to the corresponding *myo*-inositol derivatives. The results of these investigations should be viewed with caution, however, as they depend on the basis of the identification of free inositols in hydrolyzates of preparations obtained by extracting soil and plant material by Anderson's (1956) method. Such preparations are very crude and contain large amounts of polysaccharide and nitrogenous material. In order to substantiate the claims of L'Annunziata and Fuller (1971) and L'Annunziata *et al.* (1972), it would be necessary to further fractionate their preparations ("barium salts") by means of ion-exchange chromatography (Cosgrove, 1963; McKercher and Anderson, 1968) and to examine hydrolyzates of any eluted material that clearly has the properties of an inositol phosphate.

The work of Anderson (1967) showed that hydrolyzates of humic acid contained nucleic acid bases (adenine, guanine, cytosine, thymine, and traces of uracil) in proportions indicative of the presence of DNA of microbial origin. Anderson's (1967) results, however, showed that only a small proportion (ca. 2%) of soil organic phosphorus could be accounted for in this way. Anderson (1970) later isolated pyrimidine nucleotide diphosphates from soil, thus confirming the existence of characteristic structural units of DNA in soil. Although presently available information indicates that low levels of nucleic acids are normally present in soil, it is possible that improved methods of extraction may cause these values to be revised upwards in the future.

Although phospholipids enter the soil in significant quantities from plant and animal remains and by *in situ* microbial synthesis, the reported values for extractable phospholipid in soil are very low and account for no more than 1% of the total organic phosphorus present (Halstead and McKercher, 1975). It is possible that phospholipids entering the soil are rapidly degraded by microorganisms, particularly actinomycetes (Ko and Hora, 1970), but the possibility that they are incorporated into insoluble complex molecules by simple chemical reactions should not be overlooked.

Much organic phosphorus is present in soil in as yet unidentified forms. Recent work by Anderson and Malcolm (1974) has revealed the presence of a number of unusual phosphate esters in alkali extracts of soil. None of the isolated substances was completely characterized, but included among them were two esters containing glycerol, *myo*-inositol, *chiro*-inositol, and also an ester with a C : P ratio of about 7, apparently based on a polyol containing a carboxyl group. These esters accounted for only 3% of the total organic phosphorus in the soil and may have resulted from degradation during the extraction process of more complex molecules, such as phospholipids and phosphorylated polysaccharides.

Examination of soil extracts by gel filtration (Moyer and Thomas, 1970; Omotoso and Wild, 1970; Steward and Tate, 1971; Thomas and Bowman, 1966; Veinot and Thomas, 1972) has shown that a significant part of the organic phosphorus in these extracts is of high molecular weight. Partial hydrolysis of such high molecular weight material has revealed the presence in it of recognizable fragments, such as inositol polyphosphates, although much remains unidentified. The question of whether or not these macromolecular forms are naturally occurring or are merely artifacts of the extraction process has not been resolved.

In addition to organically bound phosphorus there are also present in soil forms of inorganic phosphate other than the readily available inorganic orthophosphate. Ghonsikar and Miller (1973) succeeded in extracting mixtures of linear polyphosphates from several soils using cold 0.5 M perchloric acid. Prior to this Anderson and Russell (1969) had identified small amounts of inorganic pyrophosphate in hot alkali extracts of soil. It is possible that this inorganic pyrophosphate originated by alkaline hydrolysis of linear polyphosphate (Thilo, 1962).

From a survey of the literature up to 1966 (Cosgrove, 1967), it was concluded that soils normally contain a wide range of microorganisms capable of releasing inorganic orthophosphate from all known organic phosphates and inorganic polyphosphates of microbial or plant origin. Thus, although the contribution of the microbial population to the turnover of the phosphorus constituents of the soil organic matter is not clearly understood, there is little doubt that soil microorganisms can produce a variety of phosphatases (Feder, 1973) capable of mineralizing phosphate in the system and thereby presumably affecting phosphorus availability to plants. In addition to the phosphatase activity directly attributable to living organisms, the activity due to the presence of free enzymes present in the soil must also be considered. That both types of activity are normally present was demonstrated by McLaren et al. (1962) and Skujins et al. (1962). Free enzymes are probably stabilized in soil by adsorption onto colloidal material (Skujins, 1967). As well as resulting from microbial activity, free enzymes may be produced by soil animals and may also be released from plant roots and plant residues (Skujins, 1967).

Attempts to isolate free phosphatases from soil have not apparently been successful as none is recorded in the literature. A phytase preparation (3-phytase, EC 3.1.3.8) obtained from a soil organism (*Pseudomonas* sp.) has been extensively studied, and its ability to hydrolyze various isomeric forms of inositol hexaphosphate has been demonstrated (Cosgrove, 1970; Cosgrove et al., 1970; Irving and Cosgrove, 1971). Similar studies have been made of a phytase from *Aspergillus ficuum* (Irving and Cosgrove, 1972, 1974); this fungus is known to occur in soil (Shieh and Ware, 1968).

Recent investigations of phosphatase activity of whole soil have tended to favor the use of *p*-nitrophenyl phosphate as the substrate (Cervelli et al., 1973; Speir and Ross, 1975; Tabatabai and Bremner, 1969), an understandable preference as the method is simple and rapid. It is necessary to point out, however, that this method measures the *p*-nitrophenyl-phosphatase activity of the system and that the results are not necessarily correlated with phosphatase activity towards inositol phosphates, nucleotides, etc., which are alcoholic esters, not phenolic esters. Possible objections to the use of phenolic esters as substrates for esterases have been overcome by Rammler and Parkinson (1973) and Rammler et al. (1973), who have used substrates such as the phosphate esters of 4-(*p*-nitrophenoxy)-1,2-butanediol and related diols as substrates for phosphatases. After hydrolysis of the ester the released diol is oxidized with periodate, and the resulting aldehyde is treated with a base to yield the *p*-nitrophenolate ion (as obtained by hydrolysis of *p*-nitrophenyl phosphate) (Fig. 4).

These substrates are apparently not yet available commercially, but their syntheses are not difficult (Rammler et al., 1973).

Greater numbers of organisms capable of hydrolyzing organic phosphates have been found in rhizosphere than in nonrhizosphere soils

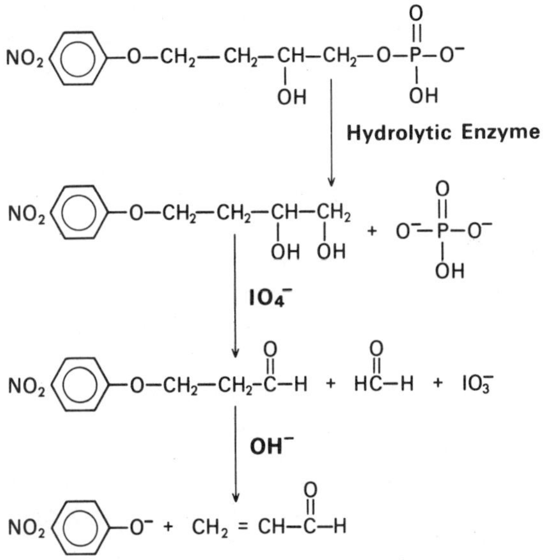

$$NO_2-\langle\bigcirc\rangle-O-CH_2-CH_2-CH-CH_2-O-\overset{\overset{O}{\|}}{P}-O^-$$
$$\underset{OH}{|}\qquad\underset{OH}{|}$$

Hydrolytic Enzyme

$$NO_2-\langle\bigcirc\rangle-O-CH_2-CH_2-CH-CH_2 \;+\; O^--\overset{\overset{O}{\|}}{P}-O^-$$
$$\underset{OH\;\;OH}{|\;\;|}\qquad\qquad\underset{OH}{|}$$

$$IO_4^-$$

$$NO_2-\langle\bigcirc\rangle-O-CH_2-CH_2-\overset{\overset{O}{\|}}{C}-H \;+\; H\overset{\overset{O}{\|}}{C}-H \;+\; IO_3^-$$

$$OH^-$$

$$NO_2-\langle\bigcirc\rangle-O^- \;+\; CH_2 = CH-\overset{\overset{O}{\|}}{C}-H$$

**Figure 4.** Assay scheme for 4-(p-nitrophenoxy)-1,2-butanediol as a substrate for a phosphomonoesterase.

(Greaves and Webley, 1965). Enzymes excreted by these organisms in intimate association with plant roots could be expected to mineralize phosphorus for plant uptake before it becomes fixed and unavailable in the soil. This deduction encouraged the practice of inoculating soils, or seeds prior to sowing, with cultures of appropriate organisms in the hope that crop yields would be increased. Owing, probably, to the lack of industrial capacity for phosphate fertilizer production at the time, the use of one such organism, *Bacillus megaterium* var. *phosphaticum* (phosphobacterin), was extensively promoted in the Soviet Union (Cooper, 1959). The claims of the Russians concerning the benefits of phosphobacterin were subsequently disputed (Smith *et al.*, 1961), and recent commentators incline to the view that any benefits that have been observed are probably due to the stimulation of growth by the production of auxins and related substances by the organisms (Brown, 1972; Rovira, 1965). Recent work (Benians and Barber, 1974) has in fact confirmed earlier suggestions (Akhromeiko and Shestako a, 1958) that inoculation with microorganisms can in some cases cause a reduction in the adsorption and translocation of phosphorus similar to that shown by plants in solution culture (Barber, 1969). Benians and Barber (1974) studied the uptake of phosphorus by barley plants growing in a basaltic loam which, although it had a large pool of labile phosphate, had a very low potential in the soil solution. Less phosphorus was taken up by plants grown under nonsterile conditions, despite a rapid turnover of the element in root-associated microorganisms. This led to a considerable reduction in the yield of plant dry matter. These microbial effects were obviated by addition of phosphate. In this soil, therefore, competition for phosphorus appears to exist between microorganisms and plants.

Martin (1973) has recently studied the influence of the rhizosphere microflora on the availability of $[^{32}P]$-*myo*-inositol hexaphosphate phosphorus to wheat. From his results Martin (1973) concluded that it is doubtful that the occurrence in the root zone of bacteria possessing phytase activity will increase the dephosphorylation of *myo*-inositol hexaphosphate at the root surface above the activity due to plant root enzymes. The dominant factor controlling the mineralization of inositol polyphosphates is most probably their concentration in soil solution.

There is ample evidence for the presence of phosphatases and phytases in ectomycorrhizas associated with various plants, particularly trees (Bartlett and Lewis, 1973; Cosgrove, 1967; Halstead and McKercher, 1975; Williamson and Alexander, 1975), but the role of these fungi in the phosphorus nutrition of the plant is still to some extent a matter for conjecture. It now seems most likely that the beneficial effects of these mycorrhizal associations are merely due to such factors as an increase in the absorbing area of mycorrhizal roots and their increased longevity (Bieleski, 1973). In the case of plants infected with vesicular–arbuscular mycorrhiza, such as *Endogone*, Mosse (1973) found that these plants took up more phosphate from phosphate-deficient soils than did uninfected plants. Sanders and Tinker (1971) and Hayman and Mosse (1972) investigated the source of the extra phosphate taken up by growing mycorrhizal and nonmycorrhizal onions in soils where the pool of "labile" soil phosphate had been labeled with $^{32}P$. The specific activity of the phosphorus in the mycorrhizal and nonmycorrhizal plants was very similar, indicating that both used the same, or a similarly labeled, fraction of the soil phosphate and that the plants with vesicular–arbuscular mycorrhiza did not utilize sources not available to the nonmycorrhizal plants, even under conditions of acute phosphorus deficiency. The most probable reason for the beneficial effect of the mycorrhiza was considered to be that the fungal hyphae outside the root were able to absorb phosphorus from the soil solution beyond the depletion zone that develops close to the root. The studies of the Rothamsted group were subsequently extended to other plants (Mosse *et al.*, 1973) and a range of *Endogone* spore types (Powell, 1975), and the results obtained confirmed those reported in the earlier papers.

As suggested by Martin (1973) and discussed above, the dominant factor controlling the mineralization of organic phosphates in soil is probably their solubility in the soil solution. For many phosphates this solubility appears to be very low, although Wild and Oke (1966) detected the presence of lower esters of inositol, including the monophosphate, in 0.005 M calcium chloride extracts of soil. Some of these substances were readily utilizable by plants. Inositol polyphosphates are usually considered to be resistant to enzymic breakdown in soil because of their ability to form very insoluble ferric iron and aluminium derivatives. It should not be overlooked, however, that ferric phytate is capable of forming soluble complexes with ammonia and amines (Nørgaard Pedersen, 1953) and that similar complexes may exist in soil as a result of combinations between nitrogenous organic matter, ferric iron, and inositol hexaphosphates. The high molecular weight organic phosphate extracted from soil by Thomas and Bowman (1966) may be a complex of this type. It contains inositol polyphosphates and releases inorganic phosphate when treated with phytase preparations obtained from the fungus *Aspergillus ficuum* (Shieh and Ware, 1968), a soil isolate. The reaction is slow, however, and rarely proceeds to completion (Thomas, 1976).

## 4. Formation of Organic Phosphorus Compounds from Available Inorganic Phosphorus in the Soil

In an earlier review of this topic (Cosgrove, 1967), it was stated that phosphate-containing organic matter is almost certainly accumulated in soil as the result of microbial activity. The implication is that the organic phosphorus in plant and animal remains is mineralized fairly rapidly and is then used for the synthesis of microbial organic phosphates. Recently, Martin and Molloy (1971) estimated that the annual amounts of humic-acid-associated organic phosphorus and inositol polyphosphate phosphorus contributed to the topsoil from sheep feces, residual plant material, and plant roots would not exceed 3% of the amount already present. This study tends to confirm the above hypothesis, but more evidence is needed to establish definitely that the stable organic phosphates in soil originate by microbial synthesis rather than by accumulation of resistant fractions of plant and animal residues.

Of the minor phosphorus constituents in soil, the phospholipids are likely to be of plant or animal origin, but their base components, choline and ethanolamine, are known to occur also in microorganisms. Kowalenko and McKercher (1971) have pointed out that an examination of the fatty acid components of the phospholipids could be informative as to their origin, as bacterial phospholipids tend to have branched or cyclic fatty acids, whereas plant phospholipids tend to have constituent unsaturated fatty acids. The nucleic acids in soils are apparently of microbial origin (Anderson, 1967; Halstead and McKercher, 1975).

The quantitatively most important identified group of phosphate esters in soil is the inositol polyphosphate mixture. Its origin in soil is still uncertain as, although *myo*-inositol hexaphosphate is a common plant constituent, especially in mature seeds, phosphates of other inositols are unknown in plants.* The inositol polyphosphate mixture is therefore assumed to be of microbial origin, but as yet no organisms capable of producing inositol polyphosphates have been isolated, let alone identified and studied. The question of the origin and transformations of soil inositol phosphate isomers has recently been discussed by L'Annunziata (1975), who considers that the available evidence, admittedly slight, favors a pathway to the soil D-*chiro*-,*scyllo*-, and *neo*-inositol hexaphosphates via an epimerization of *myo*-inositol or its hexaphosphate, rather than direct microbial synthesis of the various isomers by processes that do not involve epimerization, such as cyclization of the appropriate hexose 6-phosphate (Section 3). L'Annunziata advocates soil incubation studies with $^{14}$C-labeled *myo*-inositol, which is available, and with $^{14}$C- and $^{32}$P-labeled *myo*-inositol hexaphosphate, which can be synthesized, in order to decide between these alternatives. These suggestions are interesting, but it is likely that the addition of *myo*-inositol to soil, followed by incubation, would lead to a flush of microbial activity in which most of the added inositol would be consumed as a general carbon source. In his discussion of the cyclization of hexose 6-phosphates to inositol 1-phosphates, L'Annunziata (1975) states that Chen and Charalampous (1967) reported the cyclization of glucose to D-*chiro*-inositol by this route. The "D-inositol 1-phosphate" of Chen and Charalampous (1967) is of course a *myo*-inositol derivative.† Epimerization reactions of inositols via the acetates or via ketonic intermediates (inososes) by simple chemical reactions are well known, and it has recently been shown that epimerization of *myo*-inositol pentaphosphates to D-*chiro*- and *scyllo*-inositol pentaphosphates by oxidation followed by reduction is possible *in vitro* (Cosgrove, 1972).

By contrast with soil, the biosynthesis of *myo*-inositol polyphosphates in plants and animals is now relatively well understood. Loewus (1974) has reviewed recent work on the biosynthesis of *myo*-inositol hexaphosphate in plants. Evidence suggests a stepwise pathway from *myo*-inositol or a *myo*-inositol monophosphate to the hexaphosphate, in which intermediate di- and higher phosphate esters appear in trace amounts only. The biosynthesis of the *myo*-inositol polyphosphate found in bird blood, formerly

---

* The claims of L'Annunziata and Fuller (1971) and L'Annunziata *et al.* (1972) to have identified D-*chiro*-inositol phosphates and *muco*-inositol phosphates in plant material should be regarded as doubtful (Section 3).

† Confusion over the nomenclature of inositol derivatives has now been largely clarified as a result of the general adoption of the *1967 IUPAC/IUB Tentative Rules of Cyclitols* (Anderson, 1972).

believed to be the hexaphosphate but now known to be the 1,3,4,5,6-penta-phosphate (Johnson and Tate, 1969), has been studied by Breitenbach and Hoffmann-Ostenhof (1971).

The origin of the unidentified organic phosphate component of soil, often a large proportion of that present, can only be speculated on until such time as the problem of its chemical identity is resolved. From the nature of the fragments partially characterized by Anderson and Malcolm (1974) it would appear to consist, at least in part, of phosphorylated carbohydrate-like material of a type as yet unknown in other biological systems.

Inorganic polyphosphate was shown by Pepper and Miller (1974) to accumulate in a silt loam soil under conditions which stimulated microbial activity. Increased quantities of polyphosphate could be extracted from soil after increased glucose amendments and increased temperature of incubation. Maximum quantities of polyphosphate were found 2 days after the addition of inorganic orthophosphate to soils preincubated with glucose for 2 weeks. This accumulation of polyphosphate disappeared after a further 9-day incubation. Pepper and Miller (1974) postulated that inorganic polyphosphate may represent an important intermediate in the microbial solubilization of soil phosphorus.

## 5. Solubilization of Fixed Phosphorus and Mineral Phosphorus in Soil

When soluble inorganic orthophosphate is added to mineral soils, a proportion of it rapidly becomes "fixed" or insoluble. In acid soils the initial or fast reaction is probably dominated by polar adsorption and simple precipitation according to solubility product principles. The continuing or slow reaction, on the other hand, is probably explained by isomorphous replacement reactions involving aluminosilicates and hydrous oxides. At this stage the products resemble such naturally occurring minerals as varisite $[Al(OH)_2H_2PO_4]$ and strengite $[Fe(OH)_2H_2PO_4]$ (Kardos, 1964). In addition to the naturally occurring phosphorus-containing minerals, soils often contain finely ground phosphate rock—usually fluorapatite $[Ca_{10}(PO_4)_6F_2]$—which has been added as a fertilizer. The solubilization, and hence the return to a plant-available state, of fixed and mineral phosphorus in soil is affected by a variety of reactions. These include chelation of calcium, iron, and aluminum by organic acids, formation of phosphohumic complexes, and competition between humate ions and phosphate ions for adsorbing surfaces. Some controversy exists regarding the importance of organic acids in increasing phosphate availability; the topic has been discussed by Stevenson (1967). Undoubtedly laboratory experiments have shown that acids capable of forming stable chelate complexes with multivalent cations, e.g., citric acid and 2-ketogluconic acid, are able to release phosphate from insoluble metallic phosphates (Stevenson, 1967). It is possible that these organic acids and/or humic acids may accumulate in localized zones, as a result of organic residue decomposition by saprophytic organisms, in sufficient quantities to increase the availability of phosphate without affecting the overall phosphate solubility in the soil.

Since the publication of Stevenson's (1967) article, not a great deal of investigation in this field appears to have been attempted. Most reports deal with the phosphate-solubilizing ability of various isolates using an *in vitro* technique and various forms of calcium phosphate as the substrate. Paul and Sundara Rao (1971) isolated 12 phosphate-solubilizing bacteria from the rhizosphere of four cultivated legumes from different soil regions of India. The active organisms were identified as strains of *Bacillus subtilis*, *B. brevis*, *B. pulvifaciens*, *B. pumilis*, and *B. polymyxa*. Although phosphate solubilization by bacteria has been extensively studied (Stevenson, 1967), much less attention has been devoted to the possible role of fungi. Chhonkar and Subba-Rao (1967) investigated the solubilization of $Ca_3(PO_4)_2$ by isolates of fungi associated with legume root nodules, and Agnihotri (1970) examined the ability of soil fungi isolated from forest tree seedbeds to solubilize $Ca_3(PO_4)_2$, hydroxyapatite, and fluorapatite. Of the various isolates tested by Chhonkar and Subba-Rao (1967), *Pencillium lilacinum* and several strains of *Aspergillus* solubilized the calcium phosphates, some retaining this ability even in media containing soluble inorganic orthophosphate. Some isolates reduced the pH of the medium, others did not. Agnihotri's results shows that *Aspergillus niger* was the best of his isolates, and it also lowered the pH of the medium considerably; this fungus is known to produce oxalic acid and citric acid (Agnihotri, 1970). In his study of the properties of complexes formed in soil between organic matter and inorganic metallic phosphates, Sinha (1972) investigated the solubilizing action of fulvic acids on ferric phosphate. Formation of stable fulvic acid/iron complexes is a well-known phenomenon (Schnitzer and Khan, 1972), and this was considered by Sinha (1972) to be the mechanism responsible for the release of free inorganic orthophosphate. Sinha (1972) postulated that soluble fulvic acid/metal phosphate complexes are also formed and that these are available to plants. The presence of such complexes in the soil solution is difficult to demonstrate owing to the low concentration of soluble phosphorus present in the system. Lévesque (1969) found that there were indications that fulvic acid/metal phosphate complexes occurred in soils but later stated that fulvic acid/ferric phosphate was a poor source of phosphorus for plant growth, although apparently a good source of iron (Lévesque, 1970).

Another means for the production of free acid in soil is by the oxidation of applied elemental sulfur through the activities of organisms, particularly the autotrophic bacteria belonging to the genus *Thiobacillus*. The action of the resulting sulfuric acid in solubilizing rock phosphate *in situ* has been investigated extensively in Australia (Swaby, 1975) and is the basis for the formulation of the phosphatic fertilizer "biosuper." Experimentally produced pellets contained 16.6% of sulfur and 18.5 or 26.1% of $P_2O_5$ and were prepared from five parts by weight of ground rock phosphate, one part of powdered elemental sulfur, and a soil inoculum of thiobacilli (0.1%). Extensive trials over a number of years in North Queensland, Australia have shown that biosuper could be used economically on cattle pastures in the wetter tropics of Australia and other countries, particularly where the annual rainfall is greater than 635 mm. In temperate regions this preparation is less successful, owing to the lower level of activity of the thiobacilli and hence slower rate of release of soluble phosphate. Trials with wheat in temperate regions have been unsuccessful, as the initial rate of release of soluble phosphate is far too low to satisfy the requirements of the crop at the important early stages of growth (Swaby, 1976). Even in wetter tropical regions it is unlikely that annual crops will benefit unless the problem of slow initial release of soluble phosphate can be overcome. Technical problems of manufacture and use have been discussed by Swaby (1975).

Bromfield (1975) has recently studied the effects of a ground rock phosphate–sulfur mixture on yield and nutrient uptake of groundnuts in Northern Nigeria. The soils of this area are sulfur deficient. Bromfield (1975) found that the uptake of phosphorus by the crop was significantly better when the rock phosphate–sulfur mixture was used by comparison with ground rock phosphate alone. It is not possible to decide from Bromfield's data whether the additional phosphorus came from ground rock phosphate, dissolved by sulfuric acid produced by the oxidation of sulfur, or whether it was taken up from the soil as a result of better root systems produced by the additional sulfur supply. To distinguish between the alternatives it would be necessary to compare a ground rock phosphate–gypsum mixture with a ground rock phosphate–sulfur mixture.

The use of products such as biosuper or even rock phosphate–sulfur mixtures is an attractive possibility in countries without the industrial capacity to produce superphosphate in large amounts, provided that the soil temperatures are satisfactory. Swaby (1975) is working towards production techniques for improved forms of biosuper aiming at greater stability of pellets, higher residual effects of phosphorus and sulfur, longer viability of thiobacilli during storage of pellets, and quicker initial release of phosphorus.

## 6. Phosphorus Transformations in Wastewater Treatment Plants

The presence of phosphorus is essential to the proper functioning of all biological wastewater treatment processes, but when present in excess it can create a pollution problem in the receiving body of water. In the simplest type of treatment plant, the incoming raw sewage is screened to remove coarse solid material, and the effluent from this stage is seeded with "return sludge" (see later) before being vigorously aerated. A voluminous growth of mixed organisms takes place, and after a period of some hours the contents of the aeration tanks are pumped to settling chambers. After the "activated sludge" has settled the clear liquor is disposed of in a river, a lake, or the sea, and the sludge is removed for dumping or anaerobic digestion with the solids removed at the primary screening stage. Part of the activated sludge ("return sludge") is used to seed a new batch of liquor entering the aeration chamber. The effluent liquor is often further treated to remove phosphorus, by chemical precipitation, etc., before discharge to a natural water body. Physical, chemical, and biological treatment processes currently in use, or proposed, have been recently reviewed (Nesbitt, 1973).

The presence of excessive amounts of phosphorus in the effluent of treatment plants is likely to cause extensive growth of rooted and/or floating aquatic plants in bodies of water receiving such effluents. The undesirable effects of such growth are now too well known to require elaboration here.

The forms and sources of phosphorus found in wastewaters have been discussed by Nesbitt (1969, 1973). These forms will normally include organic phosphorus from microbial cell debris and plant remains, complex inorganic phosphates from detergents, etc., agricultural products, and finally the biologically available form, inorganic orthophosphate. Presumably most of the organic phosphates entering the biological treatment stage of the process will be hydrolyzed to inorganic phosphate before being incorporated into the tissues of active organisms. Not a great deal is known of the nature of the organically bound phosphorus present in sludge and in effluent from the process. Nishikawa and Kuriyama (1974a) have recently studied the composition of gelatinous mucilage extracted from activated sludge with ethylenediamine tetraacetic acid (EDTA) solution; they concluded that the mucilage was a complex of nucleic acids (30%) with protein, carbohydrates, and metallic ions. Both DNA and RNA were present. Further investigation of the mucilage (Nishikawa and Kuriyama, 1974b) revealed the presence of considerable amounts of inositol phosphates, mainly in the form of hexaphosphate. Nishikawa and Kuriyama reported this to be *myo*-inositol hexaphosphate, but the methods used for its identification would not have revealed whether isomers such as *scyllo*-inositol hexaphosphate were present or not. A more detailed examination of the inositol phosphates from activated sludge (Cosgrove, 1973) revealed that in the case of the two samples studied the hexaphosphate fraction was a mixture of *myo-*, *scyllo-*, *chiro-*, and *neo*-inositol esters; it therefore resembled the mixture normally present in soil organic matter (Section 3). As in the case with soil, the identity of the organisms responsible for the synthesis of inositol polyphosphates in activated sludge is not known.

In addition to organically bound phosphorus, activated sludge contains inorganic polyphosphate, probably associated with nucleic acids, lipo-proteins, etc., in the form of volutin granules. These are known to occur in *Zoogloea ramigera*, a gram-negative, floc-forming bacterium commonly found in activated sludge (Roinestad and Yall, 1970). Finally, the sludge will undoubtedly contain inorganic orthophosphate as insoluble metallic salts.

As pointed out by Pipes (1966) in an excellent review, most studies of the activated sludge process are the work of engineers or chemists, and for this reason the identification of the microorganisms in it has received limited attention. Because of the importance of achieving low levels of phosphorus in the effluent from treatment plants, some attention has recently been given to the problem of determining the optimum conditions necessary for maximum incorporation of phosphorus into the sludge during the oxidation step. Until recently, it was believed that activated sludge treatment alone was unable to remove sufficient phosphorus from its effluent to control algal blooms, etc., when phosphorus was the limiting nutrient. Jenkins and Menar (1967), as a result of pilot plant studies, expressed the opinion that under no circumstances could a phosphorus removal better than 28% be expected. In contrast to this Vacker *et al.* (1967) reported efficiencies of up to 90% at an influent concentration of about 30 mg $PO_4$/liter. The mechanisms by which sludges removed amounts of phosphorus in excess of their metabolic requirements remained for some time obscure (Yall and Sinclair, 1971). It seemed unlikely that the accumulation of phosphorus as inorganic polyphosphate in volutin granules of microorganisms was a likely important factor as this means of storing excess phosphate was believed to become operative only when a phosphate-starved culture was suddenly treated to an excess of inorganic orthophosphate (Harold, 1966; Roinestad and Yall, 1970). Wastewaters are usually relatively low in carbon sources so that growth is limited by low carbon levels rather than low phosphorus levels. Recent investigations by Yall *et al.* (1972) on sludge from Rilling Road plant (San Antonio, Texas), a well-known highly efficient phosphorus remover, have shown that its high efficiency is due to a particular coccoid, gram-negative organism. It formed clusters and abundant volutin granules on sewage and sludge media, but unlike other bacteria it did not require a nutritional imbalance to force the accumulation of phosphate-containing volutin granules. The granules are believed to contain inorganic polyphosphate, but their precise composition has yet to be determined. Yall *et al.* (1972) also showed that the performance of a low efficiency plant could be greatly improved by seeding with Rilling sludge or cultures of the isolate from it. The effect gradually wore off, however.

Another approach to the problem has been proposed by Levin *et al.* (1975). Their process, now successfully operational at Seneca Falls, New York, makes use of the observation that subjecting activated sludge to anaerobiosis induces it to release phosphorus as inorganic orthophosphate. Sludge from the aeration process is divided, part being removed for disposal and the rest stripped of phosphate by anaerobic digestion. The latter is returned to the aeration basin. The phosphorus-rich supernatant fluid is removed from the stripper as a small-volume substream and is then treated with lime to precipitate the phosphorus (Fig. 5). The plant operates at a greater than 90% efficiency, and the economics are said to be very favorable. The amount of chemical precipitant, i.e., lime, required is much less than that used in total chemical precipitation or conventional biological/chemical plants.

## 7. Phosphorus Transformations in Inland Basins

The principal storage sites or pools of organically bound phosphorus in the hydrosphere are considered by Hooper (1973) to be, (a) the organic compounds of living and dead particulate suspended matter ("seston"), (b) a variety of filterable organic compounds, usually termed "dissolved," (c) the organic compounds of rooted and excrusting plants of the bottom, (d) the phosphorus of free-swimming animals, and (e) phosphorus in sediments. Inorganic orthophosphate is the principal form of phosphorus utilized by plants and for this reason was for a long time the only component measured in water studies. There is now abundant evidence, however, that inorganic orthophosphate often makes up less than 10% of the total phosphorus in many systems (Hutchinson, 1957).

Lake sediments can be considered as special kinds of soil, differing from ordinary soils in being permanently waterlogged and being continuously regenerated by deposition. The nature of the deposit depends on such factors as the area and depth of the lake, the composition of incoming materials, and the climatic conditions. In deep lakes of extensive area, the deposits are produced mainly internally from dead organisms and unchanged organic material, and therefore the phosphorus content of the sediment will depend largely on the cycle between sediment, water, vegetation, and animals. By contrast, in shallow lakes the sediments are

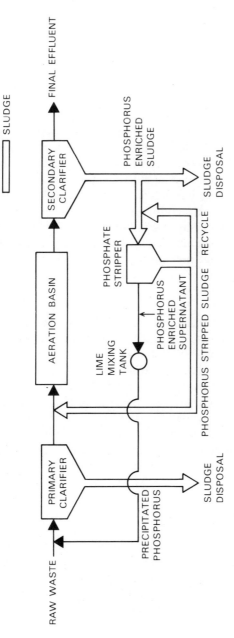

**Figure 5.** Activated sludge treatment plant, modified for phosphorus removal. (From Levin *et al.*, 1975. Reproduced with the permission of the Water Pollution Control Federation.)

formed preponderantly from material brought in from the surrounding shoreline. This constant influx of relatively large amounts of phosphorus usually means that small shallow lakes have more available phosphorus than deep, large lakes. The problems arising from increases of productivity due to increased phosphorus input into shallow lakes have received much attention in recent years (Hooper, 1973), and this in turn has generated interest in the related topic of the nature of the phosphorus in lake sediments (Hesse, 1973; Syers et al., 1973).

The meaningful study of transformations involving phosphorus that take place in water bodies and sediments is heavily dependent on the existence of reliable analytical methods for the determination of the various forms of phosphorus likely to be encountered and on methods for their isolation and identification. The molybdenum blue procedure for assaying the inorganic phosphate fraction is well known and can be quite sensitive if combined with an extraction procedure. The extraction step has the additional advantage of allowing the separation of the molybdophosphoric acid complex from contaminants. Problems arising in the direct application of the molybdenum blue method to natural waters have been discussed by Chamberlain and Shapiro (1973) and by Kimerle and Rorie (1973). An important source of error seems to be the presence in natural waters of very finely divided ferric phosphate and/or colloidal fulvic acid–ferric phosphate complexes (Chamberlain and Shapiro, 1973). These both react as soluble inorganic orthophosphate under the conditions of the assay procedure and would thus be recorded as soluble reactive phosphorus (SRP, see below). The removal of these colloids is possible by ultracentrifugation or filtration through a 0.01 $\mu$m membrane filter. Kimerle and Rorie (1973) recommended the use of an extraction with 1:1 benzene–isobutyl alcohol after a 15-sec mixing time of the molybdate reagent and the sample. The advantages of this are that interferences from labile phosphorus compounds, arsenate, and silicate are minimized and that sensitivity (1 $\mu$g/liter) is adequate for most environmental studies. Helyar and Brown (1976) have recently proposed the use of octan-1-ol (n-octyl alcohol) as the extractant. In a recent further refinement of the molybdate blue method, the extracted complex is absorbed onto diamine-treated silica gel and the molybdenum content is measured by X-ray fluorescence (Leyden et al., 1975). Another approach is to pass a large volume of water through a small anion-exchange resin column followed by a small volume of 1 M potassium chloride solution to elute the inorganic orthophosphate. Much organic phosphate, etc., is left in the column and does not interfere with the subsequent assay by the molybdenum blue method (Westland and Boisclair, 1974; Blanchar and Riego, 1975).

As yet, little progress has been made in the measurement of definite phosphorus compartments that are equivalent to morphologically or chemically distinct components of natural waters. For this reason an operational terminology is usually employed (Strickland and Parsons, 1960).

*Soluble reactive phosphorus* (SRP). This is the value obtained when membrane-filtered (0.45 µm) water is analyzed by the molybdenum blue method. Ideally, it should be the value for inorganic orthophosphate; reasons why it is often greater than this and the resulting implications for interpretation of tracer experiment results, etc., are discussed above and by Rigler (1973).

*Soluble phosphorus* (SP). This is the value obtained when membrane-filtered (0.45 µm) water is analyzed for phosphorus after digestion with an oxidizing reagent.

*Soluble unreactive phosphorus* (SUP).

$$SUP = SP - SRP$$

*Total phosphorus* (TP). The value obtained by analyzing whole water after digestion as for SP. O'Connor and Syers (1975) have recently provided evidence that the persulfate digestion is unsuitable for the determination of total phosphorus in waters containing particulate inorganic materials of soil origin and recommend the use of perchloric acid digestion for such samples.

*Particulate phosphorus* (PP).

$$PP = TP - SP$$

Methods for determination of phosphorus in lake sediments are usually similar to those used for soils. Total phosphorus and inorganic phosphorus can be determined by the method of Mehta *et al.* (1954) as used by Sommers *et al.* (1970). Organic phosphorus is then calculated by the difference. Hesse (1973) has drawn attention to certain analytical difficulties in the determination of phosphorus in sediments. The most serious of these is the change in the state of reduction likely to take place when a sample of waterlogged sediment is brought to the surface.

The literature on the quantity and distribution of organic phosphorus compounds in water is extensive. It has recently been expertly summarized and reviewed by Hooper (1973). Hooper concludes his article, however, by pointing out that despite the wealth of data on the quantity of organic phosphorus in marine and freshwater systems, only fragmentary and isolated efforts have been made toward the identification of these compounds. Until more effort is expended towards isolated and identification, a proper assessment of their role cannot be made. The situation is similar to that existing in the study of the nature and metabolism of phosphorus compounds in soil organic matter (Halstead and McKercher, 1975). Hooper's (1973) observation can be taken to include phosphorus compounds in sediments and also to apply to inorganic polyphosphates; bacterial and algal cells are known to accumulate considerable amounts of these under certain conditions of growth (Harold, 1966). The presence of inorganic polyphosphates in water/sediment systems has been demonstrated (Hooper, 1973), but rarely is an attempt made to achieve an exact identification.

Recently, some attempts have been made to identify components of the Soluble Unreactive Phosphorus (SUR) fraction in lake waters and to compare the isolated components with the soluble organic phosphorus of algal cultures. The lake water samples were treated to remove cations and then were either freeze-concentrated (Herbes et al., 1975) or freeze-dried (Minear, 1972) in order to obtain workable preparations from the dilute solutions (ca. 50 μg/liter of phosphorus). Herbes et al. (1975) tested their preparation as a substrate for (i) alkaline phosphatase, (ii) a phospho-diesterase, and (iii) wheat bran phytase. Only the latter hydrolyzed the native organic phosphorus in the water, although added phosphate esters, pyrophosphate, and inorganic tripolyphosphate were easily hydrolyzed by (i) and added RNA and DNA were easily hydrolyzed by (ii). This confirms earlier observations (Rigler, 1961; Strickland and Solorzano, 1966) that virtually no phosphatase-hydrolyzable compounds are present in solution in natural waters; degradation is presumably very rapid. Herbes et al. (1975) fractionated lake water on Bio-Gel P60 and obtained fractions of high molecular weight and low molecular weight, the latter corresponding to the elution position of myo-inositol hexaphosphate. Both were hydrolyz-able by wheat bran phytase. The higher molecular weight fraction was considered to be probably a complex of myo-inositol hexaphosphate with protein, lipid, or fulvic-acid-like material. The existence of such complexes has been postulated by earlier workers in other fields (Cosgrove, 1966; Anderson and Hance, 1963). Minear (1972) found that up to 10% of the dissolved organic phosphorus was DNA, or oligonucleotide material from DNA, capable of exclusion from Sephadex G-75 and G-100 dextran gels. The DNA material represented about 50% of the total high molecular weight fraction. The presence of DNA was confirmed by (i) a deoxyribose test, (ii) DNase breakdown of the isolated high molecular weight material, and (iii) detection of the purines adenine and guanine in hydrolyzates of the high molecular weight material.

Consideration of the results of Herbes et al. (1975) in conjunction with those of Minear allows the posing of a number of questions to which answers should be obtainable from further investigation on the lines followed by them:

(1) Is the complex of Herbes et al. composed of myo-inositol hexaphos-phate and DNA? No attempt to analyze it for bases appears to have been made. Admittedly, the complex was resistant to dephosphorylation by enzymes having RNase and DNase activity, but this could be due to such complexes being resistant to RNase and DNase. Also, the nature of the inositol present should be checked as soil inositol polyphosphates, pre-

sumably of microbial origin (Section 3), are a mixture of phosphates of several inositol isomers, whereas plant inositol polyphosphates are based on *myo*-inositol only.*

(2) Complexes of an ionic nature between inositol polyphosphates and proteins are stable at neutral pH and at low ionic strength but are easily dissociated at extremes of pH or at high ionic strength (Steward and Tate, 1969). Herbes *et al.* (1975) worked on freeze-concentrated material, and under these conditions complexes of inositol polyphosphates with high molecular weight material would be stable. Freeze-dried material may have become dissociated; also, some of Minear's observations were on alkaline extracts of freeze-dried residues. Minear (1975) has recently been in conflict with Lean (1973, 1975) over the nature of soluble organic phosphorus in lake water, particularly the contribution to this fraction due to death and decay of organisms in eutrophic lakes. Lean (1973) used kinetic studies with $[^{32}P]PO_4$ in conjunction with gel filtration to identify biological forms of phosphorus in a eutrophic lake during the summer to early autumn period. He interpreted his results as evidence for the excretion of an unidentified phosphorus compound (XP) of apparent mol. w. 250 which rapidly complexes extracellularly with colloidal material or particulate matter, either being capable of passing a 0.45 $\mu$m millipore filter. He considers also that the results of previous workers on marine algae (Antia *et al.*, 1963; Johannes, 1964; Kuenzler, 1970; Watt and Hayes, 1963) can be better explained by his hypothesis rather than one which postulates excretion of soluble organic phosphorus by marine algae during periods of rapid growth in excess inorganic phosphate, followed by utilization of this organic phosphate later on when the supply of inorganic phosphate once more becomes limiting. Lean (1973) considers also that his hypothesis is superior to any that postulate death and decay of organisms as the source of soluble organic phosphorus in natural waters. Minear (1975) criticized Lean's (1973) experimental technique in several respects and pointed out the existence of the paper referred to above on the DNA content of dissolved organic phosphorus (SUP) (Minear, 1972) of which Lean (1973) was seemingly unaware.

It is apparent that at this stage the devising of models to illustrate the exchange mechanisms between compartments of this system is an exercise of doubtful value and that Hooper's (1973) exhortation concerning the necessity for much more effort to be expended on separation and identification of organic phosphorus in natural waters has yet to be heeded.

The state of knowledge of the forms of organic phosphorus in sediments is little better than that of organic phosphorus in water. The subject has been recently reviewed (Hesse, 1973; Syers *et al.*, 1973), and it would appear that apart from traces of ATP (0.34 to 9.5 $\mu$g/g) the only other

* See first footnote on page 108.

organic phosphates positively identified are esters of inositols. Riego (1974) recently obtained inorganic pyrophosphate and tripolyphosphate from a series of 14 sediments using EDTA + ammonium fluoride followed by EDTA + sodium hydroxide as extractants. Inositol phosphates were reported to be present at levels of less than 10% by Sommers *et al.* (1972), but later determinations by Weimer (1973) indicated that up to 63% of the organic phosphorus in sediments may be inositol phosphates. Terrestial soils are reported as having up to 58% (Halstead and McKercher, 1975). In terrestial soils inositol phosphorus is most highly correlated with total aluminum, whereas in sediments inositol phosphorus is most highly correlated with acetate-extractable iron and aluminum and with organic carbon (Weimer, 1973). It is unfortunate that neither Sommers *et al.* (1972) nor Weimer (1973) attempted to characterize fully the inositol phosphates that they extracted from sediments. Plant inositol polyphosphate is based exclusively on the *myo*-isomer, whereas soil inositol polyphosphate is a mixture of isomers, predominantly based on *myo*-inositol but with a considerable proportion of *scyllo*-inositol and D-*chiro*-inositol as well as a trace of *neo*-inositol. The ratio does not appear to vary irrespective of the origin of the soil (Section 3). If sediment inositol phosphorus differs much from soil inositol phosphorus in its ratio of isomers, this could indicate a different origin; e.g., a high content of *myo*-inositol would suggest that unchanged plant material is present. Hesse (1973) considers that organic phosphorus entering a lake will be deposited eventually in the sediment largely unchanged and that subsequent mineralization of this material will be slow. The main reason for this is that plant remains usually contain sufficient inorganic phosphorus for decomposition to proceed rapidly without the necessity for mineralization of combined forms.

Inorganic orthophosphate is the form of phosphorus that is biologically mobile and is the principle state in which phosphorus is interchanged among various biological components of water systems. The group of enzymes—the phosphatases (Feder, 1973)—principally concerned with hydrolysis of organic phosphates, and inorganic polyphosphates, to inorganic orthophosphate is widespread in microorganisms (Section 3) and includes both nonspecific enzymes (e.g., acid and alkaline phosphatases) and specific enzymes (e.g., phytases, ribonucleases, phospholipases, etc.). As already stated, the level of phosphatase-hydrolyzable compounds in natural waters is usually very low (Strickland and Solorzano, 1966; Rigler, 1961; Herbes *et al.* 1975), but the measurement of phosphatase levels in such waters has received a good deal of attention.

Studies by Reichardt *et al.* (1967) on the phosphatase activity of eutrophic lake water indicated that both acid phosphatase (pH optimum 5.6) and alkaline phosphatase (pH optimum 8.7–9.1) were present in low concentrations. The activity of filtered (0.6 $\mu$m) water was much less than that of unfiltered water. These investigations were later extended by Reichardt (1971) to more detailed studies of the comparison of lake water phosphatases with the enzymic activities in culture media containing a blue-green alga such as *Anacystis nidulans*.

Jones (1972) has recently determined alkaline phosphatase levels in a series of lakes of differing degrees of eutrophication. He concluded that phosphatase activity was significantly correlated with total phosphorus and biomass and also with the degree of eutrophication of the lakes, except where the dinoflagellate *Ceratium* sp. was dominant. This organism was found to contain lower levels of the enzyme than those found in other organisms. Jones (1972) also found that there was a positive correlation between chlorophyll-*a* estimates and bacterial counts. Alkaline phosphatase activity of paper-filtered samples correlated positively with both algal and bacterial numbers, but the activity of membrane-filtered samples was significantly correlated only with numbers of phosphatase-producing bacteria.

As yet, there do not appear to have been any attempts to isolate soluble, exocellular enzymes from natural waters in order to investigate their properties by means of the standard techniques of the biochemist. The low levels of activity normally encountered should not now present insuperable difficulties.

The biological availability of sediment phosphorus has recently been discussed by Syers *et al.* (1973). The system is analogous in many ways to the terrestial soil system (Section 3), and the stabilization of organic phosphorus in such soils, through the formation of insoluble derivatives, is likely to be important in sediments also. The high levels of ferrous iron and dissolved inorganic phosphorus characteristic of deoxygenated lake bottom waters—the hypolimnion—are attributable, at least in part, to the release of insoluble inorganic phosphorus by reduction of its associated ferric iron. This is probably a similar process to the one involved in the solubilization by reduction of ferric phytate in waterlogged soils (Furukawa and Kawaguchi, 1969). Thus, this return of phosphorus to the water phosphorus cycle is an indirect and probably important effect of microbial activity. Microorganisms capable of dephosphorylating phenolpthalein phosphate, glycerophosphate, lecithin, DNA, and RNA were isolated from sediments obtained from a eutrophic lake by Strzelczyk *et al.* (1972), but none of the many isolates tested was capable of dephosphorylating phytin. This is in agreement with results of Sommers (1971), who found that *p*-nitrophenyl phosphate, nucleotides, etc., were hydrolyzed by sediment suspensions but that *myo*-inositol hexaphosphate was hydrolyzed only slightly or not at all. Results of Rodel and Armstrong (1973) were similar, and no hydrolysis of added *myo*-inositol hexaphosphate was observed even after a 16-week incubation period. Weimer (1973) found that little or no organic phosphorus was mineralized when Lake Wingra (Wisconsin) sediment samples were incubated for 12 weeks, either aerobically or anaerobically. However, Weimer (1973) admitted that analytical difficulties, with the extraction method of Mehta *et al.* (1954), made interpretation of the results uncertain. Although there was an increase in extractable inorganic phosphorus after incubation, there was an even greater increase in total extractable phosphorus. In contrast to the reports of earlier workers, Weimer (1973) showed that samples of sediment from three lakes all hydrolyzed added *myo*-inositol hexaphosphate over a 3-month period of incubation, aerobically or anaerobically. These results could possibly be explained by Weimer's use of a much higher substrate-to-sediment ratio than that used by the earlier workers but, as is so often the case with investigations involving transformations of phosphorus in biological systems, the interpretation of analytical results is fraught with difficulties.

## 8. Microbial Degradation of Phosphorus-Containing Insecticides

Since the original observations of Lange and Kreuger (1932) and of Schrader (1952) concerning their physiological effects, about 140 phosphorus compounds have been used as insecticides (Eto, 1974). A general formula for biologically active organophosphorus compounds was proposed by Schrader (1952), viz.,

$$
\begin{array}{ccc}
R^1 & & O(S) \\
 \diagdown & \diagup\!\!\diagup & \\
 & P & \\
 \diagup & \diagdown & \\
R^2 & & Acyl
\end{array}
$$

where $R^1$ and $R^2$ are alkyl, alkoxy, or amino groups and "acyl" may be an inorganic or organic acid radical, such as $Cl^-$, $F^-$, $SCN^-$, or $CH_3COO^-$. Many successful insecticides are of this type. Some examples are given in Table I.

Unlike insecticides belonging to the chlorinated hydrocarbon group, e.g., DDT, dieldrin, etc., organophosphorus pesticides are generally short-lived in the environment and in biological systems. Much of the applied pesticide finds its way into the soil unchanged, either directly or via mechanical removal from foliage through wind or rain action. Its subsequent fate depends to a great extent on the properties of the soil as, in addition to microbial modification, the pesticide may be hydrolyzed chemically, immobilized by adsorption onto soil colloids, or taken up by plant roots.

Hydrolysis rapidly converts most organophosphorus pesticides into water-soluble nontoxic compounds. Many pesticides are triesters, and in those cases enzymic hydrolysis is initiated by phosphotriesterases. These occur commonly in biological systems (Aldridge and Reiner, 1972) and

**Table I**

---

### PHOSPHOROTHIONATES

PARATHION $(C_2H_5O)_2 \overset{\overset{S}{\|}}{P}-O-\langle\bigcirc\rangle-NO_2$

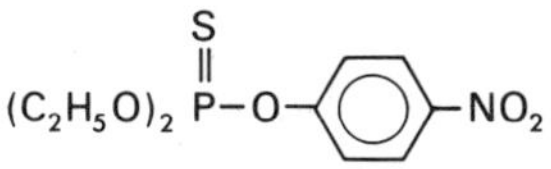

DIAZINON $(C_2H_5O)_2 \overset{\overset{S}{\|}}{P}-O-$ $-CH(CH_3)_2$

### PHOSPHATE

MEVINPHOS $(CH_3O)_2 \overset{\overset{O}{\|}}{P}-O-\overset{\overset{CH_3}{|}}{C} = CH \overset{\overset{O}{\|}}{C}O CH_3$

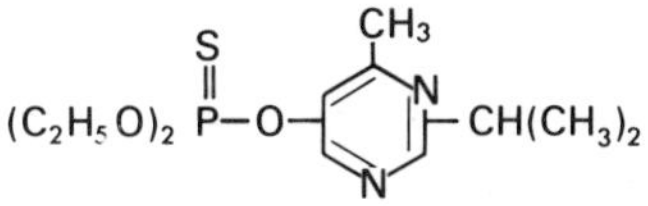

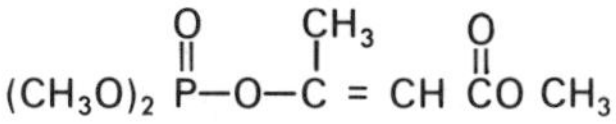

### PHOSPHORAMIDE

SCHRADAN $[(CH_3)_2N]_2 \overset{\overset{O}{\|}}{P}-O-\overset{\overset{O}{\|}}{P} [N(CH_3)_2]_2$

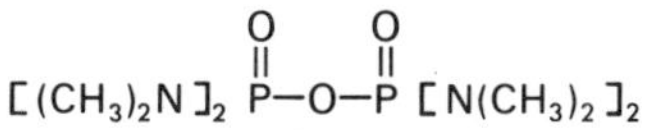

### PHOSPHOROTHIOLOTHIONATES

MALATHION $(CH_3O)_2 \overset{\overset{S}{\|}}{P}-S-\overset{\overset{}{}}{C}H-\overset{\overset{O}{\|}}{C}O C_2H_5$
$CH_2 \overset{\overset{}{}}{C}O C_2H_5$
$\overset{\overset{}{}}{\underset{O}{\|}}$

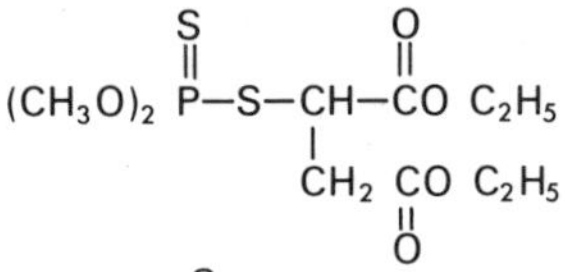

DISULPHOTON $(C_2H_5O)_2 \overset{\overset{S}{\|}}{P}-S-CH_2 CH_2 S C_2H_5$

PHORATE $(C_2H_5O)_2 \overset{\overset{S}{\|}}{P}-S-CH_2 S C_2H_5$

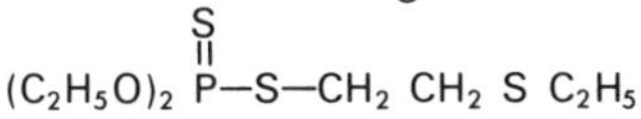

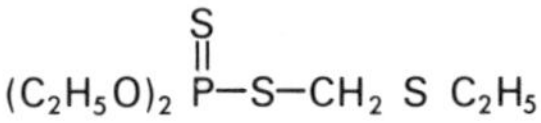

CARBOPHENOTHION $(C_2H_5O)_2 \overset{\overset{S}{\|}}{P}-S-CH_2-S-\langle\bigcirc\rangle-Cl$

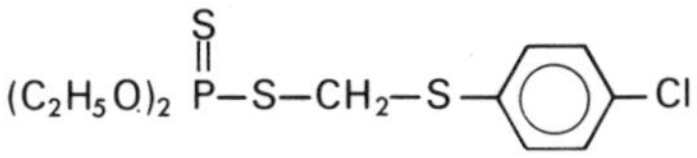

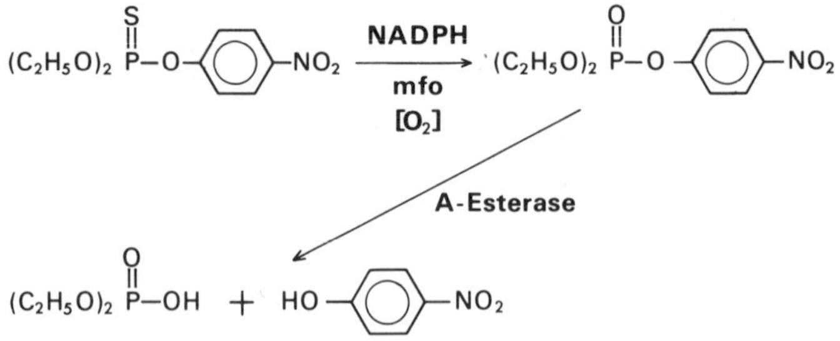

$$(C_2H_5O)_2 \overset{\overset{S}{\|}}{P}-O-\underset{}{\bigcirc}-NO_2 \xrightarrow[\substack{mfo \\ [O_2]}]{\textbf{NADPH}} (C_2H_5O)_2 \overset{\overset{O}{\|}}{P}-O-\underset{}{\bigcirc}-NO_2$$

**A-Esterase**

$$(C_2H_5O)_2 \overset{\overset{O}{\|}}{P}-OH \;+\; HO-\underset{}{\bigcirc}-NO_2$$

**Figure 6.** Oxidation followed by hydrolysis of parathion.

are usually of the arylesterase (A-esterase) type. They are capable of cleaving not only bonds such as $> P(O)-O$-aryl and $> P(O)-S$- but also such bonds as $> P(O)-O$-$P(O)<$, $> P(O)-F$, and $> P(O)-CN$. Compounds containing the thiophosphoryl group, e.g., phosphorothionates and phosphorothiolothionates (Table I) are not hydrolyzed by A-esterases. In the case of these compounds, a preliminary oxidation step is necessary before A-esterase hydrolysis becomes possible. Such oxidation of the intact pesticide molecule by so-called "mixed function oxidase" (mfo) systems (Fig. 6) has been demonstrated in mammalian and insect organs but apparently not in microorganisms.

It is likely that in soil, hydrolysis of organophosphorus compounds containing the thiophosphoryl group is a chemical reaction, probably catalyzed by metallic ions. Mortland and Raman (1967) found that diazinon was decomposed rapidly by contact with Cu (II)-montmorillonite. Products of this type of degradation are diesters and thus not only much less toxic but more susceptible to further biological degradation.

A further example of detoxication is the biotransformation of the nonphosphorus functional groups of pesticides. This is particularly important in mammals as the products, being usually more polar, are often of low toxicity. For example, malathion is rapidly hydrolyzed by mammalian carboxyesterase to give an inactive monoacid (Fig. 7).

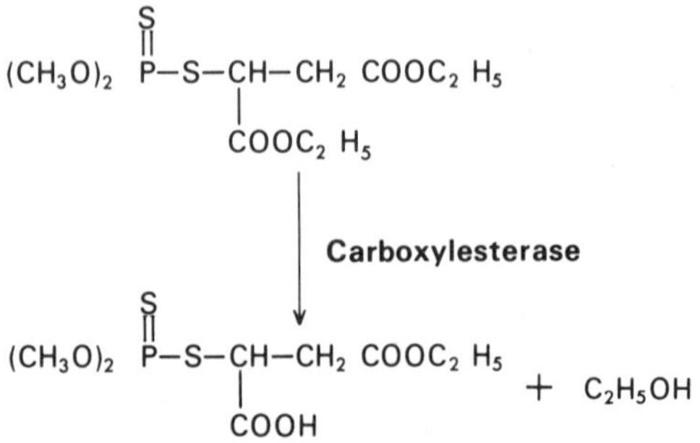

$$(CH_3O)_2 \overset{\overset{S}{\|}}{P}-S-\underset{\underset{COOC_2H_5}{|}}{CH}-CH_2\,COOC_2H_5$$

**Carboxylesterase**

$$(CH_3O)_2 \overset{\overset{S}{\|}}{P}-S-\underset{\underset{COOH}{|}}{CH}-CH_2\,COOC_2H_5 \;+\; C_2H_5OH$$

**Figure 7.** Hydrolysis of malathion.

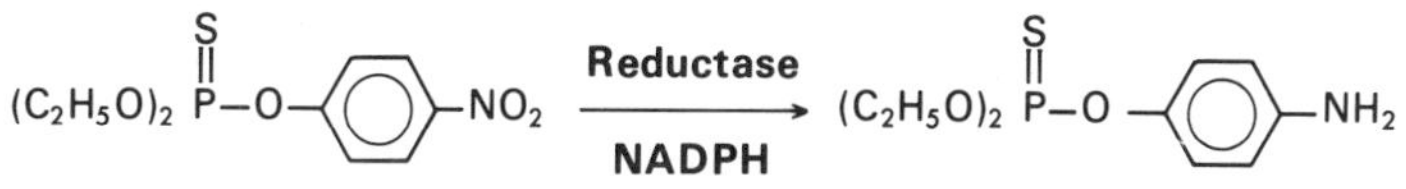

Figure 8. Reduction of the nitro group of parathion to an amino group.

An unusually stable preparation with this type of activity has recently been extracted with dilute sodium hydroxide from irradiation-sterilized soil (Getzin and Rosefield, 1971). The origin of this free enzyme is unknown; both microorganisms and higher plants contain carboxyesterases capable of degrading malathion in this manner. The common soil fungus *Trichoderma viride* was shown by Matsumura and Boush (1966) to have high malathion carboxyesterase activity, but this enzyme was not considered by Getzin and Rosefield (1971) to be identical with the soil extract. Matsumura and Boush (1966) also isolated a bacterium, *Pseudomonas* sp., with high malathion carboxyesterase activity.

Other biotransformations of pesticides mediated by microorganisms are known to occur. The nitro group of parathion and similar compounds can be reduced to an amino group, giving compounds that are inactive as insecticides or anticholinesterase agents (Ahmed *et al.*, 1958; Matsumura, 1972) (Fig. 8). Aminoparathion is a major metabolite in ruminants (Ahmed *et al.*, 1958), presumably as a result of the activity of rumen organisms. Addition of glucose to nonsterile soil containing parathion was shown by Lichtenstein (1972) to stimulate the production of aminoparathion; the reduction was attributed to the presence of a yeast in the soil. *Rhizobium* also reduces parathion rapidly to aminoparathion (Mick and Dahm, 1970).

In many instances it appears that chemical hydrolysis is necessary as a first step before microbial degradation is possible. For example, fumigation of soils with propylene oxide was shown by Getzin (1967) not to inhibit the hydrolysis of $^{14}C$-ring-labeled diazinon, but little $^{14}CO_2$ was released from the fumigated soil. Labeled 2-isopropyl-4-methyl-6-hydroxypyrimidine, a hydrolysis product of diazinon, could be extracted from the fumigated diazinon-treated soil. The predominant microbial population arising in diazinon-treated soils (*Pseudomonas, Arthobacter, Streptomyces*, etc.), appears to attack the product of chemical hydrolysis rather than the intact diazinon (Fig. 9).

In flooded paddy soils, however, soil organisms appear to be able to degrade the diazinon; in such situations the resulting 2-isopropyl-4-methyl-6-hydroxypyrimidine does not decompose further (Sethunathan and Yoshida, 1969).

Although the experimental work referred to above leaves no doubt that organophosphorus pesticides undergo microbial degradation in soil,

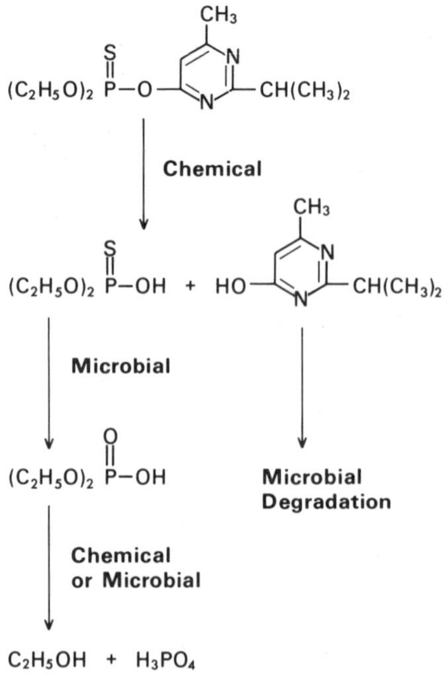

**Figure 9.** Degradation of diazinon in soil.

it does not provide answers to the crucial question, i.e., why do micro-organisms in the presence of natural nutrients degrade insecticides that are often present in minute amounts? Possible answers to this question have been discussed by Matsumura and Boush (1971). It is argued that their lipophilic properties may permit them to be selectively adsorbed on the outer surfaces of cells in a relatively polar medium such as soil and water (Ko and Lockwood, 1968). This adsorption would be followed by penetration of the cell wall. The surviving organisms would then deal with the pesticide, either by storing it unaltered, by degrading it to a less toxic, perhaps assimilable form, or by degrading it to a structure that the organism is able to reject.

## 9. Conclusion

The state of our knowledge of the microbial role in the phosphorus cycle is still uneven and, in all aspects studied, far from complete. Some progress has been made in the past 2 decades in such important fields as the chemical nature and modes of dephosphorylation of soil organic phosphates, but even here much remains to be discovered. In particular, the origin of the organic phosphorus compounds found in soil is largely unknown. The study of organic phosphorus in what are now acknowledged to be environmentally important areas, such as sewage sludges and effluents, natural water bodies, and lake sediments, has hitherto not received much attention. The application of modern methods of extraction, concentration, separation, and identification, particularly those whose success in the soil organic phosphorus field has been amply demonstrated, will no doubt be of value in these areas.

As has been repeatedly emphasized in this article, it is always necessary to check carefully the validity of any method of analysis for phosphorus when it is first applied to a new problem. The difficulties met with in the analysis of soils for organic phosphorus, in which the result is expressed as the difference between "total" and "inorganic" phosphorus, are an illustration of the magnitude of the problem when investigating changes in a complex system. In spite of considerable effort by many workers (see Anderson, 1967; Halstead and McKercher, 1975), it is still doubtful whether a method has been described that can be applied to soils indiscriminantly. Accuracy is particularly important in investigations in which it is necessary to measure the level of organic phosphorus in the presence of large amounts of inorganic phosphorus. The converse of this situation, the measurement of small amounts of inorganic phosphorus in the presence of large amounts of organic phosphorus, can also be troublesome. It has been reported by two groups (Irving and Cosgrove, 1970; Bartlett and Lewis, 1970) that *myo*-inositol hexaphosphate forms blue complexes with molybdate under conditions of the assay for inorganic phosphorus.

Although the world's reserves of phosphorus are such that serious shortages are not likely until at least well into the next century, the probability that manufactured phosphorus fertilizers will become expensive in the near future is one that cannot be overlooked. Continual increases in the cost of nitrogenous fertilizer have undoubtedly acted as part of the stimulus to research on nitrogen fixation by plants. Recent advances in the field of nitrogen fixation (Postgate, 1975) have uncovered possibilities for the extension of nitrogen-fixing capabilities to plants other than those currently able to act as hosts to rhizobia. An analogous field of research with regard to phosphorus is the possibility of increasing the ability of crop plants and pastures to make use of soil phosphorus that is not easily available. The recent work discussed in Section 3 shows that plants whose roots are infected with certain types of mycorrhizae are able to take up phosphorus present in very low concentrations in the soil solution. The extension of the ability to form such symbiotic associations to a wider variety of plants would obviously be of great value to agriculture.

# References

Agnihotri, V. P., 1970, Solubilization of insoluble phosphates by some soil fungi isolated from nursery seedbeds, *Can. J. Microbiol.* **16**:877.

Ahkromeiko, A. I., and Shestakova, V. A., 1958, The effect of rhizosphere micro-organisms upon the uptake and release of phosphorus and sulphur by the roots of arboreal seedlings, *Proc. Int. Conf. Peaceful Uses Atomic Energy, 2nd Conf., Geneva* **27**:193.

Ahmed, M. K., Casida, J. E., and Nichols, R. E., 1958, Bovine metabolism of organophosphorus insecticides: Significance of rumen fluid with with particular reference to parathion, *J. Agric. Food Chem.* **6**:740.

Aldridge, W. H., and Reiner, E., 1972, *Enzyme Inhibitors as Substrates*, North-Holland, Amsterdam.

Allison, F. E., 1973, *Soil Organic Matter and its Role in Crop Production*, Elsevier, Amsterdam.

Anderson, G., 1956, The identification and estimation of soil inositol phosphates, *J. Sci. Food Agric.* **7**:437.

Anderson, G., 1967, Nucleic acids, derivatives, and organic phosphates, in: *Soil Biochemistry*, Vol. 1 (A. D. McLaren and G. H. Peterson, eds.), pp. 67–90, Marcel Dekker, New York.

Anderson, G., 1970, The isolation of nucleoside diphosphates from alkaline extracts of soil, *J. Soil Sci.* **21**:96.

Anderson, G., and Hance, R. J., 1963, Investigation of an organic phosphorus component of fulvic acid, *Plant Soil* **19**:296.

Anderson, G., and Malcolm, R. E., 1974, The nature of alkali-soluble soil organic phosphates, *J. Soil Sci.* **25**:282.

Anderson, G., and Russell, J. D., 1969, Identification of inorganic pyrophosphate in alkaline extracts of soil, *J. Sci. Food Agric.* **20**:78.

Anderson, L., 1972, The cyclitols, in: *The Carbohydrates: Chemistry and Biochemistry*, Vol. 1A (W. Pigman and D. Horton, eds.), pp. 519–579, Academic Press, New York.

Antia, N. J., McAllister, C. D., Parsons, T. R., Stephens, K., and Strickland, J. D. H., 1963, Further measurements of primary production using a large-volume plastic sphere, *Limnol. Oceanogr.* **8**:166.

Barber, D. A., 1969, The influence of the microflora on the accumulation of ions by plants, in: *British Ecological Society Symposium 9* (I. H. Rorison, ed.), pp. 191–200, Blackwell Scientific, Oxford.

Barr, C. E., and Ulrich, A., 1963, Phosphorus fractions in high and low phosphate plants, *J. Agric. Food Chem.* **11**:313.

Bartlett, E. M., and Lewis, D. H., 1970, Spectrophotometric determination of phosphate esters in the presence and absence of orthophosphate, *Anal. Biochem.* **36**:159.

Bartlett, E. M., and Lewis, D. H., 1973, Surface phosphatase activity of mycorrhizal roots of beech, *Soil Biol. Biochem.* **5**:249.

Benians, G. J., and Barber, D. A., 1974, The uptake of phosphate by barley plants from soil under aseptic and non-sterile conditions, *Soil Biol. Biochem.* **6**:195.

Bieleski, R. L., 1973, Phosphate pools, phosphate transport, and phosphate availability, *Annu. Rev. Plant Physiol.* **24**:225.

Blanchar, R. W., and Riego, D., 1975, Phosphate determinations in waters using an anion exchange resin, *J. Environ. Qual.* **4**:45.

Breitenbach, M., and Hoffmann-Ostenhof, O., 1971, Biosynthesis of *myo*-Inositol 1,3,4,5,6-pentakisphosphate in chicken reticulocytes, *Hoppe-Seyler's Z. Physiol. Chem.* **352**:488.

Bromfield, A. R., 1975, Effects of ground rock phosphate–sulphur mixture on yield and nutrient uptake of groundnuts (*Arachis hypogaea*) in Northern Nigeria, *Expl. Agric.* **11**:265.

Bromfield, S. M., and Jones, O. L., 1970, The effect of sheep on the recycling of phosphorus in hayed-off pastures, *Aust. J. Agric. Res.* **21**:699.

Bromfield, S. M., and Jones, O. L., 1972, The initial leaching of hayed-off pasture plants in relation to the recycling of phosphorus, *Aust. J. Agric. Res.* **23**:811.

Brown, M. E., 1972, Plant growth substances produced by micro-organisms of soil and rhizosphere, *J. Appl. Bacteriol.* **35**:443.

Cervelli, S., Nannipieri, P., Ceccanti, B., and Sequi, P., 1973, Michaelis constant of soil acid phosphatase, *Soil Biol. Biochem.* **5**:841.

Chamberlain, W., and Shapiro, J., 1973, Phosphate measurements in natural waters—a critique, in: *Environmental Phosphorus Handbook* (E. J. Griffith, A. Beeton, J. M. Spencer, and D. T. Mitchell, eds.), pp. 355–366, John Wiley and Sons, New York.

Chen, I. W., and Charalampous, F. C., 1967, The mechanism of cyclization of glucose 6-phosphate to D-inositol 1-phosphate, *Biochim. Biophys. Acta* **136**:568.

Chhonkar, P. K., and Subba-Rao, N. S., 1967, Phosphate solubilization by fungi associated with legume root nodules, *Can. J. Microbiol.* **13**:749.

Cooper, R., 1959, Bacterial fertilizers in the Soviet Union, *Soils Fertilizers* **22**:327.

Cosgrove, D. J., 1963, The chemical nature of soil organic phosphorus. I. Inositol phosphates, *Aust. J. Soil Res.* **1**:203.

Cosgrove, D. J., 1966, The chemistry and biochemistry of inositol polyphosphates, *Rev. Pure Appl. Chem.* **16**:209.

Cosgrove, D. J., 1967, Metabolism of organic phosphates in soil, in: *Soil Biochemistry*, Vol. 1 (A. D. McLaren and G. H. Peterson, eds.), pp. 216–228, Marcel Dekker, New York.

Cosgrove, D. J., 1969, The chemical nature of soil organic phosphorus. II. Characterization of the supposed DL-*chiro*-inositol hexaphosphate component of soil phytate as D-*chiro*-inositol hexaphosphate, *Soil Biol. Biochem.* **1**:325.

Cosgrove, D. J., 1970, Inositol phosphate phosphatases of microbiological origin. Inositol phosphate intermediates in the dephosphorylation of the hexaphosphates of *myo*-inositol, *scyllo*-inositol, and D-*chiro*-inositol by a bacterial (*Pseudomonas* sp.) phytase, *Aust. J. Biol. Sci.* **23**:1207.

Cosgrove, D. J., 1972, The origin of inositol polyphosphates in soil. Some model experiments in aqueous systems involving the chemical phosphorylation of *myo*-inositol and the epimerization of *myo*-inositol pentaphosphates, *Soil Biol. Biochem.* **4**:387.

Cosgrove, D. J., 1973, Inositol polyphosphates in activated sludge, *J. Environ. Qual.* **2**:483.

Cosgrove, D. J., Irving, G. C. J., and Bromfield, S. M., 1970, Inositol phosphate phosphatases of microbiological origin. The isolation of soil bacteria having inositol phosphate phosphatase activity, *Aust. J. Biol. Sci.* **23**:339.

Donald, C. M., and Williams, C. H., 1954, Fertility and productivity of a podzolic soil as influenced by subterranean clover (*Trifolium subterraneum* L.) and superphosphate, *Aust. J. Agric. Res.* **5**:664.

Eto, M., 1974, *Organophosphorus Pesticides: Organic and Biological Chemistry*, C.R.C. Press, Cleveland.

Feder, J., 1973, The phosphatases, in: *Environmental Phosphorus Handbook* (E. J. Griffith, A. Beeton, J. M. Spencer, and D. T. Mitchell, eds.), pp. 475–508, John Wiley and Sons, New York.

Floate, M. J. S., 1970a, Decomposition of organic materials from hill soils and pastures. II. Comparative studies on the mineralization of carbon, nitrogen, and phosphorus from plant materials and sheep faeces, *Soil Biol. Biochem.* **2**:173.

Floate, M. J. S., 1970b, Decomposition of organic materials from hill soils and pastures. III. The effect of temperature on the mineralization of carbon, nitrogen, and phosphorus from plant materials and sheep faeces, *Soil Biol. Biochem.* **2**:187.

Floate, M. J. S., 1970c, Decomposition of organic materials from hill soils and pastures. IV. The effects of moisture content on the mineralization of carbon, nitrogen, and phosphorus from plant materials and sheep faeces, *Soil Biol. Biochem.* **2**:275.

Furukawa, H., and Kawaguchi, K., 1969, Contribution of organic phosphorus to the increase of easily soluble phosphorus in water-logged soils, especially related to phytate phosphorus, *Nippon Dojo–Hiryogako Zasshi* **40**:141. (*Chem. Abstr.* 1970, **71**:100348).

Getzin, L. W., 1967, Metabolism of diazinon and zinophos in soils, *J. Econ. Entomol.* **60**:505.

Getzin, L. W., and Rosefield, I., 1971, Partial purification and properties of a soil enzyme that degrades the insecticide malathion, *Biochim. Biophys. Acta* **235**:442.

Ghonsikar, C. P., and Miller, R. H., 1973, Soil inorganic polyphosphates of microbial origin, *Plant Soil* **38**:651.

Greaves, M. P., and Webley, D. M., 1965, A study of the breakdown of organic phosphates by micro-organisms from the root region of certain pasture grasses, *J. Appl. Bactoriol.* **28**:454.

Griffith, E. J., 1973, Environmental Phosphorus—An Editorial, in: *Environmental Phosphorus Handbook* (E. J. Griffith, A. Beeton, J. M. Spencer, and D. T. Mitchell, eds.), pp. 683–695, John Wiley and Sons, New York.

Halstead, R. L., and McKercher, R. B., 1975, Biochemistry and cycling of phosphorus, in: *Soil Biochemistry*, Vol. 4 (E. A. Paul and A. D. McLaren, eds.), pp. 31-63, Marcel Dekker, New York.

Harold, F. M., 1966, Inorganic polyphosphates in biology: Structure, metabolism, and function, *Bacteriol. Rev.* **30**:772.

Hayman, D. S., and Mosse, B., 1972, Plant growth responses to vesicular-arbuscular mycorrhiza. III. Increased uptake of labile P from soil, *New Phytol.* **71**:41.

Helyar, K. R., and Brown, A. L., 1976, Octan-1-ol extraction of molybdophosphoric acid in the colorimetric determination of orthophosphate, *Soil Sci. Soc. Amer. J.* **40**:43.

Herbes, S. E., Allen, H. E., and Mancy, K. H., 1975, Enzymatic characterization of soluble organic phosphorus in lake water, *Science* **187**:432.

Hesse, P. R., 1973, Phosphorus in lake sediments, in: *Environmental Phosphorus Handbook* (E. J. Griffith, A. Beeton, J. M. Spencer, and D. T. Mitchell, eds.), pp. 573–583, John Wiley and Sons, New York.

Hooper, F. F., 1973, Origin and fate of organic phosphorus compounds in aquatic systems, in: *Environmental Phosphorus Handbook* (E. J. Griffith, A. Beeton, J. M. Spencer, and D. T. Mitchell, eds.), pp. 179–201, John Wiley and Sons New York.

Hutchinson, G. E., 1957, *A Treatise on Limnology*, John Wiley and Sons, New York.

Huxley, A., 1928, *Point Counter Point*, Chatto and Windus, London.

Irving, G. C. J., and Cosgrove, D. J., 1970, Interference by *myo*-inositol hexaphosphate in inorganic orthophosphate determinations, *Anal. Biochem.* **36**:381.

Irving, G. C. J., and Cosgrove, D. J., 1971, Inositol phosphate phosphatases of microbiological origin. Some properties of a partially purified bacterial (*Pseudomonas* sp.) phytase, *Aust. J. Biol. Sci.* **24**:547.

Irving, G. C. J., and Cosgrove, D. J., 1972, Inositol phosphate phosphatases of microbiological origin: The inositol pentaphosphate products of *Aspergillus ficuum* phytases, *J. Bacteriol.* **112**:434.

Irving, G. C. J., and Cosgrove, D. J., 1974, Inositol phosphate phosphatases of microbiological origin. Some properties of the partially purified phosphatases of *Aspergillus ficuum* NRRL 3135, *Aust. J. Biol. Sci.* **27**:361.

Jenkins, D., and Menar, A. B., 1967, The fate of phosphorus in sewage treatment processes, I. Primary sedimentation and activated sludge, SERL Report No. 67-6, University of California, Berkeley.

Johannes, R. E. 1964, Uptake and release of dissolved organic phosphorus by representatives of a coastal marine ecosystem, *Limnol. Oceanogr.* **9**:224.

Johnson, L. F., and Tate, M. E., 1969, Structure of "phytic acids," *Can. J. Chem.* **47**:63.

Jones, J. G., 1972, Studies on freshwater micro-organisms: Phosphatase activity in lakes of differing degrees of eutrophication, *J. Ecol.* **60**:777.

Jones, O. L., and Bromfield, S. M., 1969, Phosphorus changes during the leaching and decomposition of hayed-off pasture plants, *Aust. J. Agric. Res.* **20**:653.

Kardos, L. T., 1964, Soil fixation of plant nutrients, in: *Chemistry of the Soil* (F. E. Bear, ed.), pp. 369–394, Reinhold, New York.

Kimerle, R. A., and Rorie, W., 1973, Low-level phosphorus detection methods, in: *Environmental Phosphorus Handbook* (E. J. Griffith, A. Beeton, J. M. Spencer, and D. T. Mitchell, eds.), pp. 367–379, John Wiley and Sons, New York.

Ko, W-H., and Hora, F. K., 1970, Production of phospholipases by soil micro-organisms, *Soil Sci.* **110**:355.

Ko, W-H., and Lockwood, J. L., 1968, Accumulation and concentration of chlorinated hydrocarbon pesticides by micro-organisms in soil, *Can. J. Microbiol.* **14**:1075.

Kowalenko, C. G., and McKercher, R. B., 1971, Phospholipid components extracted from Saskatchewan soils, *Can. J. Soil Sci.* **51**:19.

Kuenzler, E. J., 1970, Dissolved organic phosphorus excretion by marine phytoplankton, *J. Phycol.* **6**:7.

Lange, W., and Kreuger, B., 1932, Über Ester der Monofluorophosphorsäure, *Chem. Ber.* **65**:1598.

L'Annunziata, M. F., 1975, The origin and transformations of the soil inositol phosphate isomers, *Soil Sci. Soc. Amer. Proc.* **39**:377.

L'Annunziata, M. F., and Fuller, W. H., 1971, Soil and plant relationships of inositol phosphate stereoisomers; the identification of D-*chiro*- and *muco*-inositol phosphates in a desert soil and plant system, *Soil Sci. Soc. Amer. Proc.* **35**:587.

L'Annunziata, M. F., Fuller, W. H., and Brantley, D. S., 1972, D-*chiro*-inositol phosphate in a forest soil, *Soil Sci. Soc. Amer. Proc.* **36**:183.

Larsen, S., 1967, Soil phosphorus, in: *Advances in Agronomy*, Vol. 19 (A. G. Norman, ed.), pp. 151–210, Academic Press, New York.

Lean, D. R. S., 1973, Phosphorus dynamics in lake water, *Science* **179**:678.

Lean, D. R. S., 1975, Phosphorus dynamics in lake water: Contribution by death and decay, *Science* **187**:455.

Lévesque, M., 1969, Characterization of model and soil organic matter metal-phosphate complexes, *Can. J. Soil Sci.* **49**:365.

Lévesque, M., 1970, Contribution de l'acide fulvique et des complexes fulvo-métalliques à la nutrition minérale des plantes, *Can. J. Soil Sci.* **50**:385.

Levin, G. V., Topol, G. J., and Tarnay, A. G., 1975, Operation of full-scale biological phosphorus removal plant, *J. Water Pollut. Control Fed.* **47**:577.

Leyden, D. E., Nonidez, W. K., and Carr, P. W., 1975, Determination of parts per billion phosphate in natural waters using X-ray fluorescence spectrometry, *Anal. Chem.* **47**:1449.

Lichtenstein, E. P., 1972, Persistence and fate of pesticides in soils, water, and crops: Significance to humans, in: *Pesticide Chemistry, Proceedings 2nd International IUPAC Congress. VI* (A. S. Tahori, ed.), pp. 1–22, Gordon & Breach, London.

Loewus, F. A., 1974, The biochemistry of *myo*-inositol in plants, in: *Recent Advances in Phytochemistry 8* (V. C. Runeckles and E. E. Conn, eds.), pp. 179–207, Academic Press, New York.

Martin, J. K., 1973, The influence of rhizosphere microflora on the availability of [32]-P-*myo*-inositol hexaphosphate phosphorus to wheat, *Soil Biol. Biochem.* **5**:473.

Martin, J. K., and Cunningham, R. B., 1973, Factors controlling the release of phosphorus from decomposing wheat roots, *Aust. J. Biol. Sci.* **26**:715.

Martin, J. K., and Molloy, L. F., 1971, A comparison of the organic phosphorus compounds extracted from soil, sheep faeces, and plant material collected at a common site, *N.Z.J. Agric. Res.* **14**:329.

Matsumura, F., 1972, Metabolism of insecticides in microorganisms and insects, in: *Environmental Quality and Safety*, Vol. 1 (F. Coulston and F. Korte, eds.), p. 96, Georg Thieme, Stuttgart.

132

Dennis J. Cosgrove

Matsumura, F., and Boush, G. M., 1966, Malathion degradation by *Trichoderma viride* and a *Pseudomonas species, Science* **153**:1278.

Matsumura, F., and Boush, G. M., 1971, Metabolism of insecticides by micro-organisms, in: *Soil Biochemistry*, Vol. 2 (A. D. McLaren and J. Skujins, eds.), pp. 320–336, Marcel Dekker, New York.

McKercher, R. B., and Anderson, G., 1968, Content of inositol penta- and hexaphosphates in some Canadian soils, *J. Soil Sci.* **19**:47.

McLaren, A. D., Luse, R. A., and Skujins, J. J., 1962, Sterilization of soil by irradiation and some further observations on soil enzyme activity, *Soil Sci. Soc. Amer. Proc.* **26**:371.

Mehta, N. C., Legg, J. O., Goring, C. A. I., and Black, C. A., 1954, Determination of organic phosphorus in soils. I. Extraction method, *Soil Sci. Soc. Amer. Proc.* **18**:443.

Mick, D. L., and Dahm, P. A., 1970, Metabolism of parathion by two species of *Rhizobium, J. Econ. Entomol.* **63**:1155.

Minear, R. A., 1972, Characterization of naturally occurring dissolved organophosphorus compounds, *Environ. Sci. Technol.* **6**:431.

Minear, R. A., 1975, Phosphorus dynamics in lake water: Contribution by death and decay, *Science* **187**:454.

Mortland, M. M., and Raman, K. V., 1967, Catalytic hydrolysis of some organic phosphate pesticides by copper (II), *J. Agric. Food Chem.* **15**:163.

Mosse, B., 1973, Plant growth responses to vesicular-arbuscular mycorrhiza. IV. In soil given additional phosphate, *New Phytol.* **72**:127.

Mosse, B., Hayman, D. S., and Arnold, D. J., 1973, Plant growth responses to vesicular-arbuscular mycorrhiza. V. Phosphate uptake by three plant species from P-deficient soils labelled with $^{32}$P, *New Phytol.* **72**:809.

Moyer, J. R., and Thomas, R. L., 1970, Organic phosphorus and inositol phosphates in molecular size fractions of a soil organic matter extract, *Soil Sci. Soc. Amer. Proc.* **34**:80.

Nesbitt, J. B., 1969, Phosphorus removal—the state of the art, *J. Water Pollut. Control Fed.* **41**:701.

Nesbitt, J. B., 1973, Phosphorus in waste water treatment, in: *Environmental Phosphorus Handbook* (E. J. Griffith, A. Beeton, J. M. Spencer, and D. T. Mitchell, eds.), pp. 649–668, John Wiley and Sons, New York.

Nishikawa, S., and Kuriyama, M., 1974a, Nucleic acid as a component of mucilage in activated sludge, *J. Ferment. Technol.* **52**:335.

Nishikawa, S., and Kuriyama, M., 1974b, Phytic acid as a component of mucilage in activated sludge, *J. Ferment. Technol.* **52**:339.

Nørgaard Pedersen, E. J., 1953, On phytin phosphorus in the soil, *Plant Soil* **4**:252.

O'Connor, P. W., and Syers, J. K., 1975, Comparison of methods for the determination of total phosphorus in waters containing particulate material, *J. Environ. Qual.* **4**:347.

Omotoso, T. I., and Wild, A., 1970, Occurrence of inositol phosphates and other organic phosphate components in an organic complex, *J. Soil Sci.* **21**:224.

Paul, N. B., and Sundara Rao, W. V. B., 1971, Phosphate-dissolving bacteria in the rhizosphere of some cultivated legumes, *Plant Soil* **35**:127.

Pepper, I. L., and Miller, R. H., 1974, Factors influencing the synthesis of inorganic polyphosphates in soils, *Trans. 10th Int. Congr. Soil Sci.* **4**:290.

Pipes, W. O., 1966, The ecological approach to the study of activated sludge, in: *Advances in Applied Microbiology 8* (W. W. Umbreit, ed.), pp. 77–103, Academic Press, New York.

Postgate, J. R., 1975, Rhizobium as a free-living nitrogen fixer, *Nature (London)* **256**:363.

Powell, C. LL., 1975, Plant growth responses to vesicular-arbuscular mycorrhiza. VIII. Uptake of P by onion and clover infected with different *Endogene* spore types in $^{32}$P labelled soils, *New Phytol.* **75**:563.

Ramirez Martinez, J. R., 1968, Organic phosphorus mineralization and phosphatase activity in soils, *Folia Microbiol.* **13**:161.

Rammler, D. H., and Parkinson, C., 1973a, Hydrolytic enzyme substrates. III. Phosphomono-esterase substrates, *Anal. Biochem.* **52**:208.

Rammler, D. H., Haugland, R., and Shavitz, R., 1973b, Hydrolytic enzyme substrates. I. Chemical synthesis and characterization, *Anal. Biochem.* **52**:180.

Reichardt, W., 1971, Catalytic mobilization of phosphate in lake water and by *Cyanophyta*, *Hydrobiologia* **38**:377.

Reichardt, W., Overbeck, J., and Steubing, L., 1967, Free dissolved enzymes in lake waters, *Nature (London)* **216**:1345.

Riego, D. C., 1974, Amounts and hydrolysis of pyrophosphate and tripolyphosphate in sediments, Ph. D. Thesis, University of Missouri, Columbia.

Rigler, F. H., 1961, The uptake and release of inorganic phosphorus by *Daphnia magna* Straus, *Limnol. Oceanogr.* **6**:165.

Rigler, F. H., 1973, A dynamic view of the phosphorus cycle in lakes, in: *Environmental Phosphorus Handbook* (E. J. Griffith, A. Beeton, J. M. Spencer, and D. T. Mitchell, eds.), pp. 539–572, John Wiley and Sons, New York.

Rodel, M. G., and Armstrong, D. E., 1973, Unpublished observations cited by W. C. Weimer (1973).

Roinestad, F. A., and Yall, I., 1970, Volutin granules in *Zoogloea ramigera*, *Appl. Microbiol.* **19**:973.

Rovira, A. D., 1965, Effects of *Azotobacter*, *Bacillus*, and *Clostridium* on the growth of wheat, in: *Plant Microbe Relationships* (J. Macura and V. Vancura, eds.), pp. 193–200, Czechoslovakian Academy of Science, Prague.

Russell, E. W., 1961, *Soil Conditions and Plant Growth*, 9th Ed., Longmans, London.

Sanders, F. E., and Tinker, P. B., 1971, Mechanism of absorption of phosphate from soil by *Endogene* mycorrhizas, *Nature (London)* **233**:278.

Schnitzer, M., and Khan, S. U., 1972, *Humic Substances in the Environment*, Marcel Dekker, New York.

Schrader, G., 1952, Die Entwicklunge neuer Insektizide auf Grundlage organischer Fluor- und Phosphor-Verbindungen, Monographie Nr. 62 zu *Angewandte Chemie und Chemie-Ingenieur-Technik*, Verlag Chemie, Weinheim.

Sethunathan, N., and Yoshida, T., 1969. Fate of diazinon in submerged soil. Accumulation of hydrolysis product, *J. Agric. Food Chem.* **17**:1192.

Sherman, W. R., Goodwin, S. L., and Gunnell, K. D., 1971, *neo*-Inositol in mammalian tissues. Identification, measurement, and enzymatic synthesis from mannose 6-phosphate, *Biochemistry* **10**:3491.

Shieh, T. R., and Ware, J. H., 1968, Survey of micro-organisms for the production of extracellular phytase, *Appl. Microbiol.* **16**:1348.

Sinha, M. K., 1972, Organo-metallic phosphates. IV. The solvent action of fulvic acids on insoluble phosphates, *Plant Soil* **37**:457.

Skujins, J. J., 1967, Enzymes in soil, in: *Soil Biochemistry*, Vol. 1 (A. D. McLaren and J. J. Skujins, eds.), pp. 371–414, Marcel Dekker, New York.

Skujins, J. J., Braal, L., and McLaren, A. D., 1962, Characterization of phosphatase in a terrestial soil sterilized with an electron beam, *Enzymologia* **25**:125.

Smith, F., Fairbanks, D., Atlas, R., Delwiche, C. C., Gordon, D., Hazen, W., Hitchcock, D., Pramer, D., Skujins, J., and Stuiver, M., 1972, Cycles of elements, in: *Man in the Living Environment* (R. F. Inger, A. D. Hasler, F. H. Bormann, and W. F. Blair, eds.), pp. 48–89, University of Wisconsin Press, Madison.

Smith, J. H., Allison, F. E., and Soulides, D. A., 1961, Evaluation of phosphobacterin as a soil inoculant, *Soil Sci. Soc. Amer. Proc.* **25**:109.

Sommers, L. E., 1971, Organic phosphorus in lake sediments, Ph.D. Thesis, University of Wisconsin, Madison.

Sommers, L. E., Harris, R. F., Williams, J. D. H., Armstrong, D. E., and Syers, J. K., 1970,

Determination of total organic phosphorus in lake sediments, *Limnol. Oceanogr.* **15**:301.

Sommers, L. E., Harris, R. F., Williams, J. D. H., Armstrong, D. E., and Syers, J. K., 1972, Fractionation of organic phosphorus in lake sediments, *Soil Sci. Soc. Amer. Proc.* **36**:51.

Speir, T. W., and Ross, D. J., 1975, Effects of storage on the activities of protease, urease, phosphatase, and sulphatase in three soils under pasture, *N.Z. J. Sci.* **18**:231.

Stevenson, F. J., 1967, Organic acids in soil, in: *Soil Biochemistry*, Vol. 1 (A. D. McLaren and G. H. Peterson, eds.), pp. 119–146, Marcel Dekker, New York.

Steward, J. H., and Tate, M. E., 1969, Gel chromatography of inositol polyphosphates and the avian haemoglobin-inositol pentaphosphate complex, *J. Chromatogr.* **45**:400.

Steward, J. H., and Tate, M. E., 1971, Gel chromatography of soil organic phosphorus, *J. Chromatogr.* **60**:75.

Strickland, J. D. H., and Parsons, T. R., 1960, A manual of sea water analysis, *Bull. Fish. Res. Board Can.* **125**:1.

Strickland, J. D. H., and Solorzano, L., 1966, Determination of monoesterase hydrolysable phosphate and monoesterase activity in sea water, in: *Some Contemporary Studies in Marine Science* (H. Barnes, ed.), pp. 665–674, Allen and Unwin, London.

Strzelczyk, E., Donerski, W., and Lewosz, W., 1972, Occurrence of microorganisms capable of decomposing organic phosphorus compounds in two types of bottom sediments of eutrophic Lake Jeziorak, *Acta Microbiol. Pol., Ser. B* **4**(3):101.

Swaby, R. J., 1975, Biosuper-biological superphosphate, in: *Sulphur in Australasian Agriculture* (K. D. McLachlan, ed.), pp. 213–220, Sydney University Press, Sydney, Australia.

Swaby, R. J., 1976, Unpublished observations.

Syers, J. K., Harris, R. F., and Armstrong, D. E., 1973, Phosphate chemistry in lake sediments, *J. Environ. Qual.* **2**:1.

Tabatabai, M. A., and Bremner, J. M., 1969, Use of *p*-nitrophenyl phosphate for assay of soil phosphatase activity, *Soil Biol. Biochem.* **1**:301.

Thilo, E., 1962, Condensed phosphates and arsenates, in: *Advances in Inorganic Chemistry and Radiochemistry*, Vol. 4 (H. J. Eméleus and A. G. Sharpe, eds.), pp. 1–75, Academic Press, New York.

Thomas, R. L., 1976, Unpublished observations.

Thomas, R. L., and Bowman, B. T., 1966, The occurrence of high molecular weight organic phosphorus compounds in soil, *Soil Sci. Soc. Amer. Proc.* **30**:799.

Vacker, D., Connell, C. H., and Wells, W. N., 1967, Phosphate removal through municipal wastewater treatment at San Antonio, Texas, *J. Water Pollut. Control Fed.* **39**:750.

Veinot, R. L., and Thomas, R. L., 1972, High molecular weight organic phosphorus complexes in soil organic matter:inositol and metal content of various fractions, *Soil Sci. Soc. Amer. Proc.* **36**:71.

Watt, W. D., and Hayes, F. R., 1963, Tracer study of the phosphorus cycle in sea water, *Limnol. Oceanogr.* **8**:276.

Weimer, W. C., 1973, Inositol phosphate esters in lake sediments, Ph.D. Thesis, University of Wisconsin, Madison.

Westland, A. D., and Boisclair, I., 1974, Analytical separation of phosphate from natural water by ion exchange, *Water Res.* **8**:467.

Wild, A., and Oke, O. L., 1966, Organic phosphate compounds in calcium chloride extracts of soils:Identification and availability to plants, *J. Soil Sci.* **17**:356.

Williamson, B., and Alexander, I. J., 1975, Acid phosphatase localized in the sheath of beech mycorrhiza, *Soil Biol. Biochem.* **7**:195.

Yall, I., and Sinclair, N. A., 1971, Mechanisms of biological luxury phosphate uptake, Water Pollution Control Research Series, Project No. 17010DDQ, Report for the U.S. Environmental Protection Agency.

Yall, I., Boughton, W. H., Roinestad, F. A., and Sinclair, N. A., 1972, Logical removal of phosphorus, *Progr. Water Technol.* **1**:231.

# 4

# Biochemical Ecology of Nitrification and Denitrification

## D. D. FOCHT AND W. VERSTRAETE

## 1. Introduction

The transformations of nitrogenous compounds by soil bacteria have been studied for over a century. In terms of the global fluxes between aerial and terrestrial–aquatic systems, the simplified nitrogen cycle can be envisioned as a triangle where the only biologically reversible reaction occurs between ammonium and nitrate. The reverse reactions of dinitrogen fixation or denitrification by biological means do not occur in nature. Hence, the reductive process of denitrification, defined as the reduction of nitrate or nitrite to gaseous nitrogen (usually $N_2$), is intimately associated with the oxidative process of nitrification.

The immediate importance of these two biochemical reactions, nitrification and denitrification, was readily recognized by the nineteenth century soil microbiologists. Considerable progress has been made in elucidating the biochemical steps in the two pathways and the characteristics of the enzymes involved *in vitro*. The chemostat system employed in waste treatment facilities and the use of quasisteady state soil perfusion columns have greatly advanced our understanding on the kinetics of nitrification. On the other hand, the determination of denitrification rates has been confined almost solely to the laboratory. Because dinitrogen constitutes a very

**D. D. FOCHT** • Department of Soil Science and Agricultural Engineering, University of California, Riverside, California, U.S.A.    **W. VERSTRAETE** • Laboratory of General and Industrial Microbiology, State University of Ghent, Belgium.

large portion of the atmosphere (78% by volume), direct measurements of gaseous losses via denitrification are prohibitive *in situ*. Agronomic losses of nitrogen by denitrification are thus estimated indirectly by subtracting measured losses removed by cropping, leaching, and runoff from the added amount of fertilizer. The deficit balance is attributed to denitrification.

Many of the nitrifying and denitrifying microorganisms have been well studied *in vitro*, as have the processes brought about under nonaxenic conditions in nature. However, considerable uncertainty exists—particularly with the denitrifying bacteria—about the actual quantitative contribution from each respective strain to the overall process. The occurrence of nitrifying bacteria other than *Nitrosomonas* and *Nitrobacter* is difficult to assess in light of the competitive exclusion principle. Many genera of microorganisms have been reported to carry out heterotrophic nitrification, yet only recently has this process been shown to be of quantitative significance, as opposed to an academic curiosity, in acidic environments. The occurrence of different genera of denitrifying bacteria in response to concentrations of nitrate, nitrite, organic matter, oxygen, pH, and temperature is not always apparent or understood.

Considerable environmental attention has been directed recently to the biochemical production of nitrogen oxides. The statistically significant correlation of methemoglobinemia in infants with areas containing high nitrate concentration in the drinking water is but one example. Bacterial and/or chemical production of carcinogenic nitrosamines from nitrites and secondary amines is but another aspect of a series of reactions that are potentially favored by acidic or alkaline environments, depending on the specific mechanism and ecology of microorganisms. The concept that denitrification merely returns an innocuous product, dinitrogen, to the atmosphere is being challenged in light of the accompanying, though small, amounts of nitrous oxide that are generated and the recognition that nitrous oxide has a primary role in the natural reduction of ozone in the stratosphere. Thus, it is not merely the scope of this paper to reflect upon how the environment affects the ecology of microorganisms that reduce and oxidize nitrogenous compounds, but to consider the effects that the microorganisms may render upon the environment as well.

## 2. Chemoautotrophic Nitrification

The formation of nitrate in soil was attributed to the decay of organic matter over two centuries ago (Lipman, 1908). Conclusive proof that the production of nitrite and nitrate in soil was a biological process, and not a chemical process as envisioned by the mid-nineteenth century German chemists, was first offered by Schloesing and Müntz (1877). Warington (1884) confirmed their work and discovered that the process was a sequential two-step oxidation of ammonium to nitrite and nitrate. Winogradsky's (1890) experiments with soil perfusion columns led to the discovery and isolation of the chemoautotrophic bacteria *Nitrosomonas* and *Nitrobacter* that were responsible for the respective production of nitrite and nitrate. Not only did this discovery greatly advance the knowledge of nitrogen fertility aspects of soil, but it advanced, for the first time, the concept of chemoautotrophic growth.

## 2.1. Species Diversity

Nitrification studies in pure culture and in nature have centered primarily on the characteristics and activities of *Nitrosomonas* and *Nitrobacter*. Both organisms are gram-negative, aerobic, chemoautotrophic rods. Winogradsky and Winogradsky (1933) reported several other genera of nitrifying bacteria, which Meiklejohn (1954) and Bisset and Grace (1954) dismissed as being contaminants or as being too poorly described. Many of these objections are considered in the eighth edition of *Bergey's Manual of Determinative Bacteriology* (see Buchanan and Gibbons, 1974). Four genera of ammonium-oxidizing bacteria and three genera of nitrite-oxidizing bacteria are considered in the family *Nitrobacteraceae*. All are chemoautotrophic and have been either isolated, reisolated, or redescribed within the last decade (see Table I).

The "secondary autotrophic nitrifiers" (i.e., those other than *Nitrosomonas* and *Nitrobacter*) appear, for the most part, to be present in far smaller numbers than the primary nitrifiers and have more narrow temperature and pH ranges for growth (Table I). Watson and Waterbury (1971) obtained over 200 enrichment cultures of nitrite-oxidizing bacteria from the ocean and found that *Nitrobacter* was far and above the most significant as compared with the other two organisms, *Nitrospina* and *Nitrococcus*. However, Johnson and Sieburth (1976) cautioned about the use of the most probable number method since they were unable to culture marine species of *Nitrosomonas* and *Nitrobacter* from surface slimes which they studied extensively *in situ* by electron microscopy.

Another apparent difference between the primary and secondary nitrifiers is the ability of many species of the former to grow heterotrophically or to be stimulated by organic matter. *Nitrobacter agilis* can be grown heterotrophically (Smith and Hoare, 1968) and is stimulated in its growth by yeast extract (Delwiche and Finstein, 1965). *Nitrobacter winogradskyi* has been shown to have a much greater heterotrophic growth potential

D. D. Focht and W. Verstraete

Table I. Genera and Characteristics of the Chemoautotrophic Nitrifying Bacteria[a]

| Genus[b] | Morphology | Percentage of G + C | Growth range (pure culture) | Habitat |
|---|---|---|---|---|
| Ammonium oxidizers | | | | |
| Nitrosomonas | Straight rods; motile with one or two sub-polar flagella or nonmotile | 47.4–51.0 | 5–40°C, pH 5.8–9.5 | Soil, marine, freshwater |
| Nitrosospira | Spiral shaped; motile with peritrichous flagella or nonmotile | 54.1 | 25–30°C optimum, pH 7.5–8.0 optimum | Soil, will not grow in seawater |
| Nitrosococcus | Cocci in pairs or tetrads; motile with single or tuft of peritrichous flagella or nonmotile | 50.5–51.0 | 2–30°C, pH 6.0–8.0 | Soil, marine, freshwater |
| Nitrosolobus | Lobular, pleomorphic cells partially compartmentalized by cytomembranes; motile with peritrichous flagella | 53.6–55.1 | 15–30°C, pH 6.0–8.2 | Soil |
| Nitrite oxidizers | | | | |
| Nitrobacter | Short rods; motile with a single polar flagellum or nonmotile | 60.7–61.7 | 5–40°C, pH 5.7–10.2 | Soil, marine, freshwater |
| Nitrospina | Long, slender rods; nonmotile | 57.7 | 20–30°C, pH 7.0–8.0, grow only in 70–100% seawater | Marine |
| Nitrococcus | Spherical cells with cytomembranes forming a branched network in cytoplasm | 57.7 | 20–30°C, pH 7.0–8.0, grow only in 70–100% seawater | Marine |

[a] All information, except growth ranges for Nitrosomonas and Nitrobacter, was taken from S. D. Watson, 1974, Bergey's Manual of Determinative Bacteriology, 8th Ed., Buchanan and Gibbons (eds.), pp. 450–456.
[b] A new species, Nitrosovibrio tenuis, has recently been proposed by Harms et al. (1976).

than either *Nitrococcus* or *Nitrospina* (Watson and Waterbury, 1971). Though the rates of nitrite oxidation in the presence or absence of acetate are similar between *Nitrobacter* and *Nitrococcus*, the rates of nitrite oxidation in the presence or absence of acetate are about 100-fold greater with *Nitrobacter*. Comparative rates of nitrite oxidation and acetate oxidation between *Nitrobacter* and *Nitrospina* are about five times greater. Similarly, pyruvate and amino acids stimulate the growth of *Nitrosomonas europaea* (Clark and Schmidt, 1966, 1967a,b), while *Nitrosolobus multiformis* is unaffected by organic compounds; in some instances, a depression of oxygen uptake occurs (Watson *et al.*, 1971). The authors postulated that the strict autotrophic nature of *Nitrosolobus* may be due to the lack of α-ketoglutaric dehydrogenase and the low specific activity of succinic dehydrogenase in preventing utilization of organic compounds. In light of these observations in pure culture, it is easy to envision how natural environments, particularly soil, would selectively favor (i.e., with the presence of organic substrates) the primary nitrifiers over the secondary ones, notwithstanding the effects of temperature and pH.

Although *N. agilis* has been dropped from the eighth edition of *Bergey's Manual of Determinative Bacteriology* (Buchanan and Gibbons, 1974) on the grounds that it is the same as *N. winogradskyi*, recent evidence would suggest that they are different. Fliermans *et al.* (1974) found no cross reactions between the two when using specific fluorescent-antibody techniques for enumeration of both species in nature. Of 15 different isolates obtained, the reaction was specific with one or the other fluorescent antibody. *N. agilis* was dominant in oxidation ditches receiving solid cattle and food wastes and in caves, while *N. winogradskyi* was dominant in agricultural soils from Minnesota, Morocco, and Iceland. It is difficult to pinpoint the selective factors, though it appears that the only factors in common with the caves and waste ditches would be a high nitrate concentration, slightly alkaline pH, and high carbonate (or carbon dioxide) concentration. These last two factors would favor the faster growing species. Further studies (Fliermans and Schmidt, 1975) definitely show that the growth patterns in both pure and mixed cultures are different between the two species. This confirms earlier studies by McLaren and Skujins (1963), who also noted that *N. agilis* grew much faster than *N. winogradskyi*.

Because of the competitive exclusion principle, one would expect that the nitrifying bacteria, which compete for a highly specialized niche, would be fairly similar. Not surprisingly, there are examples of similarity between isolates of *Nitrobacter* obtained throughout the world (Silver, 1961) and within the two species, *N. agilis* and *N. winogradskyi* (Fliermans *et al.*, 1974). A strain of *Nitrosomonas* isolated from sewage sludge in England proved to be identical in its DNA base content and fine structure to one isolated from sewage in Chicago (Walker, 1975).

## 2.2. Growth and Distribution

By comparison with the growth of most heterotrophic bacteria, growth of the nitrifiers is a slow process. Generation times as short as 8 hr for the growth of *Nitrosomonas* (Skinner and Walker, 1961) and *Nitrobacter* (Morrill and Dawson, 1967) have been reported in pure culture. However, more realistic generation times in nature appear to be somewhere between 20–40 hr for both organisms (Buswell *et al.*, 1954; Knowles *et al.*, 1965). There are many factors that are overcome in pure culture which limit the growth of both organisms in nature. Also, growth in pure culture can be accelerated by use of soluble bicarbonate salts and by periodic additions of alkali to offset the drop in pH (Engel and Alexander, 1958).

On the basis of the competitive exclusion principle, McLaren (1971) suggested that chemoautotrophic processes, such as nitrification, should have more quantitative meaning than heterotrophic processes. Just as the effects of temperature, substrates, pH, and other environmental variables can be expressed with reasonable kinetic certainty, so can the amount of nitrogen oxidized and carbon dioxide fixed per unit biomass be reasonably determined in nature. *Nitrosomonas* and *Nitrobacter*, respectively, oxidize about 35 and 100 atoms of nitrogen for fixation of a molecule of carbon dioxide (Alexander, 1965). About three times as much nitrogenous substrate is thus required for the growth of nitrite-oxidizing bacteria than for ammonium-oxidizing bacteria because the resulting free energy change ($\Delta F$) is about $-65$ vs. $-20$ kcal/mol for the oxidation of ammonium and nitrite, respectively. Cell yields and oxidation rates calculated from the relationship of carbon dioxide fixation or empirically determined from bacterial counts and substrate oxidation rates give similar results: About 1 to $4 \times 10^4$ cells of *Nitrobacter*/$\mu$g N are produced from the oxidation of nitrite (Belser, 1974; Ardakani *et al.*, 1974a; Schmidt, 1974). Roughly three times this cell concentration has been shown to be supported from an equivalent amount of ammonium N oxidized by *Nitrosomonas* in soil with a continuous perfusion of ammonium (Volz *et al.*, 1975a,b).

Despite the fact that larger growth yields are evident from a mole of ammonium than a mole of nitrite, the numbers of ammonium-oxidizing bacteria do not always appear to be larger than nitrite-oxidizing bacteria in nature. Morrill and Dawson (1967) observed that the numbers of *Nitrosomonas* were slightly larger than *Nitrobacter*, though this varied depending on the pH of the soil. In one case, the maximal cell densities (slightly greater than $10^6$/g soil) were about the same for both organisms, where the soil was slightly acid (pH 6.4) and where no accumulation of nitrite was observed. Ardakani *et al.* (1974b) showed that the numbers of ammonium-oxidizing bacteria (about $10^4$/g) were highest in the surface 0–2.5 cm, while nitrite-oxidizing bacteria were 50 times higher in the same sample. These results were unlike those expected when thermodynamic considerations were taken into account, as shown by a model proposed by McLaren (1971) for the growth of nitrifiers in soil. These apparent anomalies may be resolved when one considers that ammonium, unlike nitrite, is subjected to fixation by clays and may be concentrated very close to the surface. Nitrification in soil columns or in sediments is associated with a vectoral factor that does not occur in experiments with incubation vessels. Thus, the counts obtained from a sample increment between the surface and at a depth of a few centimeters may greatly underestimate the numbers of *Nitrosomonas* per mass of soil if the effective zone of ammonium oxidation occurs in the first few millimeters. In fact, the numbers of *Nitrosomonas* decline more drastically with depth than the numbers of *Nitrobacter* in soil (Ardakani *et al.*, 1974b) and sediments (Curtis *et al.*, 1975).

The relationship of nitrite oxidation to cell density appears to be valid only for growing cultures since oxidation decreases as the cells become older (Hofman and Lees, 1952; Seeler and Engel, 1959; Gould and Lees, 1960). Maximal nitrite oxidation activity has been reported to decline 40 hr before the maximal cell density is reached with both *N. agilis* and *N. winogradskyi* (Fliermans and Schmidt, 1975). Interestingly, the pattern of activity for *N. agilis* was the same in both pure culture and mixed culture, but the maximal cell yield was reduced from $8 \times 10^7$/ml in pure culture to $1 \times 10^6$/ml in mixed culture. Growth and activity of *N. winogradskyi* were both faster in mixed culture, though the maximal cell yield in mixed culture ($10^7$/ml) was lower than in pure culture ($5 \times 10^7$/ml).

## 2.3. Biochemical Pathway

The oxidation of ammonium to nitrite by *Nitrosomonas* involves the participation of molecular oxygen, which serves in two ways: by direct incorporation into the substrate and as the terminal electron acceptor. Using $^{18}O$, Rees and Nason (1965) established that one of the oxygen atoms of nitrite originated from molecular oxygen and the other from water. The oxidation of ammonium to hydroxylamine has been demonstrated only with intact cells (Engel and Alexander, 1958; Hofman and Lees, 1952). Since this reaction is endothermic ($\Delta F = +13.4$ kcal/mol), Aleem and Nason (1963) postulated that the full intact compliments of the respiratory cytochrome system are needed to generate the energy necessary for this conversion. Thus, it would appear likely that a mixed-function oxidase would catalyze the incorporation of one atom of molecular oxygen into the substrate while using the other atom for the ultimate generation of energy through the probable oxidation of the pyridine nucleotides. One is also tempted to speculate on the similarity of this step to heterotrophic nitrification by *Arthrobacter*, which incorporates molecular oxygen into ammonium during the oxidation to hydroxylamine (Verstraete and Alexander, 1972d).

Further oxidation of hydroxylamine can be demonstrated with cell-free systems containing particulate-linked cytochromes (Nicholas and Jones, 1960; Aleem and Lees, 1963; Verstraete, 1971). A two-electron transfer to form the unstable, hypothetical nitroxyl was proposed by Aleem and Lees (1963); this then spontaneously combines with nitrite to form nitrohydroxyl-amine. The subsequent oxidation of nitrohydroxylamine was shown to yield 2 mol of nitrite, one of which could then recombine with nitroxyl in a semicyclical fashion. A ferro-c-type cytochrome was involved in the overall process, and oxygen served as the terminal electron acceptor. It is through these reactions that energy is thus obtained. In the absence of oxygen, nitrous oxide is evolved during the oxidation of hydroxylamine and the reduction of cytochrome c (Falcone et al., 1962). Anderson (1964) also noticed the production of both nitrous and nitric oxide during anaerobic incubation. The significance of these reactions, which lie off the nitrifica-tion pathway, will be discussed in a later section.

Oxidation of nitrite to nitrate by Nitrobacter also involves molecular oxygen, but its role is confined strictly to electron transport. Using $^{18}O$, Aleem et al. (1965) demonstrated that the oxygen atom in nitrate was generated from water and not from oxygen.


## 3. Heterotrophic Nitrification

### 3.1. Species and Product Diversity

A half century ago, Mishustin (1926) investigated the formation of nitrite from organic nitrogen by a Bacillus strain. Since then, numerous authors have reported other heterotrophic nitrifying organisms. Table II presents a list of micro- and macroorganisms able to effect heterotrophic nitrification. Only unequivocally established transformations, brought about by intact organisms, are reported.

Free hydroxylamine is both unstable and toxic. Traces of this com-pound have been found in the culture media of Arthrobacter globiformis, Sterigmatocystis nigra, and Aspergillus nigra (Table II). Verstraete and Alexander (1972b) isolated several Arthrobacter strains which accumulated up to 15 $\mu$g $NH_2OH$-N/ml in the liquid growth medium during the logarithmic phase of growth.

Few hydroxylamino compounds, O-alkyl hydroxylamines, and oximes occur in nature. These compounds are relatively stable to biological attack in normal environments by comparison with hydroxylamine (Smith, 1966).

The hydroxamic acids are a very important group of substituted hydroxylamine compounds. Acyclic and cyclic hydroxamic acids are formed by many microorganisms and plants. Hydroxamic acids are categorized according to the number of hydroxamate groups per molecule. They react typically with ferric iron to form a reddish complex. Hydroxamic acids, free or complexed with metal ions, are quite stable to ordinary storage.

Amine oxides are widespread throughout the plant and animal kingdom. These compounds are relatively easily reduced to the corresponding tertiary amines.

A few N-nitroso compounds formed in nature have been described. 4-Methylnitrosaminobenzaldehyde is formed by *Streptomyces achromogenes* (Herr *et al.*, 1967). Two N-nitroso hydroxylamino compounds formed by microorganisms also have been isolated and identified (Murthy *et al.*, 1966; Tamura *et al.*, 1967). So far, only two C-nitroso compounds formed by microorganisms have been reported. *Arthrobacter* sp. synthesized, in the stationary growth phase, a product tentatively identified by Verstraete and Alexander (1972b) as 1-nitrosoethanol, and Ballio *et al.* (1962) reported the formation of ferroverdin, a nitrosobenzene derivative, by a strain of *Streptomyces*. Nitrosation of aromatic substances has also been observed in mammalian metabolism (Boyland and Manson, 1966; Irving, 1964; Kiese and Rauscher, 1963). Nitroso compounds are generally susceptible to oxidation, isomerization, polymerization, and decomposition by heat and light (Smith, 1966).

A variety of natural nitro compounds has been discovered. The only acyclic compound in this group is 3-nitropropionic acid (Table II). The natural occurring cyclic and acyclic nitro compounds are chemically stable.

Nitrite is formed by a wide variety of heterotrophic organisms from various nitrogenous substrates. Accumulation of relatively high levels of nitrite by heterotrophs has only been observed for *Arthrobacter* sp. in pure culture (Laurent, 1971; Verstraete and Alexander, 1972b, 1973).

Nitrate can be formed from ammonium or amino nitrogen by a limited number of heterotrophs. Only fungi have been reported to accumulate this ion to high levels. Chemically, nitrate is stable and its major transformations in the environment are biologically catalyzed.

### 3.2. Growth and Distribution

An important difference with regard to autotrophic nitrification is that heterotrophic nitrification most often is not linked to cellular growth nor is it proportional to the total cellular biomass. Although the accumulation of intermediary metabolites, such as hydroxylamine by *Arthrobacter* sp. (Verstraete and Alexander, 1972b) and 3-nitropropionic acid by *Aspergillus flavus* (Doxtader and Alexander, 1966a), may occur during the logarithmic growth phase, most of the major nitrification end-products are formed during the stationary growth phase.

An attempt has been made to summarize some general information with regard to the rate of heterotrophic nitrification and the maximum

**Table II. Nitrification by Heterotrophic Microorganisms**

| Substrate | Product | Organism | Reference |
|---|---|---|---|
| *I. The formation of hydroxylamine and substituted hydroxylamine* | | | |
| Ammonium or amino nitrogen | (a) Free hydroxylamine (NH₂OH) | *Arthrobacter* sp. | Verstraete and Alexander (1972b) |
| | | *Arthrobacter globiformis* | Gunner (1963) |
| | | *Sterigmatocystis nigra* | Lemoigne *et al.* (1936) |
| | | *Aspergillus niger* | Steinberg (1939) |
| | (b) Hydroxylamino compounds (R—NHOH) | | |
| | N-hydroxyaspartic acid | *Azobacter* sp. | Saris and Virtanen (1957) |
| | Fluopsin | *Pseudomonas fluorescens* | Shirata *et al.* (1970) |
| | (c) Oximes (RC=NOH) | | |
| | 3-(Oximinoacetamido)-acrylamine oxime | *Streptomyces achromogenes* | Wiley *et al.* (1965) |
| | (d) Hydroxamic acids (RCONHOH) | | |
| | (1) Monohydroxamic acids | | |
| | Hadacidin | *Penicillium aurantioviolaceum* | Dulaney (1963); Kaczka *et al.* (1962) |
| | Fusarine | *Fusarium roseum* | Neilands (1967) |
| | Aspergillic acid | *Aspergillus flavus* | MacDonald (1961) |
| | Hydroxyaspergillic acid | *Aspergillus flavus* | MacDonald (1962) |
| | Mutaaspergillic acid | *Aspergillus oryzae* | Neilands (1967) |
| | Neoaspergillic acid | *Aspergillus sclerotiorum* | Micetich and Macdonald (1965) |
| | Neohydroxyaspergillic acid | *Aspergillus sclerotiorum* | Micetich and Macdonald (1965) |
| | Actinonin | *Streptomyces* sp. | Sing *et al.* (1966) |
| | Shizokinen | *Bacillus megaterium* | Neilands (1967) |
| | Cycloserine | *Streptomyces* sp. | Hidy *et al.* (1955) |
| | (2) Dihydroxamic acids | | |
| | Mycobactin P | *Mycobacterium phlei* | Snow (1970) |

| | | |
|---|---|---|
| Mycobactin T | Mycobacterium tuberculosis | Snow (1970) |
| Myceliamide | Penicillium griseofulvum | Coutts (1967) |
| Aerobactin | Aerobacter aerogenes | Gibson and Magrath (1969) |
| Arthrobactin (Terregens factor) | Arthrobacter pascens | Waid (1975) |
| Schizokinen | Bacillus megaterium | Waid (1975) |
| Rhodotorulic acid | Rhodoturula spp. and other yeasts | Waid (1975) |
| Dimeric acid | Fusarium dimerum | Waid (1975) |
| Ferribactin | Pseudomonas fluorescens | Waid (1975) |
| (3) Trihydroxamic acids | Streptomyces sp. | |
| Ferrioxamine $A_1$, $A_2$, B, C, $D_1$, $D_2$, E, F, and G | Streptomyces sp. | |
| Ferrimycins | Streptomyces olivochromogenes | |
| Succinamycin | Streptomyces albaduneus | |
| Danomycin | Ustilago sphaerogena | |
| Ferrichrome | Aspergillus melleus | Neilands (1967) |
| Ferrichrome A | Aspergillus fumigatus | |
| Ferrichrysin | Aspergillus versicolor | |
| Ferrichrocin | Penicillium variable | |
| Ferrirhodin | Actinomyces substropicus | |
| Ferrirubin | Streptomyces griseus | |
| Albomycin | Penicillium sp. | |
| Coprogen | Neurospora crassa | |
| Compound XFe | Nocardia sp. | |
| Nocardamine | | |
| Oxime nitrogen    Free hydroxylamine | Tetrahymena pyriformis | Seaman (1957) |

II. *The formation of amine oxides* ($R_3N-O$)

| | | |
|---|---|---|
| Ammonium or amino nitrogen    Pulcherriminic acid | Candida pulcherrima | Coutts (1967) |
| p-(N-Dimethyl)aminobenzaldehyde N-oxide | Streptomyces thioletus | Kawai et al. (1965) |
| N,N-Dimethylaniline N-oxide | Streptomyces thioletus | Kawai et al. (1965) |
| 4-Hydroxyquinoline-1-oxide | Pseudomonas pyocyanea | Cornforth and James (1956) |

**Table II** (*Cont.*)

| Substrate | Product | Organism | Reference |
|---|---|---|---|
| | Strychnine N-oxide | Bacillus sp., Streptomyces sp. | Bellet and Gerard (1962) |
| | Iodinine | Chromobacterium iodinum | Clemo and McIlwain (1938) |
| | Myxine | Sporangium sp. | |
| | Oxymatrine | Sophora flavescens | |
| | Seneciphylline N-oxide | Senecio sp. | |
| | Platyphylline N-oxide | Senecio sp. | Bickel (1969) |
| | Retronecine N-oxide | Senecio sp. | |
| | Senecionine N-oxide | Senecio sp. | |
| | Nupharidine | Nuphar japonicum | |
| | N,N-dimethyl tryptamine | Piptadenia sp. | |
| | Bufotenine | Piptadenia sp. | |
| | (O-Methyl-aci-nitro)  crotonic acid | Streptomyces fervens var. melrosporus | Wiley et al. (1965) |
| | Enteromycin | Streptomyces achromogenes | Wiley et al. (1965) |
| III. *The formation of nitroso compounds* | | | |
| Ammonium or amino nitrogen | (a) N-Nitroso compounds (N—NO) | | |
| | 4-Methylnitrosaminobenzaldehyde | Clitocybe suaveolens | Herman (1961) |
| | Streptozotocin | Streptomyces achromogenes | Harr et al. (1967) |
| | (b) N-Nitroso hydroxylamino compounds (—NOH—NO) | | |
| | Fragin | Pseudomonas fragi | Tamura et al. (1967) |
| | Alanosine | Streptomyces alanosinus | Murthy et al. (1966) |
| | (c) C-Nitroso compounds (C—NO) | | |
| | Ferroverdin | Streptomyces sp. | Ballio et al. (1962) |
| | 1-Nitrosoethanol | Arthrobacter sp. | Verstraete and Alexander (1972b) |
| Nitrite | (a) N-Nitroso compounds | Many bacteria | Ayanaba et al. (1973); Ayanaba and Alexander (1973) |

IV. *The formation of nitrite and nitro compounds*

| Substrate | Product | Organism | Reference |
|---|---|---|---|
| Ammonium or amino nitrogen | (a) Nitrite (NO$_2^-$) | Many bacteria, actinomycetes and fungi | Hirsch et al. (1961); Gode and Overbeck (1972); Odu and Adeoye (1970) |
| | (b) Nitro compounds (C—NO$_2^-$) | | |
| | 3-Nitropropionic acid | Aspergillus flavus | Marshall and Alexander (1962) |
| | | Penicillium atrovenetum | Birkinshaw and Dryland (1964) |
| | 1-Amino-2-nitrocyclopentane-1-carboxylic acid | Aspergillus wentii | Brian et al. (1965) |
| | Aureothin | Streptomyces thioletus | Kawai et al. (1965) |
| | Azomycin | Streptomyces sp., Nocardia sp. | Waksman and Lechevalier (1962) |
| | p-Nitrobenzoic acid | Streptomyces thioletus | Kawai et al. (1965) |
| | Ilamycin, ilamycin B | Streptomyces sp. | Takita et al. (1962) |
| | Pyrrolnitrin | Pseudomonas sp. | Arima et al. (1964) |
| | 2-Hydroxy-3-nitrophenylacetic acid | Penicillium chrysogenum | Isono (1954) |
| Free hydroxylamine | Nitrite | Pseudomonas sp. | Amarger and Alexander (1968) |
| | | Proteus sp. | Castell and Mapplebeck (1956) |
| | | Microbacterium sp. | Castell and Mapplebeck (1956) |
| | | Arthrobacter sp. | Verstraete and Alexander (1972c) |
| Hydroxylamino compounds | Nitrite | Various bacteria and fungi | Doxtader and Alexander (1966b) |

**Table II** (*Cont.*)

| Substrate | Product | Organism | Reference |
|---|---|---|---|
| Oximes | Nitrate | *Pseudomonas aeruginosa* | Obaton et al. (1968) |
| | | *Fusarium* sp. | Doxtader and Alexander (1966b) |
| Hydroxamic acids | Nitrate | Various bacteria and fungi | |
| Nitro compounds | | | |
| Aliphatic | Nitrate | *Arthrobacter* sp. | Verstraete and Alexander (1972c) |
| | | *Neurospora crassa* | Little (1951) |
| | | *Pseudomonas aeruginosa* | Obaton et al. (1968) |
| | | *Aspergillus flavus* | Doxtader (1965) |
| | | Various yeasts and fungi | Germanier and Wuhrmann (1963); Kido et al. (1975); Jensen and Lautrap-Larsen (1967) |
| Aromatic | Nitrate | *Arthrobacter* sp. | Raymond (1970) |
| | | *Flavobacterium* sp. | |
| | | *Pseudomonas* sp. | |
| | | *Nocardia* sp. | Germanier and Wuhrmann (1963) |

| Substrate | Product | Organism | Reference |
|---|---|---|---|
| Nitrate | Nitro compound (C—NO$_2$) Chloramphenicol | Streptomyces venezuelae | Gottlieb (1967) |
| **V. The formation of nitrate** | | | |
| Ammonium or amino nitrogen | Nitrate (NO$_3^-$) | Arthrobacter sp. | Verstraete and Alexander (1972b) |
| | | Arthrobacter globiformis | Gunner (1963) |
| | | Aspergillus flavus | Saris and Virtanen (1957) |
| | | Aspergillus parasiticus | Shih et al. (1974) |
| | | Cephalosporium sp. | Eylar and Schmidt (1959) |
| | | Penicillium sp. | Hora and Ivengar (1960) |
| | | Chlorella sp. | } Kessler and Oesterfield (1970) |
| | | Ankistrodesmus sp. | |
| Nitrite | Nitrate | Aspergillus wentii | } Malavolta et al. (1955) |
| | | Penicillium sp. | |
| Nitro compounds aliphatic | Nitrate | Aspergillus flavus | Doxtader (1965) |
| | | Arthrobacter sp. | Verstraete and Alexander (1972c) |
| Aromatic | Nitrate | Pseudomonas sp. | Germanier and Wuhrmann (1963) |

concentration of products formed by some well-studied heterotrophic nitrifiers. These data are brought together in Table III and, as a comparison, the data for *Nitrosomonas* and *Nitrobacter* are also included. They relate only to axenic cultures grown under optimal circumstances. The data indicate that the nitrification rate of the heterotrophs is $10^3$–$10^4$ times smaller than that of the autotrophs. In addition, the carrying capacities of the heterotrophs towards their nitrification products remain $10^2$–$10^3$ times smaller than those of their autotrophic counterparts.

A plausible explanation for heterotrophic nitrification can be sought in the identity of some of the nitrification products. First of all, an important group of these products, namely the hydroxamic acids, includes a number of well-known microbial growth factors. Secondly, a large number of the products summarized in Table II have strong biocidal properties. With regard to the first group, typical examples are the mycobactins which are essential for the growth of *Mycobacterium* sp. and the so-called Terregens Factor, a dihydroxamic acid, which is required for the growth of *Arthrobacter pascens*. Neilands (1967) and Haydon *et al.* (1973) have shown that hydroxamic acids are involved in the process of iron uptake by microorganisms. When little iron is available, the organism synthesizes relatively large amounts of the hydroxamate, which then serves as an iron chelating agent for the organism. When iron is abundant, little or no chelating agent is needed, and only small amounts of the hydroxamate are formed. Such a definite inverse relationship between the amount of hydroxamic acid formed on the one hand and the iron concentration of the growth medium on the other hand has been reported for various bacterial and fungal species (Garribaldi, 1971; Neilands, 1967). In particular, the nitrification process by *Arthrobacter* sp. (Verstraete and Alexander, 1972b) turned out to be governed by this principle since the unidentified hydroxamic acid became the dominant product of nitrification in iron-deficient media, whereas hydroxylamine, nitrite, and 1-nitrosoethanol formation were favored in iron-rich solutions. From the latter perspective, it is tempting to assume that the other products of heterotrophic nitrification, for which no apparent physiological role can be visualized, are formed as secondary or end metabolites of constitutive nitrification pathways.

As to the group of the toxic compounds, the antibiotic activity of hydroxamic acids such as actionin, mycelianamide, hadacidin, the aspergillic acids, etc., has been established beyond doubt. In addition, the toxic and even mutagenic properties of hydroxylamine, nitrite, and the *C*-nitroso and the *N*-nitroso compounds have been well documented. As a consequence, it is tempting to assume that these classes of heterotrophic nitrification products are generated to combat or destroy competitors, predators, or parasites. Unfortunately, clearcut evidence to support this last hypothesis is not available.

**Table III. Nitrification Rate and Maximum Concentration of Products Formed by Heterotrophic and Autotrophic Nitrifiers**

| Organism | Substrate | Nitrification product | Rate of formation of product ($\mu$g N/day/g dry cells) | Maximum concentration of product[a] ($\mu$g N/ml) | Reference |
|---|---|---|---|---|---|
| **Heterotrophs** | | | | | |
| *Arthrobacter* sp. | $NH_4^+$ + succinate | Hydroxylamine Nitrite Nitrate | 12 375 250 | 0.2 0.2 4.5 | Gunner (1963) |
| *Arthrobacter* sp. | $NH_4^+$ + organic N | Nitrite Nitrate | — — | 1.6 14.1 | Laurent (1971) |
| *Arthrobacter* sp. | $NH_4^+$ + organic N + acetate | Nitrite +Nitrate | 2800 | 1.0 | Gode (1970) |
| *Arthrobacter* sp. | $NH_4^+$ + acetate | Hydroxylamine Nitrite Nitrate Hydroxamic acid 1-Nitrosoethanol | 4500 9000 650 600 300 | 15.0 18.0 2.0 6.0 10.0 | Verstraete and Alexander (1972a) |
| *Pseudomonas aeruginosa* | Acetaldoxime | Nitrite | 2000 | 150.0 | Obaton et al. (1968) |
| *Hansenula mrakii* | Nitroethane 1-Nitropropane 2-Nitropropane | Nitrite | 1680 | 2.8 | Kido et al. (1975) |
| *Aspergillus flavus* | $NH_4^+$ + sucrose | 3-Nitropropionic acid Nitrate | 1400 1350 | 45.0 75.0 | Doxtader and Alexander (1966b) |
| **Autotrophs** | | | | | |
| *Nitrosomonas* sp. | $NH_4^+$ | Nitrite | 1,000,000–30,000,000 | 2000–4000 | Painter (1970); Wong-Chong and Loehr (1975) |
| *Nitrobacter* sp. | $NO_2^-$ | Nitrate | 5,000,000–70,000,000 | 2000–4000 | |

[a] Growing culture studies; resting cells may give rise to higher accumulations.

The evidence available so far about the occurrence and significance of heterotrophic nitrification in waters and soils is still tenuous. Where nitrification is known to occur, some investigators have shown that the numbers of autotrophic cells are far lower than what would be expected from theoretical considerations based on pure culture studies. Yet, these results are not wholly convincing. First of all, the most probable number (MPN) techniques to recover and enumerate the autotrophic nitrifiers lack accuracy and may recover only a fraction of the *in situ* nitrifying populations. Secondly, the growth and energy efficiency of the autotrophs under anexic culture conditions may not resemble that under field circumstances. Indeed, microbial interactions may reduce cell yields in natural ecosystems below those in axenic cultures (Kao *et al.*, 1973). Furthermore, it is conceivable that, within a microbial community with a limited influx of nitrogen, nitrification supplies energy at a rate just rapid enough to maintain the nitrifying populations. Hence, the amount of nitrate formed over a period of time cannot be converted to the generation of a fixed number of microbial cells. Although the occurrence of large populations of heterotrophic nitrifiers with low populations of autotrophic nitrifiers (Alexander *et al.*, 1960; Gode and Overbeck, 1972; Laurent, 1971; Odu and Adeoye, 1970; Remacle and Froment, 1972) suggests the potential occurrence of heterotrophic nitrification, it remains to be shown that those organisms which can oxidize nitrogen under axenic culture conditions can do so *in situ*. However, recent findings suggest that nitrification by heterotrophs might be qualitatively, as well as quantitatively, of major importance in two types of environments where autotrophic nitrification is not observed to occur. These environments are the acidic soils and the highly alkaline, nitrogen-rich aqueous environments, which are discussed in later sections.

### 3.3. Biochemical Pathway

The biochemistry of heterotrophic nitrification is all but unraveled. A crucial problem still to be resolved is whether the pathway of nitrogen oxidation is organic or inorganic. Indeed, the oxidation of nitrogen can proceed by an inorganic pathway, an organic pathway, or a combination of both (Fig. 1). Possible intermediates in the inorganic pathway are ammonium, hydroxylamine, nitroxyl, and nitrite. The organic pathway can be visualized in general terms to proceed from an amine or an amide to a substituted hydroxylamine. The latter compound might be oxidized to a nitroso compound and further to a nitro compound. Cleavage of the nitro group from the carbon moiety may give rise to nitrite and/or nitrate. Closely associated with this problem is the question of whether all or some of the steps in the oxidation of nitrogen are catalyzed by dehydrogenases or oxygenases.

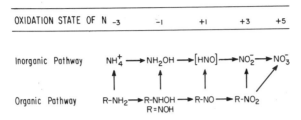

**Figure 1.** Pathways for nitrification.

First, let us review the possibility of an organic pathway of nitrogen oxidation. Doxtader (1965) has suggested that the biosynthesis of 3-nitropropionic acid by *Aspergillus flavus* proceeds from $\beta$-alanine and gives rise to $\beta$-hydroxaminopropionic acid and $\beta$-oximinopropionic acid. Similarly, Birkinshaw and Dryland (1964) suggested that the oxidation of the amino group of aspartic acid might be a prelude to the biosynthesis of 3-nitropropionic acid. On the basis of the nitrification pattern of *Arthrobacter* sp., Verstraete and Alexander (1972c) postulated that, in the early stages of nitrification by this organism, ammonium is converted to an organic compound, possibly an amide, which is then oxidized to yield acetohydroxamic acid. The latter is rapidly converted by a reversible, rapid reaction to free hydroxylamine, but it is also further oxidized to 1-nitroso-1-oxoethane during the stationary growth phase. Reduction of the latter compound gives rise to the typical metabolite, 1-nitrosoethanol, while enzymatic hydrolysis of 1-nitroso-1-oxoethane results in the formation of nitrite and nitrate. These organic pathways of nitrification in *Aspergillus flavus* and *Arthrobacter* sp. are quite plausible in view of the available experimental evidence, but unequivocal proof of their occurrence in these organisms is still lacking.

The study of the biosynthesis of hydroxamic acids has been more successful. Emery (1967) determined the stimulatory effect of various potential precursors of ferrichrome compounds on their formation by growing cells. From his observations, he postulated that the biosynthetic pathway of the ferrichrome compounds proceeds by oxygenation of the $\delta$-amino group of ornithine to yield $\delta$-$N$-hydroxyornithine. The hydroxylamino group is then $N$-acylated to form ornithine $\delta$-hydroxamate. Similar studies by Tateson (1970) on the formation of mycobactins, however, indicate that lysine is first acylated to $\delta$-$N$-acyllysine. Hydroxylation of the resulting amide nitrogen would then give rise to the hydroxamic acid. Strong evidence for the oxidation of amino nitrogen was obtained by Stevens and Emery (1966) in their study of the biosynthesis of hadacidin ($N$-formyl-$N$-hydroxyglycine). Their experimental results were consistent with the hypothesis that the biosynthesis of hadacidin occurs by $N$-oxygenation of glycine to yield N-hydroxyglycine, followed by N-formylation of the product to yield the hydroxamic acid. In addition, their studies with $^{18}O$ showed that the hydroxylamino oxygen of hadacidin is derived from $O_2$ rather than from water.

N-hydroxylation of an amide nitrogen has also been shown by MacDonald (1965). He demonstrated that *Candida pulcherrima* forms pulcherrimic acid by N-hydroxylation of cyclo-L-leucyl-L-leucine. Similarly, an oxygenation of the amide nitrogen of flavocol is involved in the biosynthesis of neoaspergillic acid (Micetich and MacDonald, 1965).

N-Hydroxylation of primary and secondary aromatic amines by higher animals has been studied intensively (Cramer *et al.*, 1960; Das and Ziegler, 1970; Irving, 1964; Kiese and Rauscher, 1963). These studies with microsomal fractions of liver indicate that the enzyme responsible for the reaction is an NADPH-dependent oxygenase.

N-Oxygenation of both aliphatic and aromatic tertiary amines has been shown in microorganisms, plants, and animals (Baker *et al.*, 1963; Bellet and Gerard, 1962). Studies with the microsomal fraction of animal liver revealed that this N-oxygenation reaction, too, was carried out by an NADPH-dependent oxygenase (Baker *et al.*, 1963).

A study by Kawai *et al.* (1965) showed that *Streptomyces thioletus* converted p-aminobenzoate to p-nitrobenzoate, and by use of $[^{18}O]$dioxygen, they showed that both oxygen atoms in the nitro group of p-nitrobenzoate were derived from molecular oxygen. In a similar experiment using $[^{18}O]$dioxygen, Verstraete and Alexander (1972d) showed that the oxidation of ammonium to hydroxylamine by cells of *Arthrobacter* sp. involved the incorporation of molecular oxygen. These findings clearly demonstrate that, in the latter organism, the initial oxidation step is brought about by an oxygenase–enzyme system. The fact that this bacterium needed an external supply of organic substrate for the formation of hydroxylamine suggests that the oxygenation process requires a continuous supply of reduced pyridine nucleotides. Thus, it is clear that both micro- and macro-organisms oxidize various organic nitrogenous compounds. The oxidation seems to involve an NADPH-dependent oxygenation of the nitrogen.

Indications of an inorganic nitrification pathway in the heterotrophs are found essentially in three reports. First of all, Amarger and Alexander (1968) reported that *Pseudomonas aeruginosa* forms nitrite from hydroxylamine and several oximes. Enzymatic studies proved that nitroalkanes were not intermediates in the conversion of oximes to nitrite by the bacterium. Apparently, the oximes function as a source of hydroxylamine. Hydroxylamine is then converted stoichiometrically to nitrite by extracts of the bacterium. The authors found that NADP was a cofactor for the nitrifying enzymes but they found no indication that hydroxylamine was linked to a NADP-requiring dehydrogenase.

Verstraete and Alexander (1972c), in their study of the mechanism of nitrification by *Arthrobacter* sp., noted that this organism excreted enzymes into the culture medium which converted hydroxylamine to nitrite. It was found that NAD, NADP, or cytochrome c, alone or in combination, did not enhance the rate of hydroxylamine oxidation by these enzymes. Since the culture fluid contained small amounts of catalase and peroxidase activities, the authors postulated that a peroxidase or catalase enzyme might be responsible for the extracellular oxidation of hydroxylamine to nitrite.

Finally, it should be mentioned that Aleem *et al.* (1964) have proposed that ammonium oxidation by *Aspergillus flavus* proceeds by an inorganic pathway, since their cell-free preparations of this fungus oxidized ammonium to hydroxylamine, nitrite, and nitrate. However, subsequent investigators could not repeat or support these studies (Doxtader and Alexander, 1966a).

If heterotrophic nitrifying organisms generate oxidized inorganic nitrogenous compounds by an inorganic pathway, how do they incorporate these inorganic compounds into organic molecules? There are several mechanisms by which hydroxylamine can be incorporated into organic complexes. Glutamine synthetase forms $\delta$-glutamylhydroxamic acid from glutamic acid and hydroxylamine in the presence of $Mg^{2+}$ and ATP (Waelsch, 1952). A similar phenomenon has been observed with aspartic acid. In addition, bacteria, plants, and mammalian tissue contain transferases which catalyze the formation of a hydroxamic acid and ammonium from an amide and free hydroxylamine (Waelsch, 1952). Finally, aspartase catalyzes the formation of *N*-hydroxyaspartic acid from hydroxylamine and fumarate (Emery, 1963).

These reactions have tempted a number of authors to suggest that a similar incorporation of hydroxylamine might be involved in the synthesis of naturally occurring hydroxamic acids or nitroso or nitro compounds (Birkinshaw and Dryland, 1964; Hylin and Matsumoto, 1960). In this regard, Verstraete and Alexander (1972c) showed that cell-free extracts of *Arthrobacter* sp. contained enzymes capable of synthesizing 1-nitrosoethanol from hydroxylamine and acetate. Incorporation into organic complexes of nitrite and/or nitrate has also been suggested as a mechanism for synthesis of nitro compounds (Gottlieb, 1967), though no evidence is available to substantiate this view.

## 4. Denitrification

### 4.1. Species Diversity

Schoenbein (1868) suggested that the reduction of nitrate to gaseous products, considered to be purely chemical at the time, might be performed by soil microorganisms. Gayon and Dupetit (1886) made the first conclusive observation that the production of nitrite, nitrous oxide, and dinitrogen in sand columns perfused with nitrate, in the absence of molecular oxygen, was a biological process. Wagner (1895) expressed considerable concern that this process would lead to significant depletion of nitrogen from soil, and he went so far as to suggest that manure, which he thought to be the source of denitrifying bacteria, be acidified before addition to soil. Deherain (1897) showed that denitrifying bacteria were ubiquitous in soil and that the process of denitrification was accelerated by the addition of manure. Weissenberg (1902) noted that the process was effected by normal aerobic bacteria that used nitrate in lieu of molecular oxygen. The coupled reduction of nitrate with the oxidation of carbonaceous compounds was recognized by Giltay and Alberson (Lipman, 1908), who devised methods for the isolation and culturing of many different species of denitrifying bacteria. Depending on the medium used in enumeration by the most probable number method, denitrifying bacteria range from decimal fractions of a percent to half of the natural soil heterotrophic bacterial flora (Valera and Alexander, 1961; Focht and Joseph, 1973; Ardakani et al., 1975; Vives and Parés, 1975).

It is presently common knowledge that denitrification is brought about by the same respiratory electron transport chain present in many aerobic bacteria. Nevertheless, much of the current literature incorrectly attributes denitrification as being due to facultative anaerobes. By definition, facultative anaerobes utilize the full compliment of a respiratory cytochrome system with oxygen (when it is present) as the terminal electron acceptor (i.e., oxidative phosphorylation) but use organic compounds as terminal electron acceptors (fermentation) in the absence of oxygen (i.e., substrate-level phosphorylation). Aerobic and denitrifying bacteria are incapable of growing anaerobically by fermentative means. Many facultative anaerobes, however, are nitrate respiring; that is, they utilize nitrate as a terminal electron acceptor but cannot further reduce the nitrite which is formed. The respiratory reduction of nitrate to nitrite by both denitrifying and nitrate-respiring bacteria is specifically referred to as dissimilatory nitrate reduction, as distinct from assimilatory nitrate reduction. The latter process does not involve electron transport and is common to eucaryotes, including green plants.

In his review, Payne (1973) compiled a comprehensive list of genera of nitrate-respiring and denitrifying bacteria. Since publication of the eighth edition of Bergey's Manual of Determinative Bacteriology (Buchanan and Gibbons, 1974), many generic names have been dropped. Also, the description of supposed denitrifying bacteria in both the seventh and eighth editions is quite equivocal, and one cannot be certain if "reduction of nitrates" means to nitrite or gaseous products. It is doubtful that microaerophiles, anaerobes, and facultative anaerobes for the most part are anything more than nitrate-respiring organisms.

Table IV. Reported Genera of Denitrifying Bacteria

| Genus | Reference[a] |
|---|---|
| *Acinetobacter* | Focht and Joseph (1974) |
| *Gluconobacter* (*Acetomonas*) | Focht and Joseph (1974) |
| *Alcaligenes* (*Achromobacter*) | Chatelain (1969); Matsubara and Iwasaki (1971) |
| *Bacillus* | Several; see text |
| *Halobacterium* | Buchanan and Gibbons (1974) |
| *Hyphomicrobium* | Sperl and Hoare (1971) |
| *Micrococcus* | Several; see text |
| *Moraxella* | Hollis *et al.* (1972) |
| *Paracoccus* | Buchanan and Gibbons (1974) |
| *Pseudomonas* | Many; see text |
| *Rhodopseudomonas* | Satoh *et al.* (1974) |
| *Spirillum* | Buchanan and Gibbons (1974); Mechsner and Wuhrmann (1963) |
| *Thiobacillus* | Several; see text |
| *Xanthomonas* | Mechsner and Wuhrmann (1963) |

[a] The most recent references are generally given. Additional references can be found in the text or in the review by Payne (1973).
*Note Added in Proof*: In a recent study, Kessel (1976) isolated and identified denitrifying bacteria that comprised the genera *Cytophaga*, *Flavobacterium*, *Propionobacterium*, *Vibrio*, and *Xanthomonas*.

The quantitative significance of the different denitrifying genera in nature is not known. Only two of the 14 genera listed in Table IV are gram positive. The available evidence suggests that the gram-negative bacteria of the genera *Alcaligenes* (including *Achromobacter*) and *Pseudomonas* are the most ubiquitous in soil (Valera and Alexander, 1961; Vives and Parés, 1975). However, enumeration by the most probable number method will select for those organisms that grow faster and are less exacting nutritionally, conditions that would tend to favor pseudomonads and other gram-negative rods. Vives and Parés (1975) found only one soil sample out of several dozen that yielded a denitrifying species of *Bacillus*. Nonetheless, Nommik's (1956) studies with Scandinavian soils show that thermophilic organisms, presumably strains of *Bacillus*, are present and may be significant at elevated temperatures.

## 4.2. Biochemical Pathway

Dissimilatory nitrate reductases are particle bound, competitively inhibited by azide, and inhibited by oxygen, and they reduce chlorate. By contrast, assimilatory nitrate reductases are soluble enzymes that are inhibited by chlorate but not by oxygen. However, the two reductases of

*E. coli* were found by Murray and Sanwal (1963) to be immunologically the same, and Van't Riet *et al.* (1972) found that the two reductases of *Aerobacter aerogenes* had similar kinetic properties. Since there are also some denitrifying and nitrate-respiring bacteria that do not appear to contain a distinguishable assimilatory nitrate reductase yet are able to utilize nitrate for nutritional purposes, Payne (1973) suggests that these apparent anomalies can be resolved if one considers that the two enzymes may contain the same protein but may combine differently in response to the regulatory control of the cell.

The branch point in the electron transport chain occurs at cytochrome c (Payne, 1973). Electrons are transferred to the terminal cytochrome oxidase $a_3$ and then to oxygen. In the absence of oxygen, electrons are transferred directly to nitrate from cytochrome c, a process which involves the participation of an iron–molybdenum complex (Fewson and Nicholas, 1961; Forget and Dervartanian, 1972). Because dissimilatory nitrate reductases are particle bound, it is not surprising that phosphorylation is associated with this step (Payne, 1973; Koike and Hattori, 1975a,b; Terai and Mori, 1975).

Though it has been well established that the initial step involves the reduction of nitrate to nitrite, considerable conflicting reports exist in much of the earlier literature concerning the intermediates between nitrite and dinitrogen. Many of these studies lacked refined equipment that did not permit detailed enzymological studies or rapid and specific identification of intermediates. Many recent studies (Cox and Payne, 1973; Ishaque and Aleem, 1973; Baldensperger and Garcia, 1975; Koike and Hattori, 1975a,b; Terai and Mori, 1975) have conclusively established the following pathway, one originally proposed by Payne (1973):

$$NO_3^- \rightarrow NO_2^- \rightarrow NO \rightarrow N_2O \rightarrow N_2$$

The nitrite and nitric oxide reductases of *Pseudomonas perfectomarinus* were found to be soluble (Cox *et al.*, 1971; Cox and Payne, 1973). On this basis, Payne (1973) postulated that phosphorylation was not associated with either nitrite or nitric oxide reduction. More recent studies with *Pseudomonas denitrificans* (Koike and Hattori, 1975a,b; Terai and Mori, 1975), *Micrococcus denitrificans* (Terai and Mori, 1975), and *Thiobacillus denitrificans* (Ishaque and Aleem, 1973) conclusively show that phosphorylation is particle-linked with nitric oxide reductases but not with nitrite reductases in these organisms. Many of the common cytochrome inhibitors block every step in the above reaction series except the reduction of nitrite. Thus, nitrite reduction appears to be coupled near the flavoprotein level (Ishaque and Aleem, 1973), and phosphorylation activity associated with nitrite reduction occurs only during the reduction of nitric oxide to nitrous oxide (Terai and Mori, 1975).

The reduction of nitrous oxide to dinitrogen is also effected by particulate enzymes and is coupled with phosphorylation through the respiratory cytochrome system (Cox and Payne, 1973; Ishaque and Aleem, 1973; Payne, 1973; Terai and Mori, 1975; Koike and Hattori, 1975a,b; Baldensperger and Garcia, 1975). In only a few reports (Hart et al., 1965; Renner and Becker, 1970) has nitrous oxide been shown not to be reduced by intact cells.

Nitrous oxide can be produced chemically from nitrite and hydroxylamine or phenylene diamines. No doubt this has led to much confusion in attempting to establish definitely the pathways of both nitrification and denitrification. Payne (1973) and Bollag and Tung (1972) questioned whether in fact the production of nitrous oxide with hydroxylamine as substrate was truly biochemically mediated in Renner and Becker's (1970) study. Terai and Mori (1975) similarly observed a large increase in the generation of nitrous oxide from nitrite upon the addition of hydroxylamine, yet they detected no increase in phosphorylation that normally accompanied the reduction from nitric oxide. The evolution of nitrous oxide by Nitrosomonas europaea (Yoshida and Alexander, 1970) may involve a similar mechanism since the organism produces both hydroxylamine and nitrite. Payne (1973), however, has suggested that Nitrosomonas may also be a denitrifying bacterium that is able to survive brief anaerobic periods in soil by this means. Bollag and Tung (1972), though noting that nitrous oxide was generated in pure culture by the fungus Fusarium, concluded that such a process was insignificant in the generation of nitrous oxide in nature and that the products, though formed biologically, reacted chemically to give nitrous oxide. Studies by Hart et al. (1965) did not take the necessary precautions used by Bollag and Tung (1972) in removing carbon dioxide by gas chromatography prior to mass spectral analysis. They assumed, incorrectly, that the alkali well in the incubation vessel removed all interfering carbon dioxide and that the $m/e$ 44 peak formed in the mass spectrometer was solely nitrous oxide.

Several reported cases of the dissimilatory reduction of nitrogen oxides exist that differ from the normal standard pathway. Prakash and Sadana (1973) showed that anaerobic reduction of nitrate to nitrite by Vibrio fischeri (Achromobacter fischeri) was coupled with energy. However, no energy was generated though the coupled oxidation of substrate and reduction of nitrite to ammonia. Furthermore, nitrite was shown to be toxic to the bacterium. They concluded that the reduction of nitrite was a detoxication mechanism. A recent report by Stanford et al. (1975c) shows evidence for a similar pathway in soil; when $^{15}NO_3^-$ and glucose were added, over one-third of the nitrogen was recovered as $^{15}NH_4^+$. However, it is doubtful that their studies represent a detoxication mechanism since the amount of nitrate converted to ammonium was directly proportional to the amount of glucose added, which simply suggests the assimilatory reduction of nitrate.

Youatt (1954) reported on an organism, described as an *Achromobacter*, that reduced nitrite but not nitrate by dissimilatory means. Similarly, Vagnai and Klein (1974) isolated several cultures of gram-negative, polar flagellated bacteria (described as *Pseudomonas*) that were unable to reduce nitrate but were able to reduce nitrite to nitrous oxide and dinitrogen. Nitrous but not nitric oxide supported growth as the terminal electron acceptor. Dinitrogen production ranged from 0 to 99.8% when KCN, $NaN_3$, and 2,4-dinitrophenol were used as inhibitors for nitrite and nitrous oxide reduction by four different strains. The activity of some strains was reduced by a given inhibitor in the reduction of nitrous oxide but not nitrite; the reverse was also true. It is hard to assess the variability of these results in explaining how nitrous oxide reduction but not nitrite reduction to dinitrogen could be blocked in the same organism if nitrous oxide is a precursor. Furthermore, the partial inhibition of nitrite reduction by cytochrome inhibitors suggests that nitrite reduction in these isolates may be linked with phosphorylation, though the variability among strains, inhibitors, and substrates makes this conclusion tenuous. Nevertheless, the nitrite dissimilating organisms which fail to dissimilate nitrate are probably of minimal importance in most terrestrial and aquatic systems where nitrite concentrations are generally low. These organisms may be of far greater importance in nitrogen-rich environments, specifically waste treatment operations (Prakasam and Loehr, 1972; Voets *et al.*, 1975) utilizing the incomplete nitrification–denitrification sequence.

Payne (1973) and McCarty (1972) calculate that the energy efficiency of denitrification is about 60% of that realized when oxygen is used as the terminal electron acceptor. Their calculations, however, are based on the voltage differentials between the $NAD^+/NADH + H^+$ and $NO_3^-/NO_2^-$ couples. These calculations would appear to be supported by empirical observations of John and Whatley (1970) and Koike and Hattori (1975a,b), aside from the fact that these studies showed complete reduction to dinitrogen, not nitrite. An apparent anomaly exists since no energy is coupled with nitrite reduction. Koike and Hattori (1975b) report identical cell yields of *P. denitrificans* per electron transferred with nitrate and nitrite as terminal electron acceptors despite the fact that there is a greater amount of "phosphorylating-available electrons" with nitrate (four out of five) vs. nitrite (two out of three). These apparent anomalies can be resolved only if the free-energy change per electron transferred is greater with nitric oxide or nitrous oxide reduction than with nitrate reduction. The free-energy change, $\Delta F$, for each reaction can be calculated from the data of Koike and Hattori (1975b). The molar growth yields with nitrate, nitrite, and nitrous oxide as terminal electron acceptors were found to be

28.6, 16.9, and 8.8 g/mol, respectively. Since the reaction steps from the most oxidized species are identical to the sequential reduced species, any differences in entropy losses would be confined to the preceding step of the reductive pathway. The ratio of the molar growth yields, $Y$, is equal to the ratio of the total free energy change, $\Delta F$, from each step in the reduction by the respective equations, where all species are denoted by the subscripts below:

$$\frac{Y_{NO_3^-}}{Y_{NO_2^-}} = \frac{\Delta F_{NO_3^-} + \Delta F_{NO_2^-} + \Delta F_{NO} + \frac{1}{2}\Delta F_{N_2O}}{\Delta F_{NO_2^-} + \Delta F_{NO} + \frac{1}{2}\Delta F_{N_2O}} \tag{1a}$$

$$\frac{Y_{NO_2^-}}{Y_{N_2O}} = \frac{\Delta F_{NO_2^-} + \Delta F_{NO} + \frac{1}{2}\Delta F_{N_2O}}{\Delta F_{N_2O}} \tag{1b}$$

Assuming the voltage differential $(\Delta E_0')$ is initiated at the pyridine nucleotide level $(E_h = -320 \text{ mV})$, $\Delta F$ for the reduction of nitrate to nitrite is $-34$ kcal/mol, and $\Delta F_{NO_2^-}$ is assumed to be zero. Substituting these numerical values and those from Koike and Hattori (1975b) into equation (1b) yields

$$\Delta F_{NO} = 1.42 \Delta F_{N_2O} \tag{2}$$

Substituting equation (2), along with the respective numerical values, into equation (1a) yields the following numerical solution: $\Delta F_{NO} = 35.57$, $\Delta F_{N_2O} = 26$ kcal/mol. The respective redox couples $(E_h)$ can then be calculated to give 1230 and 250 mV, respectively. Since there are no published $E_h$ values for either $NO/N_2O$ or $N_2O/N_2$ couples in aqueous solutions, there is no alternative source to verify these values. Pourbaix (1966) gives $E_h$ values for $NO/N_2O$ and $N_2O/N_2$ of 1352 and 1168 mV, respectively, for gaseous substances.

Nevertheless, these calculations from available evidence in the literature point to the fact that nitric oxide must be a more efficient terminal electron acceptor than oxygen. This would suggest the presence of another cytochrome above the $E_h$ scale of cytochrome $a_3$ oxidase that transfers electrons to nitric oxide. Though other studies have shown the participation of a c-type cytochrome in nitric oxide reduction that is distinct from the cytochromes linked to nitrate and nitrous oxide reductases (Cox and Payne, 1973; Payne, 1973; Terai and Mori, 1975), the redox potential of this cytochrome has never been measured. Krul (1976) recently showed that nitric oxide was rapidly reduced by activated sludge in the presence of dissolved oxygen. Furthermore, he found that nitric oxide actually inhibited oxygen uptake by *Alcaligenes*. These observations support the calculations which show that, on strictly thermodynamic grounds, nitric oxide would be a preferential electron acceptor to molecular oxygen.

## 5. Environmental Factors

Where the species diversity is narrow, as in the case of nitrification, a single environmental determinant may have a measurable and predictable effect upon the microorganisms and the rate process. Yet the physical interaction of two or more environmental factors frequently tends to yield results between independent studies that appear to be contradictory. For example, the effect of temperature on biological processes alone is also a factor that influences the solubility of oxygen or the oxidation–reduction potential ($E_h$). Adding to this complexity is the effect of available organic matter, which may serve as both a reductant for oxygen and nitrogen oxides, with the reduction of the latter being highly dependent upon the rate at which oxygen is removed from this system. The rate of oxygen removal is temperature dependent from a biochemical standpoint (oxidation of substrate) and a physical standpoint (solubility and diffusion). Though the effects of important environmental factors upon the microorganisms and the rate process related to nitrification and denitrification are treated separately in the following section, interactions between these factors will also be considered.

### 5.1. Substrates and Products

Because of the narrow species diversity and the relative simplicity of the pathway of substrate oxidation, the effect of inorganic nitrogen compounds upon the growth and metabolism of the autotrophic nitrifiers is the most predictable aspect of the nitrogen cycle. $K_m$ for ammonium oxidation ranges between 1 and 10 mg N/liter between 20 and 30°C (Hofman and Lees, 1953; Ulken, 1963; Loveless and Painter, 1968), while $K_m$ for nitrite oxidation falls between 5 and 9 mg/liter in this same temperature range (Lees and Simpson, 1957; Gould and Lees, 1960; Laudelout and van Tichelen, 1960; Ulken, 1963). Values of 0.2 and 29 mg/liter have been reported for temperatures of 8°C (Knowles et al., 1965) and 49°C (Laudelout and van Tichelen, 1960), respectively. The higher values agree reasonably well with those observed in soil (Ardakani et al., 1974a,b) and those calculated from turnover rates in lakes as a function of concentration (Daubner and Ritter, 1973).

Because $K_m$ is usually equal to or larger than substrate concentrations in nature, the rates of ammonium and nitrite oxidation tend to be dependent on the substrate concentration. The nitrifiers are most populous in the surface soils (0–10 cm), where the total nitrogen (both organic and ammonium) is highest (Ardakani et al., 1974b; McLaren, 1971). These studies show that populations of Nitrosomonas may increase from several hundred to several hundred million per gram of soil upon the addition of ammonium fertilizer and that the growth and oxidation of ammonium follow the standard Monod logistics. Similar results can be shown in the growth of *Nitrobacter* upon the addition of nitrite to soil (Ardakani *et al.*, 1973).

Substrate inhibition of the nitrifying bacteria has been observed at relatively high levels (Meyerhof, 1916a,b; Boon and Laudelout, 1962; Watson, 1971; Prakasam and Loehr, 1972) and is pH dependent since the inhibition is effected by free ammonia or undissociated nitrous acid. End-product inhibition by nitrite to *Nitrosomonas* is an apparently more complex mechanism since lower concentrations (500 mg N/liter) are more toxic during the lag phase of growth than the logarithmic phase (2500 mg N/liter) (Painter, 1970). Nitrite concentrations of 1400 and 4700 $\mu$g N/ml effect a 26 and 100% inhibition of oxygen uptake by washed cell suspensions of *Nitrosomonas* (Meyerhof, 1916a,b). Watson *et al.* (1971) observed inhibition of nitrite to *Nitrosolobus* at 1400 but not at 700 $\mu$g N/ml. High concentrations of nitrate appear to noncompetitively inhibit the oxidation of nitrite by *Nitrobacter* (Boon and Laudelout, 1962). Gould and Lees (1960) found that greater cell densities could be maintained by continuous removal of nitrate by dialysis or by continuous culture. The intimate association of nitrification with carbon dioxide fixation may account for the greater rates of ammonium or nitrite oxidation and the lesser sensitivity to end-product inhibition observed by growing cells. Resting cells of *Nitrobacter*, for example, lose the ability to oxidize nitrite, but by supplying nitrite, oxygen, and carbon dioxide (all are necessary), the oxidative activity returns. Similarly, the rate of nitrite oxidation is drastically reduced when the stationary growth phase is reached (Fliermans and Schmidt, 1975; Fliermans *et al.*, 1974).

Autotrophic nitrifiers depend entirely on the oxidation of nitrogen for their energy supply. It is highly unlikely that heterotrophs gain a significant amount of their overall metabolic energy from their nitrification processes. Indeed, no link between nitrogen oxidation and ATP-generating systems has as yet been found for heterotrophic nitrifiers. Furthermore, even if this would be the case, the net amounts of energy trapped from these nitrogen oxidations would be small compared with those of the normal organic carbon oxidations since these organisms oxidize only small amounts of nitrogen (Table III). However, the heterotrophs can obtain energy from the carbon moiety of certain nitrogenous compounds which they metabolize. Kido *et al.* (1975) report that a variety of microorganisms, in particular *Hansenula mrakii*, grow with alkyl nitro compounds as a carbon source. These organisms liberate nitrite from the nitro compounds. They appear to reduce and subsequently assimilate the nitrite during their logarithmic growth phase, but they accumulate the liberated nitrite in the stationary phase.

The C/N ratio of the growth medium, rather than the carbon or nitrogen substrate concentration alone, appears to be of major importance for heterotrophic nitrifiers. Generally, most media for growing heterotrophic nitrifiers have a C/N ratio ranging from 3–5. Verstraete and Alexander (1972b) studied this parameter more closely and found that for *Arthrobacter* sp. the formation of hydroxylamine per cell increased with decreasing C/N ratio in the range 10/1 to 1/10. The formation of nitrite per bacterial cell, however, was relatively independent of the C/N ratio for this organism.

For many years, the rate of denitrification was thought to be independent of nitrate concentration (Wiljer and Delwiche, 1954; Nommik, 1956; Bremner and Shaw, 1958; Cooper and Smith, 1963; Broadbent and Clark, 1965). First-order kinetics for denitrification in waste water was observed by Balakrishnan and Eckenfelder (1969), Johnson (1969), and Mulbarger (1971), while Wuhrmann (1963) and Dawson and Murphy (1972) reported zero-order kinetics. Stanford et al. (1975c) used lower concentrations (below 32 mg $NO_3^-$–N/liter) and subsequently found the reaction to be first order. Starr and Parlange (1975) similarly found that first-order kinetics applied at concentrations less than 40 $\mu$g N/ml in soil columns. Bowman and Focht (1974) showed that the kinetics of denitrification fitted the standard Michaelis–Menten equation, providing the reductant (carbonaceous substrate) was not limiting. Thus, the discrepancies between the first-order or zero-order kinetics reported in the literature are resolved when one considers the nitrate concentration ranges used in the experimental procedure and whether the system is carbon or nitrogen limiting.

Since the rate of denitrification is also dependent on the available carbon concentration, kinetic constants have little quantitative meaning on their own. The $K_m$ values derived from carbon-limiting systems are far lower than those derived from systems that are not carbon limiting. Moore and Schroeder (1971) reported a $K_m$ value of 0.1 $\mu$g N/ml in sewage low in carbon (chemical oxygen demand of 7.5 $\mu$g/ml), while Bowman and Focht reported a value of 170 $\mu$g N/ml in soils amended with 0.6% glucose-C. Similarly, Kohl et al. (1976) showed a difference between soil under 12 yr of bromegrass ($V_{max} = 77.7$ $\mu$g N/ml/day, $K_m = 9.6$ $\mu$g N/ml) and soil under a yearly corn–soybean rotation ($V_{max} = 48.7$ $\mu$g N/ml/day, $K_m = 4.1$ $\mu$g N/ml), which they attributed to differences in the available carbon between the two soils, even though both were amended with glucose. Without glucose, there were 312 and 1467% decreases in the denitrification rates of the soils under bromegrass and corn–soybean rotation, respectively.

Bowman and Focht (1974) explained the kinetics of denitrification by Bray and White's (1966) equation for the product of two simultaneous Michaelis–Menten expressions such that

$$-\frac{dN}{dt} = \frac{V_{max} N C}{(N + K_N)(C + K_C)} \tag{3}$$

where $t$ is time, $N$ is the concentration of nitrate, $C$ the concentration of reductant, $K_N$ and $K_C$ are the respective Michaelis constants, and $V_{max}$ is the maximum rate. This model assumes simplistically that both oxidant and reductant combine specifically with the same enzyme. In a strict sense, the above equation would apply to the terminal oxidase and assumes that electron transport from the primary dehydrogenase is not limiting. Bowman and Focht (1974) reported a similar $V_{max}$ for nitrate-N in a desert soil supplemented with glucose, when the system was carbon (glucose) limiting and nitrogen (nitrate) saturated (135 $\mu$g N/ml) and when it was nitrogen limiting and carbon saturated (150 $\mu$g N/ml). The $K_C$ and $K_N$ values were 500 and 170 $\mu$g N/ml, respectively. In another soil, an apparent $V_{max}$ of 630 $\mu$g N/ml/day was observed in both treatments. However, because of the failure to achieve saturation with either carbon or nitrogen, while the concentration of the other was varied, the typical Michaelis–Menten curve was not achieved. Instead, an abrupt transition between pseudo first-order and zero-order states was noted in both cases. When both carbon and nitrogen were increased proportionally, the true $V_{max}$ under these conditions was found to be 1500 $\mu$g N/ml/day.

Doner et al. (1974) proposed a three-step sequence in which the enzyme first reacted with nitrate, and then the resulting complex reacted with reductant and finally dissociated into products and enzyme. The equation describing this reaction is

$$-\frac{dN}{dt} = \frac{V_{max}\,NC}{K_m + C} \tag{4}$$

where $K_m = (K_2 + K_3 C)/K_1$ instead of the usual $(K_2 + K_3)/K_1$, $V_{max}$ is defined as $K_3 E_0$ ($E_0$ is the enzyme concentration), and each $K$, as designated by the subscripts, is the reaction constant for the three catalytic sequences. These investigators were unable to determine the $K_m$ because the process was catalyzed by soil organic matter, and they had no effective way of measuring the available carbon concentration, unlike Bowman and Focht (1974), who added measured levels of glucose to a soil having a low indigenous "background carbon" content. Kohl et al. (1976) showed that Bowman and Focht's (1974) data could also be fitted to a series of non-linear differential equations that were derived from the kinetics of a more complex reaction sequence that assumed a limitation of reductant (available carbon) at the primary dehydrogenase level.

Carbon per se lacks quantitative meaning in the present context since the stoichiometric C/N relationship is dependent upon the electrons supplied per mole of carbonaceous substrate and the extent of completion of the reduction of nitrate to dinitrogen. Nevertheless, the C/N ratio for the combined optimal reduction of nitrate and oxidation of carbon is between 2 and 3 in waste water (Dawson and Murphy, 1973; Wuhrmann and

Mechsner, 1973) and in soil when exogenous carbon is added (Bremner and Shaw, 1958; Bowman and Focht, 1974). Since the C/N ratio of most soils is generally higher than 10, it would appear that available carbon is limiting in soil and that all of it is used for respiratory rather than assimilatory purposes during denitrification. Excellent correlations between the rates of denitrification and available carbon as assessed by "glucose equivalent" (Stanford et al., 1975c) and by soluble carbon and mineralizable carbon (Burford and Bremner, 1975) have been reported. In the latter study, the amounts of total, mineralizable, and soluble carbon were 1.93, 1.36, and 0.72 times the amount needed for the observed production of nitrous oxide and dinitrogen. Thus, if all the water-soluble carbon was used by denitrifiers, the authors concluded that at least 28 % would be derived from the mineralizable insoluble organic carbon.

Since other microorganisms compete for a vast array of carbonaceous compounds, it is unlikely that all of the available carbon would be used strictly by denitrifiers. A considerable portion of substrate would also be used in normal oxidative respiration by the denitrifiers and other organisms until the system became anoxic. The kinetics of conversion of total soil organic matter to mineralizable organic matter to soluble organic matter by coupled denitrification are unknown. According to Doner (1975), transient kinetics of denitrification are primarily determined by some solubilization of organic matter during wetting of air-dry soil, and subsequent steady states are dependent on less available organic matter.

Wetting and drying cycles in soil have a pronounced effect upon all microbial processes, presumably due to the physical breaking of bonds to liberate smaller, more labile organic molecules (Birch, 1958). Thus, increasing the severity of the wetting and drying cycles liberates more labile organic matter, as noted by the rapid decrease in $E_h$ (Yamane and Sato, 1968) and the rapid loss of nitrate (MacGregor, 1972; Cawse and Sheldon, 1972) following the wetting of air-dried soils. Similarly, Paul and Myers (1971) found that the highest losses of indigenous organic nitrogen from soil through nitrification–denitrification were greater where drying was more severe.

The length and frequency of wetting and drying also govern the liberation of available organic nitrogen, which can then be subjected to nitrification and denitrification. Reddy and Patrick (1975) found that the losses of total indigenous and applied nitrogen—as high as 24%—occurred in the treatment with the maximum number of cycles (2 days aerobic, 2 days anaerobic) and that losses were proportionally less as the cycles lengthened for the 128-day incubation period.

In nature, the transformations of nitrogen are represented by a vectoral factor that does not exist in most laboratory studies. Thus, the observation that shorter cycles effect greater losses of nitrogen in contained systems does not wholly apply to the field. Much greater removal of nitrogen from secondary waste effluent occurs with longer cycles (Bouwer, 1970; Lance, 1972) or with reduced infiltration rates. Leaching of nitrate produced in the mineralization of peat soils could be prevented in a similar manner by controlling the irrigation rate (Avnimelech, 1971; Avnimelech and Raveh, 1974,1976). It would appear that the failure to induce significant denitrification following or occurring concomitantly with nitrification during shorter vs. longer irrigation periods (or shallow vs. deep irrigation) would be due to the insufficient time required to create anoxic conditions since the growth, activity, and numbers of denitrifying bacteria far exceed the nitrifiers.

The concentration of nitrate may possibly play a role in the selection of different denitrifying floras. Verhoeven (1950) reported that pseudomonads were the predominate bacterial flora with low (1–2%) $KNO_3$ concentrations in cured ham, while spore-forming bacilli were more prevalent at higher concentrations (5–12%). Whether this selection factor is due to differences in response to toxicity of nitrate or the competitive utilization for nitrate as an electron acceptor between the two groups is not known. If applicable to soil and aquatic systems, these results would suggest that pseudomonads would be predominant in the denitrifying microflora since nitrate concentrations are normally below 1%. Woldendorp (1968) compared the two groups and observed that *Bacillus* was initially the dominant denitrifier, but it was rapidly displaced by gram-negative rods (probably pseudomonads) after incubation with nitrate. He concluded that soil conditions in his study were less favorable for the growth of the *Bacillus* spp. than the gram-negative rods. Doner (1975) observed the establishment of different steady-state rates of denitrification in three soil columns, which could not be attributed to different flow rates in soils with an initial nitrate concentration of 0, 100, and 1000 $\mu$g N/ml. When the two columns containing 0 and 1000 $\mu$g N/ml were switched to 100 $\mu$g N/ml, they both showed transitions to steady-state removal of nitrate identical to the third.

When the reduction of nitrate and nitrite are compared, some apparent anomalies are observed at different concentrations. At lower concentrations of nitrite, nitrate, and reductant, nitrate appears to be preferentially reduced. At higher concentrations of reductants and oxidants (where zero-order kinetics tend to occur), nitrite is preferentially reduced, and at a faster rate, when the two nitrogen oxides are compared separately. The first pattern is more characteristic of aquatic and terrestrial environments, while the latter pattern is observed in highly nitrogenous waste waters and in pure cultures.

The $E_h$ for the reduction of nitrite has not been clearly established. Though studies in pure culture suggest that the reduction occurs near the voltage of the primary hydrogenases, Kefauver and Allison (1957) noted a continual drop in the $E_h$ from 350 to 50 mV during anaerobic reduction of nitrite, and Bailey and Beauchamp (1973a,b) noted a poising effect near 200 mV. Based on the previously considered thermodynamic calculations (Section 4.2), it appears more likely that this poising was due to nitrous oxide and not nitrite.

Nitrite is a more commonly observed product of denitrification in oceans than in other environments, including soil. Goering and Cline (1970) found that the appearance of nitrite from nitrate in ocean waters coincided with a depletion of oxygen, and at these low concentrations of nitrogenous oxide, nitrate was reduced faster than nitrite. Richards and Broenkow (1971) similarly found that gaseous nitrogen was not formed until the ratio of nitrate to nitrite was less than 1.00. The active zones of nitrate reduction and denitrification in the deep ocean are largely dependent upon the oxygen consumption rates in the euphotic zone, mixing of thermoclines, and diffusion of oxygen in the liquid. In bays or estuaries, the level of organic matter is higher than in the deep ocean so that active zones tend to be associated with the benthos in oxygen-depleted zones below 40 m (Richards and Broenkow, 1971). In the open ocean, organic matter rarely exceeds 0.5 mg C/liter (Carlucci and Schubert, 1969), so that anoxic zones tend to be deeper and nitrite tends to accumulate.

The nitrite-rich zone in the open sea was found by Hattori and Wada (1971) to occur at about 150 m. The mean residual time of nitrite was estimated to be about 30 days. The numbers of denitrifying and ammonium-oxidizing bacteria in the immediate subsurface samples were very small by contrast to the numbers of nitrate-respiring organisms. No denitrifiers or nitrifiers could be isolated from the deeper, nitrite-rich waters of the South Pacific. Carlucci and Schubert (1969) similarly observed that reduction of nitrate to nitrite was the primary reductive process in samples obtained in the deep nitrite-rich waters off the coast of Peru, though they did observe some gas production and assimilation. Richards and Broenkow (1971), noting that denitrification did occur, nevertheless concluded that most of the organic matter produced *in situ* was oxidized by nitrate-respiring bacteria.

Zones of nitrite accumulation in soil profiles have also been noted along with higher population densities of nitrate-respiring bacteria. Volz *et al.* (1975b) perfused a lysimeter containing a sandy loam soil low in organic matter ($<0.5\%$ carbon) with a nitrate solution and observed a peak nitrite concentration in the top 60 cm that corresponded to a low nitrate concentration. They showed that nitrate was first reduced to nitrite by nitrate-reducing bacteria, and the nitrite was subsequently reoxidized by nitrite-oxidizing bacteria as it moved below the nitrite zone. A similar study (Volz *et al.*, 1975a) showed that the denitrifying bacteria (range 10–29,000/g) were generally less numerous than nitrate-respiring bacteria (range 10–690,000/g) at comparable depths and time. McGarity and Meyers

(1968) attributed the accumulation of nitrite in the B horizon of soil to the depletion of available carbon during nitrate reduction.

The selection of nitrate-respiring bacteria over denitrifying bacteria at low reductant concentrations is not clearly understood. Possibly, the nitrate-respiring organisms have a lower $K_m$ for oxidation of carbonaceous substrates than denitrifiers, which would make them better competitors at low substrate concentrations. A second possibility is that the nitrate reductase and/or cytochrome c may have a greater affinity (i.e., lower $K_m$) for nitrate in nitrate-respiring bacteria than in denitrifying bacteria. The latter explanation appears to be the least likely in view of studies that show increased rates of nitrate reduction but no observed denitrification when nitrate is added to ocean samples taken from the nitrite-maximum zones (Hattori and Wada, 1971). Furthermore, the addition of glucose by Ardakani et al. (1975) to the same soil used in previous studies where nitrate-respiring populations were high (Volz et al., 1975a,b) resulted in an almost complete conversion to dinitrogen and an increased number of denitrifiers. Thus, it appears that the organic matter concentration is the primary selective factor in determining whether nitrate-respiring or denitrifying bacteria will be dominant.

The rate of nitrous oxide reduction has been shown to follow Michaelis–Menten type kinetics in rice paddy soils (Garcia, 1974). The $K_m$ determined from Garcia's (1974) data is high (310 $\mu$g N/ml) by comparison to concentrations normally observed in soil, so that the reaction can be considered to be practically first order. Nitrous oxide is also a very soluble gas (1300 mg/liter at 20°C and 1 atm), but the concentration in solution is dependent on its partial pressure in accordance with Henry's law. Thus, the lag period in the production of dinitrogen that is frequently observed in incubation studies can be attributed to a dependence of rate on the concentration of nitrous oxide in solution. When soil atmospheres contain 100% nitrous oxide, there is no lag in the production of dinitrogen (Nommik, 1956; Van der Staay and Focht, 1977).

Proportionally less nitrous oxide is evolved from gaseous products when exogenous substrate is added to soil (Nommik, 1956; Wiljer and Delwiche, 1954), which is due to sufficient reductant being present to promote completion of the reductive process. The effect of plant roots in accelerating denitrification has been well established. The mechanism is twofold: The excretion of available organic matter lowers the oxygen tension, and this carbon serves as a substrate for reduction of nitrate. Stefanson (1972a) found that more dinitrogen was evolved in incubation chambers containing plants, while more nitrous oxide was evolved from containers not containing plants. Thus, the amount of labile organic matter affects the extent to which denitrification goes to completion. Nitrous oxide is rarely observed in sediments or in sewage (Goering, 1968; Dawson and Murphy, 1972; Terry and Nelson, 1975; Goering and Cline, 1970; Goering and Dugdale, 1966a,b) during denitrification. Again, it is not easy to determine whether this is due to abundant reductant *per se* or to limiting concentrations of terminal-electron acceptors (e.g., $O_2$, $NO_3^-$, $N_2O$) because of more anoxic conditions.

Recent independent studies (Yoshinari and Knowles, 1976; Balderston *et al.*, 1976) show that acetylene reversibly inhibits the reduction of nitrous oxide; when the acetylene is removed, the activity is restored. This observation may provide a useful method for measuring the rates of nitrous oxide production and reduction *in situ*. Though nitrous oxide and acetylene are both substrates for nitrogenase activity (Postgate, 1974), it is not known whether or not the mechanism is similar. It does not appear that acetylene functions in any way as an electron acceptor since it is not reduced by denitrifying bacteria.

Though some denitrifying bacteria can use molecular hydrogen (*Micrococcus denitrificans*) or sulfur (*Thiobacillus denitrificans*) as an energy source, it is difficult to assess the significance of this activity in nature. However, Martin and Ervin (1953) observed that amendments of sulfur to citrus groves resulted in nitrogen deficiency, which they attributed to autotrophic denitrification. This explanation seems reasonable in view of the low organic matter content of those soils, the use of excessive nitrate fertilizer at that time (see Ayers and Branson, 1973), and the type of irrigation practices that would promote periods of anoxia. Using soils from this same geographic area, Mann *et al.* (1972) demonstrated that denitrification rates were greatly accelerated when sulfur was added to soil columns continuously perfused with an $NO_3^-$ solution.

## 5.2. Aeration

For the most part, microbial respiration is independent of oxygen concentrations because of the low Michaelis constant ($K_m$), which is about $10^{-6}$ M (Greenwood, 1961; Wimpenny, 1969; Painter, 1970). Surprisingly, the $K_m$ for oxygen consumption by *Nitrosomonas* and *Nitrobacter* is significantly higher, ranging from 0.3–1.0 mg $O_2$/liter (Scholberl and Engel, 1964; Loveless and Painter, 1968; Boon and Laudelout, 1962). Reduced rates of nitrification in sewage occur at oxygen levels of 1 mg/liter, and the concentration remains steady (i.e., zero-order) above 2 mg/liter (Wuhrmann, 1963). The higher respiratory $K_m$ of the autotrophic nitrifiers vs. heterotrophic bacteria means that the former would be poorer competitors for oxygen at lower concentrations. Heterotrophic respiration would be increased with a possible reduction of soluble oxygen upon the addition of an available carbon source. This may explain the fallacy that persisted for over half a century that organic matter *per se* was toxic to the autotrophic nitrifiers.

Studies of axenic culture stress the importance of adequate aeration for active nitrification. Analysis of redox potentials on the one hand and nitrite and nitrate formation on the other led Laurent (1971) to speculate that in muds, ammonia is oxidized through the action of autotrophic and heterotrophic bacteria in microaerophilic conditions and solely by the action of heterotrophs in slightly anaerobic conditions ($-85$ mV). This is in agreement with ZoBell's (1935) studies, which showed that the formation of nitrite by autotrophic marine nitrifiers did not occur below 250 mV, even though autotrophic nitrifiers could be isolated from marine muds where the $E_h$ was quite low, between $-100$ and $-300$ mV.

Since the early discoveries that the nitrification and denitrification processes occurred optimally under aerobic and anaerobic conditions, respectively, frequent controversy has arisen as to whether or not the latter process was of any significance in seemingly well-aerated aquatic and terrestrial habitats. Waksman (1927) concluded that losses of nitrogen from well-aerated agricultural soils were insignificant, though Allison (1955), in an extensive review of nitrogen balance studies, showed that generally 10–30% of the nitrogen could not be accounted for, a deficiency which he attributed to denitrification.

Though denitrification is an anaerobic process, it can be demonstrated that the reaction occurs under what may appear to be well-aerated conditions. Nitrogen losses can occur with concurrent nitrification in aerated activated sludge tanks (Wong-Chong and Loehr, 1975; Wuhrmann, 1963), oxidation ditches (Murray et al., 1975), soils (Starr et al., 1974; Greenland, 1962; Misra et al., 1974), and incubation chambers in which plants are actively growing (Stefanson, 1972a,b,c). Losses from aerated shake-flask cultures have been reported (Collins, 1955; Sacks and Barker, 1949). The occurrence of both processes in close proximity is due to the existence of many anaerobic microsites, which are frequently too small to measure with an oxygen probe or a platinum electrode. The creation of an anoxic microsite is dependent on three factors: (1) the oxygen consumption rate, (2) the oxygen diffusion rate, and (3) the geometry. Greenwood (1961) showed that anoxic microsites could exist in a well-aerated soil crumb having a 200 $\mu$m diameter. Similarly, Wuhrmann (1964) showed that flocs in an activated sludge tank could contain anoxic microsites. Using an oxygen diffusion coefficient of $5 \times 10^6$ cm$^2$/sec and an oxygen consumption rate of $10^{-4}$ mg/cm$^3$, he showed that a soluble oxygen concentration of 2 mg/liter on the outside of a sphere would drop to the "critical level" at a 500 $\mu$m radius. Considering these same rates, only a 145 $\mu$m distance would be required to effect anoxic conditions in a one-faced film. The geometry of gaseous–aqueous diffusion was also shown by Collins (1955), who found that the shape of the culture flask influenced the rate of denitrification by Pseudomonas aeruginosa under apparent aerobic conditions. Painter (1970) reviewed much of the conflicting reports on anaerobic vs. "aerobic" denitrification, and he concluded that many studies reporting that denitrification occurred under aerobic conditions did not contain any or sufficient data on the soluble oxygen concentrations. Thus, the critical factor governing whether aerobic or anaerobic metabolism occurs at a definite site is the soluble oxygen concentration at that site and not the average gaseous concentration of the surrounding milieu.

Concentrations of gaseous oxygen in soil profiles are very poor and misleading indications as to the anoxia status in microsites. Amer and Bartholomew (1951) found that there was very little difference in the rate of nitrification with soils subjected to continuous aeration at 11 vs. 20% oxygen and that the rate was only halved at 2%. On the other hand, oxygen concentrations of 17% and higher are frequently observed in soil profiles where denitrification is known to occur (Meek et al., 1969; Focht et al., 1975; Stefanson, 1972a,b,c; Burford and Millington, 1968; Dowdell and Smith, 1974). In a comprehensive review of the literature, Wesseling and van Wijk (1957) showed that oxygen diffusion in soil became critical at about 85% water saturation of the total pore space. Gaseous diffusion equations for conditions wetter than this are not applicable since water films greatly impede the flux of gas (Flühler, 1973). Denitrification appears to cease when soils become drier than "field capacity" (roughly equivalent to −0.3 bar of tension or about 50% of saturation) in most incubation studies (Pilot and Patrick, 1972; Bremner and Shaw, 1958; Mahedrappa and Smith, 1967; Abd-el-Malek et al., 1975). Similarly, soil profiles do not contain nitrous oxide concentrations significantly above ambient at tensions more negative than −0.3 bar (Focht et al., 1975; Dowdell and Smith, 1974).

The moisture content of soils is important to nitrification and denitrification insofar as it relates to aeration—except at extreme moisture stress, where desiccation reduces all microbial activity. Cooper and Smith (1963) found that denitrification proceeded rapidly at −1/3 bar in argon atmospheres. Mahedrappa and Smith (1967) showed that in a closed anaerobic system, the rate of denitrification was unaffected by increasing the moisture content, but in an open system, the rate was accelerated in all cases when the soil was wetted beyond field capacity. Thus, the effects of moisture in the ranges between saturation and field capacity are mostly directed towards the diffusion of oxygen to microsites where microbial activity occurs.

Oxygen diffusion rates in cropped soils may also be restricted in the rhizosphere where oxygen consumption and available carbon levels are high. Stefanson (1972a) showed that the pore space relationships at the end of the study were more restrictive, the oxygen demand was greater, and the amount of available carbon was higher in a cropped vs. pasture soil. Significantly more dinitrogen was evolved from the cropped soil. The greater potential for denitrification to occur in the presence of roots than in nonrhizosphere soil has been fairly conclusively demonstrated (Brar, 1972; Bailey and Beauchamp, 1973b; Valera and Alexander, 1961; Woldendorp, 1962).

Since oxygen is always utilized in preference to nitrate by denitrifying and nitrate-respiring bacteria, reduction of nitrate would not be expected to occur until oxygen was limiting. The point at which utilization of oxygen shifts from an apparently zero-order to first-order reaction is between 0.1 to 0.2 mg/liter (Wuhrmann, 1964; Chance, 1957). Denitrification has been reported to occur at this range in pure culture (Skerman and MacRae, 1957), sewage (Dawson and Murphy, 1973), and ocean waters (Richards and Broenkow, 1971). However, denitrification has also been observed at much higher oxygen concentrations, such as 0.7 mg/liter in the ocean (Goering and Cline, 1970) and as high as 2.0 mg/liter in sewage (Wheatland et al., 1959), though the rate was reduced by 90% in the latter case. It is not definitely known whether denitrification at high oxygen concentrations truly occurs at the measured site or within a microsite that cannot be measured. Alternatively, the differences in reported "critical oxygen concentration" above which denitrification does not occur may reflect the selection of bacteria which possess different affinity constants for cytochrome c or nitrate (Focht and Chang, 1975). Avnimelech and Raveh (1976) suggested that high nitrate concentrations overcome the inhibition by oxygen—possibly through some type of competition. White and Sinclair (1971) noted that the critical oxygen concentration is increased when cellular levels of cytochrome c are elevated by previous anaerobic incubation in the presence of nitrate or nitrite as electron acceptors. They suggested that the rate-limiting step in substrate oxidation is not at the primary dehydrogenase level but at the terminal-electron acceptor. A greater cytochrome content would ensure faster rates of denitrification where respiration might be limited by the amount of cytochrome $a_3$. Furthermore, cytochrome $a_3$ is independent of either nitrate or oxygen concentrations, while cytochrome c content is elevated in response to nitrate (Downey and Kiszkiss, 1969; Sapshead and Wimpenny, 1972; Daniel and Appleby, 1972). Thus, any limitations in electron flow through cytochrome $a_3$ could be offset by shunting off more electrons at the branch point from cytochrome c.

It has been shown that the kinetics of oxidative decomposition of domestic wastes is similar with oxygen and nitrate as oxidants (Mulbarger, 1971; Stensel, 1973). Yet, there is no evidence which suggests that these kinetics would be applicable to soil or benthic environments, in which the proportion of labile organic matter is very low. The oxidation of aromatic rings, which are significant structural units of humus, involves the participation of molecular oxygen through the action of dioxygenases. Thus, a significant fraction of the soil organic matter would be unavailable under continued anoxia, a condition common to the formation of peat. Though an isolated report by Taylor and Heeb (1972) showed that nitrate could be used in place of oxygen during the metabolism of simple aromatic compounds, the ecological significance of this reaction is obscure.

Aeration, though an appropriate term for comparing the preference of oxygen to other terminal-electron acceptors, is best expressed as $E_h$ when comparing the utilization of terminal-electron acceptors other than oxygen or when the soluble oxygen level is too low to measure. Ideally, the $E_h$ for the $NO_3^-/NO_2^-$ equilibrium is 421 mV (Latimer, 1952; Pourbaix, 1966), though nitrate appears to be reduced at potentials slightly lower than this (300–400 mV) in biological systems (Patrick, 1960; Grass et al., 1973; Pearsall and Mortimer, 1939; Bailey and Beauchamp, 1973a; Gambrell et al., 1975; Kefauver and Allison, 1957). Asghar and Kanehiro (1976) similarly observed no reduction of nitrate above 400 mV and found that the rate of denitrification was highly correlated with $E_h$ and increased by 23 % for every 100 mV decrease in $E_h$.

The more rapid reduction of nitrite over nitrate is important in the treatment of highly nitrogenous waste waters where the truncated nitrification–denitrification sequence is used. Voets et al. (1975) showed that the rates of denitrification were greater under aerobic conditions when nitrification was blocked at the nitrite stage by perchlorate than when the reaction proceeded to completion and nitrate was the product. Chemodenitrification could be ruled out in these investigations because the process did not occur in autoclaved and filter-sterilized samples. In view of the fact that Voets et al. (1975) had meticulously measured the soluble oxygen concentration throughout the experiment, their conclusion that denitrification can occur under aerobic conditions with nitrite as substrate is unequivocal. Nevertheless, they still observed greater rates of reduction in the daily cycles that had a 4-hr anaerobic incubation period than in the reactors that were subjected to continuous aeration. A similar observation was made by Vagnai and Klein (1974), who noted that four isolates of nitrite-dissimilating bacteria reduced nitrite to nitrogen gas much more rapidly under anaerobic conditions. Thus, the term "oxydenitrification" (Pasveer, 1971) should be used judiciously, and it should not be taken to mean that the reaction is favored under aerobic conditions.

The specific mechanism for oxydenitrification is not clearly understood. Voets et al. (1975), using McCarty's (1972) published values for $E_h$, noted that the $E_h$ for the $NO_2^-/N_2$ couple (996 mV) was higher than for $O_2/H_2O$ (818 mV), while that of the $NO_3^-/NO_2^-$ couple (408 mV) was lower. $E_h$ does not appear to be directly involved, however, because the $NO_3^-/N_2$ couple under these same conditions is also higher, i.e., 1.024 mV (Pourbaix, 1966), than the oxygen couple, and nitrite is more rapidly reduced under anaerobic than aerobic conditions. This suggests that the limiting factor in the oxidation of reductant is the terminal-electron acceptor. It is, therefore, not surprising that oxydenitrification has been observed only in conditions where cell densities and available organic matter levels are very high, specifically in activated sludge and in culture. Thus, the observation of Murray et al. (1975) that 57% of the total nitrogen entering a waste oxidation tank was lost despite high dissolved oxygen concentrations may also have been due to oxydenitrification since they used the truncated nitrification–denitrification process.

Nitrous oxide is commonly found as a final product or as a transient intermediate in soils, though it is rarely found in sediment or in sewage. Nommik (1956) and Wiljer and Delwiche (1954) observed that proportionally less nitrous oxide than dinitrogen was evolved under reduced oxygen tensions. Burford and Stefanson (1973) observed that the proportion of the two gases was greatly dependent on the moisture-aeration status. Stefanson (1972a) found that the reaction tended to go more to completion in cropped soils, in which the oxygen demand was greater and the diffusion was more restricted, than in pasture soils. The highest concentrations of nitrous oxide observed in soils appears to be in the $-50$ to $-200$ mbar of tension range, with concentrations generally much lower under saturated conditions (Dowdell and Smith, 1974; Focht et al., 1975). Focht (1974) suggested that this was due to a rapid use of nitrous oxide as a terminal-electron acceptor when conditions became more anaerobic. The $E_h$ calculated for the reduction of nitrous oxide to dinitrogen in the previous section suggests that nitrous oxide would be more likely to accumulate where the redox potential does not fall below 200–240 mV.

### 5.3. Temperature

Rate processes are exponentially affected by temperature, and the standard Arrhenius equation can be used within moderate temperature limits (15–35°C) to describe the effect of temperature on the rate of nitrification, while a much broader range (15–75°C) can be used for characterizing the denitrification process. Lower temperatures (generally below 12–15°C) have a much more drastic effect on rate processes, a phenomenon characteristic of all biochemical systems (Ingraham, 1962) that cannot be characterized as well by the Arrhenius equation.

The optimal temperature for nitrification in pure culture ranges from 25–35°C (Buswell et al., 1954; Meiklejohn, 1954; Deppe and Engel, 1960; see also Table I), although higher optimum temperatures have been reported for Nitrobacter winogradskyi (Laudelout and van Tichelen, 1960). The failure of the autotrophic nitrifiers to grow above 40°C is generally believed to explain why nitrate or nitrite is not found at all, or in trace quantities, in thermophilic composts.

There are some indications that heterotrophic nitrification might be of quantitative significance in high-temperature environments such as desert soils and composting systems, especially because activity of the autotrophic nitrifiers ceases above 40°C. Thus, the formation of nitrate in desert soils at temperatures above 40°C has been reported by Etinger-Tulczynska (1969), who noted that the formation of nitrate from ammonium was optimal at 28°C and that it was inhibited by $KClO_3$ and chloromycetin. In contrast, nitrate formation from indigenous soil nitrogen proceeded better at 37–40°C and was not susceptible to those inhibitors. Also, the addition of ammonium did not enhance nitrification. Similar studies by Ishaque and Cornfield (1974) of an acid soil lacking autotrophic nitrifiers revealed that nitrification was fairly rapid at 40°C and continued at temperatures of 50 and 60°C, though the rate was reduced to a few mg/kg soil/month. Myers (1975) also was able to measure the production of nitrate at 60°C and, in fact, found that the rate at 50°C was greater than at 20°C.

These findings point in the direction of heterotrophic nitrification, particularly because the thermophilic process apparently uses organic nitrogen as its initial substrate. With regard to the composting systems, it is interesting to note that nitrate formation seems to occur during the thermophilic composting (40–55°C) of solid wastes (Knuth, 1970). Detailed studies with regard to the nature of these nitrifying organisms are not known to us but would certainly be worthwhile in view of the restricted temperature tolerance of the classic autotrophic nitrifiers.

It must be added that, in contrast to solid composting, no nitrate formation has so far been reported for liquid thermophilic composting. Thaer et al. (1975) noted that the formation of oxidized nitrogenous products could not be detected in animal wastewaters treated aerobically at temperatures above 40°C. In these liquid environments, only bacteria were active. Thermophilic actinomycetes might be responsible for the formation of nitrate in high-temperature environments where the autotrophic nitrifiers are inhibited.

Temperatures above 35°C are generally of minimal significance to most aquatic and terrestrial systems. Of far greater importance is the effect of lower temperature (0–15°C), which is fairly common and difficult to assess quantitatively. It has been well established that, upon thawing of frozen soils, organic matter is readily mineralized. A similar seasonal change in aquatic systems occurs during the warming periods in the spring when the organic and biomass nitrogen are rapidly converted to mineral forms. The subsequent release of ammonium through ammonification stimulates the growth and metabolism of nitrifiers. Nitrogen availability is also promoted by freezing and thawing. Sabey et al. (1956) showed that nitrification was more rapid under field conditions at low average spring temperatures than in the laboratory, the explanation being that diurnal fluctuations in temperature affected greater amounts of available nitrogen. Frederick (1956) similarly showed that soils subjected to fluctuating square-wave temperatures below 15.5°C supported more nitrification than soils held at the constant mean temperature. Campbell et al. (1971) found that the ammonium levels present in sterile soils subjected to fluctuating temperatures were also higher than in soils held at the corresponding mean temperature, thereby suggesting a nonbiological mechanism as well. Campbell et al. (1974) found sine-wave incubation studies to be comparable with field studies of nitrification in enclosed plastic bags. About half of the soil nitrogen was mineralized in 7 weeks.

Autotrophic nitrifying bacteria apparently become acclimated to the temperature regime of their habitat and do not appear to vary in their adaptability to low temperatures. Anderson et al. (1971) found that soils from colder mountainous areas in Georgia had the highest nitrifying activity, soils from the warmer coastal plain had the lowest, and soils from the intermediate Piedmont plateau were in between when all were incubated at 6°C. When the two soils from the coldest and warmest climates were mixed in varying proportions and incubated at 6°C, it was found that the lag period in the production of nitrate was shortened accordingly as the proportion of mountain soil was increased at the expense of the coastal soil. Upon inoculation of sterilized soil with the two soils, a greater lag period in nitrate production was noted with the coastal plain soil than with the mountain soil. In all sets of experiments, the rate of nitrification appeared to be the same, with the chief effect being the length of the lag period. They noted no difference in the rate of nitrification at the optimum temperature (35°C) for both soils. Mahendrappa et al. (1966), however, found that nitrification was faster at 20 and 25°C than at 35 and 45°C with western soils in northern climates, while the reverse was true with southern soils. The highest concentrations of nitrite were found at 35 and 40°C in the northern soils and at 20 and 25°C in the southern soils. They attributed this to the environment exerting a more adverse effect upon *Nitrobacter* than upon *Nitrosomonas*.

It is not definite which of the two bacteria, the ammonium oxidizers or the nitrite oxidizers, are more greatly affected by temperature. Knowles et al. (1965) found that the maximum growth rates of ammonium-oxidizing and nitrite-oxidizing bacteria in river water were both 2.0/day, and the respective $Q_{10}$ values were 2.7 and 1.9, which suggests that the former organisms are more drastically affected by temperature changes. Hence, nitrite would not tend to accumulate. A $Q_{10}$ of 1.9 for the maximum oxidation rate (1.5 ml $O_2$/day) for *Nitrobacter* in pure culture (Laudelout and van Tichelen, 1960) agrees with this finding. On the other hand, a lower $Q_{10}$ of 1.7 is reported for the maximum growth rate (2.2/day) of *Nitroso-*

*monas* in pure culture (Buswell *et al.*, 1954). These apparent anomalies are resolved when one considers the kinetic studies of nitrification by Wong-Chong and Loehr (1975). They showed that at the optimal pH for nitrite oxidation (7.3) the activation energy ($E_a = 6.6$ kcal/mol) was lowest by comparison with pH 8.5 (39.6 kcal/mol) and pH 6.5 (14.0 kcal/mol). The activation energy for ammonium oxidation was lowest at pH 7.5 (16.0 kcal/mol) and was not greatly affected by a pH shift to 8.5 (20.0 kcal/mol) or to 6.0 (19.8 kcal/mol). Thus, at optimal pH, changes in temperature would be more pronounced for the activity of ammonium oxidizers than nitrite oxidizers, while at an acid or alkaline pH, the reverse would be true. Hence, the conclusion of Mahedrappa *et al.* (1966) would be valid for the acidic soils used in their study.

Wong-Chong and Loehr (1975) also noted that the optimal rate of nitrite oxidation at pH 8.5 and 6.5 was about 25°C, while at pH 7.5 it was 34°C. Thus, the combined effects of pH and temperature must be considered when comparing the metabolism and growth of the ammonium- and nitrite-oxidizing bacteria.

Since the species diversity affecting denitrification is far greater than that affecting nitrification, it is not surprising that the rate is optimal at much higher temperatures, namely at about 65°C (Nommik, 1956), due no doubt, to the existence of thermophilic species of *Bacillus*. Similar observations were made by Bremner and Shaw (1958), though they observed no change in the rate beyond 35°C, probably because their rates were calculated on the basis of a 3-day incubation period, during which 90% or more of the added nitrate was gone with higher incubation temperatures. Stanford *et al.* (1975a) found essentially no change in the rate from 35 to 45°C, but no higher temperatures were tested. Failure to observe higher rates at higher temperatures may be due to an insufficient incubation period to allow for the succession of thermophilic species of *Bacillus*. This would be important if *Bacillus* does indeed constitute a small fraction of the species diversity of denitrifying bacteria, as suggested by Vives and Parés (1975) and Woldendorp (1968).

In general, the kinetic effect for temperatures between 12 and 35°C can be related to the Arrhenius equation. Dawson and Murphy (1972) showed that the Arrhenius equation held for temperatures as low as 3°C, though most other reports show a profound break in the curve below the 12–15°C range (Nommik, 1956; Bremner and Shaw, 1958; McCarty *et al.*, 1969; Bailey and Beauchamp, 1973a; Stanford *et al.*, 1975). $Q_{10}$ values determined for the process are quite variable, ranging from 1.4 to 3.4 in the moderate temperature range and from 5.0 to 16.0 in the lower temperature ranges (see Focht and Chang, 1975).

Several interacting factors are undoubtedly responsible for the wide range of $Q_{10}$ values reported in the literature. One of the most important is the reductant–substrate interaction. Since various carbonaceous materials are known to have different activation energies and since denitrification involves essentially the same respiratory mechanisms of substrate oxidation, it is probably not surprising that the quantitative effect of temperature varies greatly among different environments. Novak (1974) showed that $Q_{10}$ was influenced by substrate concentration, particularly at lower concentrations. Complex wastes appeared to have little effect on $Q_{10}$, while acetic acid and synthetic waste material increased the $Q_{10}$ exponentially at the lower concentrations. It is not known if the diversity of organic substrates, the selection with temperature of different bacterial species, or both mechanisms account for the variable quantitative effects of temperature upon denitrification.

A significant finding by Misra et al. (1974) was that the interaction of temperature and aeration greatly affected the rate of denitrification (see Fig. 2). The activation energy (or $Q_{10}$) was far greater at 20% gaseous $O_2$ concentration than at 0.2%. Stated another way, the rate of denitrification was much more greatly affected by the gaseous oxygen concentration at 19°C than at 34°C. This leads one to extrapolate the results to higher temperatures (high temperature composting) and poses the following question when viewed in the light of recent reports on high-temperature heterotrophic nitrification: Does the failure to find significant quantities of nitrate under such conditions truly prove that nitrification does not occur, or does the diffusion rate of nitrate over a very short distance to anoxic microsites enable it to be denitrified as fast as it is formed? The loss of mineral nitrogen during high temperature composting has classically

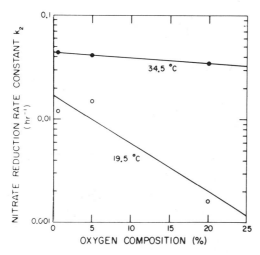

**Figure 2.** Combined effects of temperature and gaseous oxygen concentration upon denitrification. From Misra et al. (1974).

been attributed to deamination and subsequent volatilization of ammonia. Though this mechanism of nitrogen loss has been well established, the possibility of heterotrophic nitrification–denitrification should not be excluded.

Patterns of product formation from denitrification as influenced by temperature are not clearly understood, though there does not appear to be any significant differences in the gaseous composition of nitrous oxide or dinitrogen (Nommik, 1956; Bailey and Beauchamp, 1973a; Focht, 1974). However, nitric oxide appears to be more frequently observed as a product at lower temperatures; i.e., at 3–5°C (Nommik, 1956; Bailey and Beauchamp, 1973a). Citing the studies of Nommik (1956) and Bailey and Beauchamp (1973a), Focht and Chang (1975) suggested that the $Q_{10}$ for nitrate reduction at lower temperatures was much higher than the $Q_{10}$ for nitrite reduction. Bailey and Beauchamp (1973a) found no reduction of nitrate at 5°C, though their incubation period (22 days) may have not been long enough. They found considerable production, however, of nitric oxide from nitrite. They speculated that the nitric oxide may have come from chemodenitrification, but they also suggested that nitrite might have repressed further reduction to dinitrogen. Their latter explanation would appear to be more realistic since the pH of their soil (7.1) was rather high for chemodenitrification to occur and since high nitrite concentrations are known to have the greatest inhibitory effects on the particulate-linked enzymes (Payne, 1973). Thus, the direct effect of temperature on product formation would pertain only to environments that would favor the accumulation of nitrite rather than nitrate.

### 5.4. pH

It is well established that the growth and metabolism of the autotrophic nitrifying bacteria are optimal in the neutral to slightly alkaline range (pH 7–8). The pH range for complete nitrification is, however, very restricted, due largely to the toxicity of free ammonia (at alkaline pH) and nitrous acid (at acid pH) to *Nitrobacter*. The concentration of free ammonia or undissociated nitrous acid is pH dependent. (The $pK_a$ for $NH_3/NH_4^+$ is 9.3, and the $pK_a$ for $HNO_2/NO_2^-$ is 3.4). *Nitrobacter* will grow in pure culture up to a pH of 10.2 (Meyerhof, 1916a,b) when $NO_2^-$ is the only nitrogen source present. However, in nature, this situation is not likely to exist since nitrite is formed from the oxidation of ammonium by *Nitrosomonas*. The effective pH range for autotrophic nitrification is depicted in the nitrification tolerance graph (Fig. 3) developed by Anthonisen *et al.* (1976). This graph indicates that, at ammonia levels surpassing 150 mg $NH_3$-N/liter, autotrophic nitrification is totally inhibited. Such levels of ammonia have been found so far only in highly nitrogenous waste waters.

Prakasam and Loehr (1972) found that nitrification was unaffected up

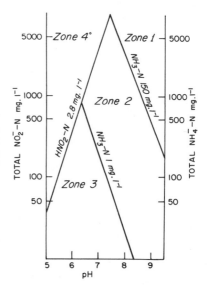

**Figure 3.** Nitrification tolerance graph redrawn from Anthonisen *et al.* (1976). Zone 1: Inhibition of *Nitrosomonas* and *Nitrobacter* by $NH_3$. Zone 2: Inhibition of *Nitrobacter* by $NH_3$. Zone 3: Complete nitrification. Zone 4: Inhibition of *Nitrobacter* by $HNO_2$.

to pH 11.2 as long as the free ammonia concentration was less than 0.02 $\mu$g N/ml. Rather than attempting to control the concentration of free ammonia, they devised a system for the treatment of highly nitrogenous waste water by allowing the oxidation to stop at nitrite. They termed this incomplete nitrification process "nitritification." The accumulation of nitrite in alkaline soils is frequently observed after the addition of ammonium fertilizers. The effect of pH upon the $NH_3/NH_4^+$ equilibrium is more complex in soil because of its cation-exchange capacity. Ammonia inhibition of *Nitrobacter* in soil is thus abated with increasing clay content (Broadbent *et al.*, 1957).

In a study encompassing a wide diversity of soils, Morrill and Dawson (1967) found that there were basically four types of nitrification patterns, which were directly dependent on the initial pH. The first type was characterized by a pH of 7.9, and such soils effected the rapid oxidation of ammonium with nitrite accumulating; nitrate was not oxidized until most of the ammonium was gone. This type of pattern exhibited a long lag phase in the growth of *Nitrobacter* that coincided with the accumulation of nitrite. In the second type (pH 6.4), the oxidation of both ammonium and nitrite was rapid, with no accumulation of nitrite. Type 3 (pH 5.4) was similar to type 2 except that the rates of both processes were markedly reduced. In the fourth type (pH 5.1), no detectable oxidation was observed. The numbers of *Nitrosomonas* were generally higher than *Nitrobacter*, except in the second type, where they were both about the same.

Indications that heterotrophic nitrification occurs during the aerobic treatment of highly nitrogenous, alkaline waste waters are not known to us.

However, in a specific attempt to verify whether heterotrophic nitrification could take place under nonaxenic circumstances, Verstraete and Alexander (1973) enriched soil suspensions as well as samples of sewage and of river and lake water with ammonium up to a level of 1000 mg N/liter. Upon aerobic incubation of the latter suspensions, the pH rose to values of 8.0–9.0. According to Fig. 3, autotrophic nitrification under those conditions is highly unlikely and can only be expected after a prolonged and gradual adaptation phase. Nevertheless, the authors observed within a matter of days the formation of hydroxylamine, 1-nitrosoethanol, nitrite, and nitrate. These nitrification products, as well as the patterns in which they were formed, resembled exactly those of *Arthrobacter* sp. when grown under axenic culture conditions (Verstraete and Alexander, 1972b). This kind of evidence strongly points in the direction of heterotrophic nitrification. The accumulation of trace amounts of hydroxylamine in aerobic, highly alkaline aquatic environments has recently also been reported by Baxter *et al.* (1973) in samples from an Ethiopian lake. However, the latter report does not permit one to speculate on the autotrophic or heterotrophic origin of the hydroxylamine.

The hypothesis that in acidic soils (pH ≤ 4.5) heterotrophic nitrifiers are responsible for the formation of nitrate has been advanced by numerous authors. There are three principal characteristics of the nitrification process in these soils. First, autotrophic nitrifiers cannot be isolated from such soils in significant numbers (Lemee, 1967; Overrein, 1971; Herlichy, 1973; Ayanaba and Omayuili, 1975; Ishaque and Cornfield, 1974; Cooper, 1975; Verstraete and Voets, 1976). As mentioned earlier, this point on its own is equivocal because of enumeration problems. The second point, that the pH optimum is 4.5, reinforces the first point. Ishaque and Cornfield (1972) observed in this regard that the addition of lime to an acid Pakistan "tea" soil inhibited nitrification. Verstraete and Voets (1976) also noted a pH optimum of 4.5–5.0 for acid forest soils. The third and perhaps most convincing point is that nitrate formation is related to the amount of organic nitrogen present, and the addition of ammonium inhibits or has no effect on the nitrification process. Such phenomena have been observed in acid soils by Weber and Gainey (1962) and recently by Verstraete and Voets (1976). In the latter case, nitrate was formed from indigenous soil nitrogen in an acid forest soil. Addition of peptone (100 mg N/kg soil) doubled the rate of nitrate formation, although during the ammonification of the peptone, the pH never rose above 4.5. However, when ammonium instead of peptone was added as a nitrogen source at a dose of 100 mg N/kg soil, nitrification was completely inhibited. Since this inhibition occurred with ammonium chloride, ammonium sulfate, and ammonium phosphate as well, it appears that the inhibitory activity is exhibited by the cation rather than by the anion. A quite similar observation was made by Ishaque and

Cornfield (1974), who reported that urea and oxamide stimulate the formation of nitrate, but not ammonium sulfate in acid Bangladesh soils lacking autotrophic nitrifying organisms. Cooper (1975) similarly observed that nitrogen from pig slurry was nitrified to a greater extent than ammonium sulfate in a slightly acid soil (pH 5.8). Probably because of the higher pH in this study as opposed to the previously cited studies, autotrophic nitrification still occurred, though the production of nitrate from nitrite was attributed to heterotrophic action because numbers of *Nitrobacter*, unlike *Nitrosomonas*, were too low ($<10/g$ vs. $>10^4/g$). These facts, when taken together, suggest that in acidic soils some form of heterotrophic nitrification occurs which proceeds optimally at acidic pH values. The organisms which bring about the formation of nitrate in these soils are, as yet, unknown.

The distribution of denitrifying bacteria covers a much wider range of the pH spectrum than the autotrophic nitrifying bacteria. Though the process is favored at slightly alkaline pH (Wiljer and Delwiche, 1954; Nommik, 1956; Bremner and Shaw, 1958; Dawson and Murphy, 1972), it proceeds to a pH of 3.5 (Nommik, 1956; Cady and Bartholomew, 1960). The pH at which the reaction completely ceases is not known, though Prakasam and Loehr (1972) report the highest value (pH 11.2) found in the literature.

Since most of the common soil and aquatic denitrifying bacteria have optima for growth similar to those of the general bacterial flora (at least in the pH 5–9 range), it would appear that the effect of pH on population shifts would be minimal in this range. More extreme ranges at the higher and lower end of the pH spectrum might favor a more narrow species diversity, though there are no reports that indicate whether or not this is true.

The pH in nature has a profound effect upon not only the rate process but upon the composition of the gaseous products released during denitrification. It is well established that the process tends more towards completion (i.e., the relative amounts of $N_2$ vs. $N_2O$ are greater) as the pH increases (Wiljer and Delwiche, 1954; Hauck and Melstead, 1956; Nommik, 1956; Matsubara and Mori, 1968). If it is assumed that the same bacterial species are effecting the conversion at all the pH ranges tested, then the nitrous oxide reductase enzyme would appear to be more sensitive to low pH. It is questionable, however, whether the species diversity of denitrifying bacteria is relatively uniform among different soils. Nommik's (1956) data show that the rates of $N_2O$ production and reduction are fairly linear up to pH 8.0 (Focht, 1974). On the other hand, Wiljer and Delwiche (1954) showed that $N_2O$ makes up most of the denitrification gases up to pH 6.0, and then its relative abundance falls off rapidly. Both of these studies suggest that the initial microflora were different, since both adjusted various samples of a single soil to different pH ranges without prior incubation before the addition of nitrate. Nommik (1956) used a soil that was initially more acidic than the one used by Wiljer and Delwiche (1954). It would be expected that the former soil would contain bacteria better acclimated to acid conditions and more effective in reducing $N_2O$ under these conditions than the bacteria present in the latter soil. Cady and Bartholomew (1960) showed that nitrous oxide was reduced to dinitrogen at pH values below those reported in the two studies above, though it took over 40 days. Possibly the incubation periods in the other two investigations were not long enough to allow for the development of an adaptation of more acid-tolerant denitrifiers. While studying the decomposition of a hydrocarbon, Focht and Joseph (1974) noted that there was a difference in acid tolerance among several denitrifying pseudomonads. It is thus possible that the failure to note significant conversion of nitrous oxide to dinitrogen in short-term incubations in acid soils may be attributed to insufficient time for acclimation of more acid-tolerant denitrifiers.

Nitric oxide also is a gaseous product of denitrification, though it is rarely found with intact cell suspension (in lieu of high nitrite concentrations) or in neutral to alkaline soils. Although considerable quantities of nitric oxide are frequently generated in acid soils (Nommik, 1956; Wiljer and Delwiche, 1954; Bollag et al., 1973; Cady and Bartholomew, 1960; Garcia, 1973), this reaction appears to be primarily nonbiological since sterilized soils usually evolve as much nitric oxide as controls when nitrite is added (Bollag et al., 1973). Though the rate of chemodenitrification of nitrite are favored by metallic ions and organic matter (Nelson and Bremner, 1970a,b) and may exceed that of biological denitrification in acid environments, nitrite does not usually accumulate in large quantities during denitrification when there is sufficient reductant present. Van Cleemput and Patrick (1974) concluded that the reduction of nitrite in acid and alkaline soils that had been sterilized by irradiation was a biochemical process mediated by enzymes of nonviable, though intact, cells.

## 5.5. Surfaces

In nutrient-poor environments, surfaces enhance the establishment of the microbial biomass by providing attachment sites and by concentrating nutrients from dilute solutions (Heukelekian and Heller, 1940; ZoBell and Anderson, 1936; ZoBell, 1943). Heukelekian and Heller (1940), for example, found that the total cell mass of E. coli incubated with glass beads was increased 50 times over the control containing no beads and was equal to the cell mass of a culture without beads containing a substrate concentration 100 times greater. It is therefore not surprising that the overall numbers and activity of heterotrophic benthic bacteria exceed those of planktonic bacteria by several orders of magnitude (Potter, 1963; Keeney et al., 1971). The greater abundance of organic matter and the consequent higher population of denitrifying bacteria in sediments undoubtedly account for the greater denitrification activity in the sediments (Chen *et al.*, 1972a,b; Terry and Nelson, 1975; Goering and Dugdale, 1966).

Most of the nitrifying activity is also associated with the sediments in fresh waters and estuaries, where the numbers of *Nitrosomonas* and *Nitrobacter* are about a thousand times higher in the benthos ($10^6$–$10^7$) than the water ($10^3$–$10^4$) (Tuffey and Matulewich, 1974; Curtis *et al.*, 1975). Tuffey and Matulewich (1974) showed that nitrification does not occur along the entire length of a polluted river but occurs in identifiable zones because of oxygen content and residence time. One zone where nitrification occurs is in the shallow headwater streams and tributaries. Shallow rivers are more affected by nitrifying slimes on rocks than deeper rivers because of the greater contact proportion of biomass to water volume at equal flow rates. The other zone where nitrification would be significant is in estuaries and rivers of long detention time due to growth of planktonic bacteria. Between these zones, significant nitrification will not occur.

In the deep ocean, the effect of benthic activity on the nitrification process appears to be minimal since the nitrate-maximum zone is not above the sediments but is located at a depth of about 250–800 m below the surface (Watson, 1963). Rittenberg (1963) calculated that the inputs of nitrogen into the ocean could not be balanced through nitrification and denitrification because the numbers of planktonic organisms were too low. Webb and Wiebe (1975) provide evidence which shows that significant inputs of nitrate to the ocean can be obtained through nitrogen fixation and subsequent nitrification by symbiotic associations and attachment of nitrifying bacteria to masses of blue-green algae in algal flat regions. They found that only 20 out of 30 and 4 out of 30 enrichment cultures for *Nitrosomonas* and *Nitrobacter*, respectively, were positive. However, they obtained counts of *Nitrobacter agilis* in the algal flat of $1.4 \times 10^6$ cells/cm$^2$ by the fluorescent antibody method. Using these numbers, they calculated nitrification rates of 17 kg N/ha/yr; this figure was in agreement with their experimental observations and would resolve the dilemma presented by Rittenberg (1963). Paerl *et al.* (1975) similarly found that attached bacteria accounted for greater than 80% of the total bacteria in an oligotrophic lake. They postulated that the nitrifying activity in the lake was associated with the attached rather than the free-floating forms.

Soil particles influence the nitrification process in several ways. The strong affinity of ammonium through cation retention is but one means by which the numbers of *Nitrosomonas* are maintained very close to the surface of soils and sediments, where the ammonium concentration is highest. The rates of nitrification also appear to be greater with increasing clay content (Broadbent *et al.*, 1957; Justice and Smith, 1962), probably because less free ammonia is available through exchange equilibria.

The $K_m$ for nitrification by *N. agilis* has been shown to be 0.5 pH units higher in soil than in liquid suspension, though McLaren and Skujins (1963) attributed this to repulsion of nitrite from soil as an additional possibility to greater saturation of clays by $H^+$. The greater rates of nitrification in finer-textured soils may also be attributed to higher numbers of nitrifiers and greater amounts of nitrogen in such soils, the finer-textured soils normally containing more organic matter than sandier soils.

The higher rates of nitrification that are generally observed as clay content increases has led to some misconceptions concerning the supposed preference for adsorbed ammonium over free ammonium by *Nitrosomonas*. Lees and Quastel (1946a,b) concluded that only adsorbed ammonium was oxidized. Exchange capacity, however, was shown to be insignificant by Faurie *et al.* (1975). Furthermore, displacement of adsorbed ammonium by a flowing solution of KCl enhances nitrification in a soil column (Ardakani and McLaren, unpublished data). Lees and Quastel's observation that nitrification was increased by the addition of sterile soil but not sterile sand was attributed by Doner and McLaren (1976) to the likely increase in nutrients from the soil leading to an increase in the biomass. Furthermore, the origin of the ammonium ions once they diffuse into the bacterial cell is not significant. Finally, they argue that such an ion-exchange phenomenon has never been suggested for nitrite oxidation by *Nitrobacter* in soil. It appears that the preferential oxidation of exchangeable ammonium seems unrealistic except in very dilute solutions.

Higher rates of denitrification have also been observed in finer-textured soils (Wagner and Smith, 1960; Lund *et al.*, 1974), though Van der Staay and Focht (1977) showed that this was not related to surface area since they were unable to demonstrate any differences in denitrification rates between different clays, different glass bead sizes, and different minerals after removal of indigenous organic matter and subsequent addition of substrate and inoculum. Thus, it appears as though the effect of soil texture upon the denitrification process is related to physical properties (e.g., structure, aggregation, and water infiltration rate) that affect aeration.

## 6. Applied and Environmental Aspects

### 6.1. Agriculture

Reactions involving the cycling of nitrogen in nature have, for the most part, been most extensively studied by agricultural scientists since the fate of nitrogen in soil is profoundly associated with food and fiber production. Since nitrate is the form of nitrogen that is most readily assimilated by plants, the nitrification process is a crucial aspect of soil fertility. Though the growth and metabolism of the nitrifiers are comparably slow when compared with the heterotrophic flora, the rate of nitrate formation may be proceeding faster than the plant can assimilate the anion, and the excess nitrate may be leached to the ground water. The loss of nitrate through leaching is undesirable from both soil fertility and water quality standpoints. Agronomic practices which slow the rate of nitrate formation involve the use of slow release fertilizers, the addition of fertilizers in split applications, or the use of nitrification inhibitors such as 2-chloro-6-(trichloromethyl)-pyridine. The first two procedures minimize the rate of nitrification by effectively lowering the substrate concentration. The third procedure works directly by inhibiting the initial oxidation of ammonium (Campbell and Aleem, 1965a). With time, the inhibitor is degraded by other soil microorganisms, and nitrification commences. Ideally, degradation would occur and nitrification would start when the plant is actively growing; however, the success of this method is related to the rate at which the microflora decomposes the inhibitor.

Though losses of nitrogen from rice paddies are minimized by using ammonium rather than nitrate fertilizers, losses do occur through nitrification and denitrification. The concentration and thickness of the aerobic zone at the top of a flooded soil appear to be the determining factors influencing nitrogen losses from rice paddies (Broadbent and Tusneem, 1971; Patrick and Gotoh, 1974). Patrick and Gotoh (1974) also found that the thickness of the oxidized layer was reduced from 22 to 2 mm when the atmospheric oxygen concentration was reduced from 20 to 10%.

It is fairly clear that nitrification occurs near the soil surface, but there is considerable variation where the zone of denitrification occurs. The depth at which denitrification is most likely to occur in a soil profile is largely dependent on several factors: available organic matter, a suitable denitrifying population, nitrate concentration, and anoxic zones. The first three factors are almost always greatest near the surface, which suggests that most of the nitrate produced by either mineralization of organic or fertilizer nitrogen is denitrified before it reaches the plant roots if anoxic zones are present at the surface. Using different soils, Starr et al. (1974) found that denitrification occurred close to the surface in the soil having a higher organic matter content. They showed this relationship to be directly related to the oxygen consumption rate, with the accompanying depletion of oxygen and appearance of [15]N-labeled nitrogenous gas from added [$^{15}$N]NO$_3$. The soil that was poorer in organic matter required a longer residence time and a greater depth to effect reduction of the oxygen concentration. Rolston et al. (1976) confirmed these findings. They also found that by increasing the flux of water through a soil column, the active zone of denitrification was localized at a greater depth than in a column receiving a smaller flux of water.

Avnimelech (1971) studied the mineralization and release of nitrate from peats. He concluded that denitrification occurred near the surface, where the nitrate concentrations were greater, and that by regulating the irrigation regime, the depth and duration of anoxic zones could be controlled to remove all nitrate through denitrification. Similarly, removal of nitrogen from waste waters applied to highly porous soils is greatly dependent upon the residence time (Bouwer, 1970; Lance, 1972; Lance et al., 1976). Thus, denitrification in soils that are freely drained apparently does not occur at great depths in the profile below the root zone. Where shallow water tables or restrictive zones occur, denitrification may take place at greater depths. Studies by Gambrell et al. (1975) showed that certain soils were too aerobic in the upper profile to effect denitrification; only in the depths below the plant roots (i.e., near the water table) were conditions anoxic. In the two soils that were tested, they calculated that more than sufficient soluble organic matter was leached with the nitrate to this zone to effect complete reduction of the nitrate. Similarly, Meek et al. (1970) presented evidence of a continual drop in the redox potential from the surface to a 2 m depth at the tile drain. They showed that sufficient organic matter was leached with the drainage water to eventually reduce almost all of the nitrate throughout the profile.

Considerable controversy currently exists as to whether or not the use of nitrogenous fertilizers per se contributes to increased nitrate concentrations in ground waters (Viets, 1971; National Academy of Sciences, 1971). Kohl et al. (1971), by comparing natural variations in $^{15}N$, claimed that over half of the nitrate in drainage waters could originate from fertilizer, although others (Hauck et al., 1972; Focht, 1973) have questioned the validity of the approach. The increase in numbers of nitrifiers and production of nitrate certainly occurrs in response to substrate concentration, and if nitrogen is supplied beyond the needs of the plant, the excess nitrate must go somewhere if it is not immobilized into organic nitrogen or denitrified.

Since denitrification cannot be measured in situ without the use of $^{15}N$, agronomic losses of nitrogen through this process are generally calculated as the amount unaccounted for after crop and leaching removals are subtracted from the fertilizer inputs. In his review of the enigma of nitrogen balance studies, Allison (1955) stated that about 10–30% of the nitrogen lost from soils annually was through denitrification. Because of the errors in spatial variability, the transit times in nitrate leaching, and the mixing of fertilizer and organic nitrogen pools, denitrification losses by the difference method cannot be accurately calculated over one growing season even when $^{15}N$ is used. Long-term studies yield much better results even without the use of isotopes. Many of these studies support the contention that excessive use of fertilizer nitrogen over periods of several years will lead to an increase in the amount leached and/or denitrified.

In a 10-yr study with citrus, Pratt *et al.* (1972) found that nitrate not used by the plant either leached through soil profiles or was denitrified. The former fate was characteristic of the two sandy porous soils, while the latter fate was confined to the two soils containing argillic horizons. Broadbent (1975) found very little significant increase in the quantity of nitrate moving beyond the root zone until the fertilizer was used in excess of plant needs. Even with excessive additions of $^{15}NH_4^+$, Broadbent (1975) showed that over half of the nitrate in ground water was derived from mineralization of soil organic nitrogen. Excessive fertilization (448 kg N/ha/yr) resulted in a 720 kg N/ha increase in nitrate below corn roots to the ground water in a 3-yr study by Schuman *et al.* (1975).

Considerable nitrate leaching can also result from the high additions of manure to soils. Studies by Adriano *et al.* (1971) showed that high nitrate concentrations (some exceeding 1000 mg N/liter) in deep soil profiles were characteristic of soils receiving heavy applications of dairy wastes. Other studies have shown that the nitrate concentration in the drainage waters is low regardless of the amount of ammonium or manure added to soil (Gambrell *et al.*, 1975; Meek *et al.*, 1973; Elliot *et al.*, 1972). In all these instances, the water table was fairly high and the redox potential was below the couple for nitrate reduction at zones where the nitrate concentration declined. Therefore, much of the nitrate not removed by crops was denitrified. Thus, the drainage characteristics of the soil are very important in determining whether or not nitrate not utilized by the plant is denitrified. Another point to consider is that concentration is not always related to flux; in fact, the two were found to be inversely related by Devitt *et al.* (1976). Letey (unpublished data) found no correlation between application rates of fertilizer nitrogen and the concentration of nitrate in the tile drain, but he did observe a positive correlation between applied nitrogen and total nitrogen leaving the drains. Thus, he concludes that the amount of water used for irrigation must be considered.

Minimizing denitrification to prevent reduction in the nitrogen level of soil is also an important agronomic concern. Deherain (1897) recommended that manure and nitrate should never be added together, since the manure was observed to stimulate gaseous loss of nitrogen. His recommendation for adding manure in the fall and nitrate in the spring to minimize nitrogen losses is still good advice to this day. The abundance of zones of anoxia in rice paddies is but another reason for not using nitrate fertilizers and suggests instead the use of ammonium or reduced-nitrogen fertilizers.

The principal characteristics of the nitrification processes in acidic soils have been discussed before. Additional information supporting the hypothesis of heterotrophic nitrification can be deduced from the studies of Verstraete and Voets (unpublished data), who found that thiram at application rates of 250 and 1000 mg/kg soil completely inhibited nitrate

**Table V. Rate of Nitrate Formation in Acid Soils under Laboratory Conditions**

| Biotope | pH | Milligrams $NO_3^-$-N per kilogram soil dry weight per month | Reference |
|---------|-----|------------------------------------------------------------|-----------|
| Crop land | 5.4 | 1–5 | Pang *et al.* (1975) |
| Crop land | 4.2 | 5–10 | Ishaque and Cornfield (1972, 1974) |
| Forest soil | 3.0–5.0 | 15–100 | Runge (1974) |
| Forest soil | 3.9–4.4 | 5–50 | Lemee (1967, 1975) |
| Forest soil | 3.9–4.6 | 12–24 | Verstraete and Voets (1976) |

formation in acid podzolic forest soils. Mainwright and Pugh (1973) similarly reported that thiram at a dose of 250 mg/kg strongly reduced autotrophic nitrification. Chinn (1973) furthermore indicated that thiram at a concentration of 1000 mg/kg significantly reduced the growth of fungi and actinomycetes for a period of more than 32 weeks but only slightly affected the growth and metabolism of the soil bacteria. Hence, these facts suggest that nitrate formation in the acidic forest soils is either due to autotrophs or fungi. Verstraete and Voets (unpublished data) also found that the formation of nitrate in these soils was not inhibited by 2-chloro-6-(trichloromethyl)pyridine at a concentration of 4 mg/kg. From these inhibition studies and more particularly from the experiments revealing that organic nitrogen but not ammonium can serve as initial substrate, it appears that fungi probably are responsible for the formation of nitrate in these acidic soil environments.

The quantitative importance of such heterotrophic nitrification in the overall nitrogen cycle is presently not well documented. Typical rates of nitrate formation in these acidic soils are summarized in Table V. Although these rates are relatively low compared to those reported for autotrophic nitrifiers, they are nevertheless not without environmental significance since, on a hectare basis, even the lowest rate reported represents the formation of some 20 kg of nitrate N/yr.

## 6.2. Waste Treatment

Reduction of the biochemical oxygen demand of treated waste effluent was, until a few years ago, the primary waste treatment objective. In fact, the appearance of nitrate was an indication of the completion of biooxidation. In view of increasing concern about the premature eutrophication of receiving lakes and rivers, considerable attention has been directed to a tertiary treatment process designed for the removal of nitrogen from secondary waste effluent. After completion of secondary waste treatment, about 85–90% of the nitrogen is in the form of ammonium (5–50 mg N/liter) with the remainder present as organic nitrogen; only trace amounts of nitrate or nitrite are found (McCarty and Haug, 1971). The most successful procedure for removal of nitrogen from domestic plants is the sequential nitrification–denitrification processes. Reviews on one or both of these processes, as related to waste water treatment, have been written by Painter (1970), Focht and Chang (1975), and Francis and Callahan (1975).

Rapid removal of nitrogen from sewage involves separate treatments for reduction of BOD, nitrification, and denitrification. Though all processes can occur simultaneously in a trickling filter bed or in an activated sludge tank (Wuhrmann, 1964), it is the nitrification process that is rate limiting and thus necessitates a long retention time. Separate carbon and nitrogen oxidation processes minimize wash-out of the nitrifiers and can be operated at shorter detention time, lower MLSS (mixed liquids suspended solids), and sludge age (Rimer and Woodward, 1972). A greater improvement still is the three-stage process involving carbon removal, nitrification, and denitrification. In terms of nitrogen removal, operation, and costs, this procedure is clearly the best (Mulbarger, 1971). Population stability is also not a problem, unlike the switching over from aerobic to anaerobic cycles within the same treatment unit.

Eckenfelder (1967) calculated a sludge retention time of 2 days for adequate nitrification of domestic waste water (10–50 mg N/liter). Batch studies by Wong-Chong and Loehr (1975) showed that ammonium oxidation was the rate-limiting step, with a maximum removal rate of about 80 mg N/liter/hr. In mixed culture, nitrite oxidation by *Nitrobacter* was double this rate, and nitrite did not accumulate. Complete oxidation of urea and casein from a 1000 mg N/liter concentration was shown to occur in 40 hr at pH 7.0 and in 45 hr at pH 8.0. The slightly longer time in the latter case was apparently due to the toxicity of ammonia to *Nitrobacter*, as noted by the reduced rate of nitrite oxidation, the corresponding increase in nitrite, and the extended lag period.

Prakasam and Loehr (1972) suggested that the nitrite oxidation step could be eliminated, thereby also bypassing the reduction of nitrate to nitrite during the anaerobic stage. These two incomplete oxidation and reduction processes they described as "nitritification and denitritification" (note the additional "ti") to distinguish them from complete nitrification and denitrification. Use of the incomplete nitrification method obviates the problem of ammonia toxicity so that the pH in highly nitrogenous waste waters (e.g., from chicken manure) would not have to be intensively controlled or lowered to yield suboptimal rates of treatment. This method also has two distinct advantages during the denitrification stage: Nitrite is reduced faster than nitrate, and less reductant (organic matter) is required.

Voets *et al.* (1975) were concerned about highly carcinogenic *N*-nitrosamines, which are formed chemically and biochemically by condensation of a secondary amine with nitrite. Both of these substances might be found in highly nitrogenous wastes employing incomplete nitrification. Despite studies with sewage and lake water amended with secondary amines and nitrite, which showed that nitrosamines were formed (Ayanaba et al., 1973; Ayanaba and Alexander, 1974), Voets et al. (1975) were unable to detect the presence of nitrosamines.

Despite apparently highly oxidative conditions, incidental losses of nitrogen through denitrification occur as a result of anoxic microsites, a topic discussed earlier. Wong-Chong and Loehr (1975) found that a 10% loss of nitrogen occurred during the early incubation when an organic nitrogen substrate was used. Since no such loss of inorganic nitrogen occurred with the use of ammonium as initial substrate at comparable pH regimes, they concluded that this loss was due to denitrification in flocs, with the organic matter furnishing the reducing power. Lowered dissolved oxygen concentrations (0.5 mg/liter) measured during this period would confirm their conclusions.

Generally the rate-limiting process in waste treatment is the reductant concentration. Frequently the C/N ratio may be below the amount of reductant necessary to complete the process. McCarty et al. (1969) proposed the use of methanol to offset carbon-limiting wastes. Wuhrmann (1964), however, pointed out that there is sufficient carbon in the sludge biomass, and with longer retentions, all of the nitrate could be denitrified. Johnson and Schroepfer (1964) diverted a portion of the primary effluent to the nitrified effluent in the anaerobic reactor. Though nitrate was denitrified, the BOD (biochemical oxygen demand) of the effluent increased.

Molasses, humus (Finsen and Sampson, 1959), cellulose (Skinner, 1972), and bakery waste sugars (Adams et al., 1970) have also been used as reductants, with the result being a higher biomass and corresponding BOD. Wuhrmann and Mechsner (1973) illustrated why additions of exogenous carbon to carbon-limiting waste waters should not exceed the dissimilatory nitrogen demand; otherwise the effluent BOD would increase. It appears that the most efficient C/N ratio for denitrification is between 2 and 3 (Dawson and Murphy, 1973; Wuhrmann and Mechsner, 1973). Since this is higher than the C/N ratio needed for complete denitrification, it appears that nondenitrifying microorganisms are also competing for available carbon. When sufficient reductant is present, denitrification is extremely rapid, requiring less than an hour retention time (McHarness and McCarty, 1973; Finsen and Sampson, 1959).

Addition of secondary waste to land followed by nitrification-denitrification has been successful in warm, arid areas where open land is relatively plentiful. The Flushing Meadows Project in Arizona has been claimed to render cheaper renovated water of the same quality as tertiary sewage treatment (Bouwer et al., 1974). Shorter cycles (2 days wet, 3 days dry) resulted in complete conversion to nitrate with no denitrification, while longer cycles (17 days wet, 7 days dry) led to 90% removal of the added nitrogen. Cycles of 15 days wet, 15 days dry resulted in monthly nitrate peaks which ranged from 5–50 mg N/liter in one well. Lance (1972) observed that soil columns with 10-day wet, 10-day dry periods removed about 67% of the applied nitrogen. These studies indicate that the length of the wet period is the critical aspect of the cycle. A dry period, however, is essential to effect the nitrification process and to obviate the problem of clogging of soil pores by continuous application. When denitrification is limited by available carbon, the nitrogen removal rate can be increased by reducing the infiltration rate or by mixing a minimal amount of primary effluent with the secondary effluent (Lance, 1972; Lance et al., 1976).

## 6.3. Global Fluxes

Considerable uncertainty exists with respect to the cycling of terrestrial, aquatic, and atmospheric forms of nitrogen. The only unequivocal numbers that are presently available are the inputs of chemically fixed nitrogen fertilizers to cultivated lands because of commercial records. Delwiche (1970) gives a value of 30 million metric tons/ha/yr for chemical fixation. This is probably a conservative estimate in view of increasing fertilizer usage, which is expected to reach 57 million metric tons by 1980 (Nelson, 1975). About 43 million tons/yr of nitrogen are returned to the atmosphere by denitrification from terrestrial sources, about 40 million tons from marine sources, and an insignificant 0.2 million from sediments to give a total global denitrification flux of 83 million metric tons. This is about 9 million tons short of what was calculated to be the total nitrogen input from industrial, volcanic, legume, historic, and marine sources (Delwiche, 1970).

Applying Delwiche's (1970) calculations of terrestrial denitrification to the total land area gives a yearly surface flux of about 3 kg N/ha. This is considerably lower than estimated losses of denitrification on cropped lands and lysimeters, which range from 10–40 kg N/ha (Pratt et al., 1972; Allison, 1955, 1966). About 17 kg N/ha/yr is the estimated loss from denitrification in the Santa Ana Basin of southern California (Ayers and Branson, 1973), a figure which includes both agricultural and nonagricultural land together. Thus, it appears that denitrification losses are more prominent from agricultural soils.

Direct attempts to measure denitrification by use of isotopes and nitrous oxide fluxes (Rolston et al., 1976; Arnold, 1954; Stefanson, 1973; Burford and Stefanson, 1973) or sealed argon/oxygen growth chambers (Stefanson, 1972a,b,c) have given highly variable results. One problem is that rates higher than normal occur when nitrate is added in excess of what normally is found in the field. Also, it is not known whether denitrification in soil occurs rapidly in a short period of time or slowly for a long period of time. Broadbent and Clark (1965) believe the latter to be a more accurate account. Arnold's (1954) classic studies show that both cases can occur and are directly related to the extent and time that the soil is wet. Thus, the observation of rapid denitrification for a short time, if calculated to a yearly flux, can give very large and unrealistic numbers. Arnold's (1954) estimate of the nitrogen available for denitrification in soil to match the troposphere nitrous oxide half-life of 70 yr (Bates and Hays, 1967) would yield an annual flux between 0.6 and 5.21 kg $N_2O$-N/ha. Schütz et al. (1970) similarly calculated a nitrous oxide flux from soil of 1–10 kg $N_2O$-N/ha/yr. Delwiche (1970) stated that land areas would be a more likely source of nitrogen oxides than marine sources, but he did not estimate the relative proportions of dinitrogen and nitrous oxide.

Because of the tremendous variation in the proportion of nitrous oxide and dinitrogen, one cannot unequivocally estimate, at this time, how much of the latter would be evolved if the evolution rate of the former were known. However, it appears that proportionally less nitrous oxide is evolved in wetter, more anoxic soils (Jones, 1951; Arnold, 1954; Wiljer and Delwiche, 1954; Nommik, 1956; Focht, 1974) and sediments (Goering and Dugdale, 1966a; Chen et al., 1972b), or from marine sources (Goering and Dugdale, 1966b; Goering, 1968; Goering and Cline, 1970). Though these observations are consistent with thermodynamic considerations that would favor the further reduction of nitrous oxide only after the other nitrogenous oxides have been depleted, questions arise as to whether closed incubation vessels tend to overestimate the amounts of nitrous oxide that is reduced, since the diffusion of the gas out of the soil in a closed vessel would be slower than in an open system because of the smaller concentration gradient in the former. Thus, the depth at which denitrification occurs in soil may be an important function relative to concentration gradients and to contact time for further reduction of nitrous oxide as it diffuses to the surface. Intuitively, one would expect that more nitrous oxide would escape the closer the zone of denitrification is to the surface. Thus, Goering's studies (as cited previously) which show that nitrous oxide is an insignificant gaseous product in marine denitrification and in sediments would be consistent with this principle, as would studies by Rolston et al. (1976) and Gilliam and Dasberg (unpublished results), which show that more nitrous oxide is evolved from soil the closer the zone of denitrification is to the surface.

Delwiche's (1970) estimate that about 40 million tons is denitrified from marine sources per annum gives a yearly flux of 0.92 kg N/ha/yr. Rittenberg (1963) assumed that denitrification in the open ocean is small in magnitude, and he calculated a denitrification rate of 8.6 million tons/yr. Though the rates of denitrification in littoral and coastal sediment are much higher than in the open ocean, it is not known how much of the total flux generated from marine denitrification comes from sediments in the light of more recent studies. Richards and Broenkow (1971) calculated that the rates of denitrification over a sediment–water interface of Darwin Bay amounted to 182 kg N/ha/yr. From Delwiche's (1970) calculations, this means that about 0.5% of the marine surface area, as represented by littoral and coastal waters, could account for all denitrification from marine sources. This equatorial, oxygen-poor environment, however, would represent optimal zones and would not be representative of all littoral and coastal areas. Goering and Dugdale (1966) calculated a depth-integrated rate of 11 kg N/ha/yr in the oxygen-poor, nitrite-rich zone of this same bay, although they acknowledged that the addition of nitrate to samples on the ship's deck may have stimulated the rates. Thus, using these same calculations, all marine denitrification could occur in about 8% of the ocean surface area if their sample site was representative.

   Nitrite-rich zones can be due to both nitrification and nitrite reduction. Wada and Hattori (1971) found that the appearance of nitrite in the euphotic zone of the open sea was due mainly to nitrifying bacteria. A second zone of nitrite-rich water usually occurring below 50 m in the open sea, however, is due primarily to nitrate reduction by nitrate-respiring bacteria (Miyazaki et al., 1973). The case with bays, however, is different: The shallower depth and greater mixing frequently give rise to nitrite from both means. Miyazaki et al. (1973) showed with $^{15}N$ that half of the nitrite came equally from nitrification and nitrate reduction.

   Information in a recent comprehensive paper by Wada et al. (1975) may unravel some of the discrepancies. Using differences in natural abundance of $^{15}N$ that result from isotope fractionation, they offer some novel concepts about this enigma. They propose four geochemical mechanisms for marine denitrification. Type 1 represents complete anoxia whereby total conversion to $N_2$ occurs. In this situation, the per mil excess enrichment of $^{15}N$ $(\delta N^{15})$ of the gas is identical to the original nitrate (usually about 2.9). Type 2 represents oxygen-deficient, nitrate-rich waters that border the high nitrate zones and contain high concentrations of nitrate, where the reaction rate is slow. Type 3 represents the inhibition of denitrification by high oxygen concentrations. Here denitrification occurs only in the microsite or in the transient aggregate, and the gaseous nitrogen is enriched to the same degree as in Type 1 $(\delta N^{15}$ for soil and marine nitrate is $-1.7$ and $+5.8$, respectively). Type 4 represents deep sediments like the Cariaco Trench and Norwegian Fjords where $\delta N^{15}$ is $+34$. By studying the naturally occurring abundance of $^{15}N$ in nitrate and dinitrogen, Wada et al. 1975) calculated that most of the denitrification in the marine environment (88%) was due to conditions resembling Type 3, 12% was due to conditions representing Type 2, and only negligible amounts were due to the other two systems. When contributions from terrestrial sources are considered, 66 % of the total nitrogen evolved from denitrification is calculated to come from soil representing a Type 3 environment and 34% from the sea (30% from Type 3 and 4% from Type 2). The conclusions of Wada *et al.* (1975) are also in agreement with those proposed by others that terrestrial sources account for the great majority of gases produced by denitrification (Arnold, 1954; Albrecht *et al.*, 1970; Schütz, 1970; McElroy *et al.*, 1976). Furthermore, the findings that nitrous oxide production is almost always observed from soil but rarely from marine environments suggest that terrestrial sources account for almost all of the nitrous oxide that is produced in nature.

### 6.4. Toxicological Aspects

The microbiology and biochemistry of nitrification and denitrification are directly related to terrestrial and aquatic ecosystems. Yet the presence or absence of the bacteria affecting these processes sometimes has a direct effect upon humans and other mammals. The nitrite ion is a potential health hazard; as a direct toxicant, it is much more poisonous to adults than nitrate (3 mg vs. 3 g/kg body wt.) (Lee, 1970).

The reduction of nitrate to nitrite by microorganisms is the causative factor of methemoglobinemia. Fixation of nitrite to hemoglobin prevents oxygen transport, and this results in the characteristic "blue baby syndrome." The disease is common only to infants of about 6 months of age or less. The nitrate-respiring bacteria are apparently unable to survive the highly acid gut of older children and adults. Thus, most of the nitrate is absorbed in the adult gut before it reaches the duodenum, where the more moderate pH favors the growth of these organisms (Lee, 1970).

The ruminant gut contains a rich flora of obligate and facultative anaerobic bacteria, many of which reduce nitrate to nitrite. Consequently, the ingestion of nitrate-rich water can be fatal to livestock (Lee, 1970). Despite the fact that the $E_h$ of the ruminant gut would favor further reduction of nitrite to dinitrogen, it appears that denitrifying bacteria are absent from the rumen or present in such small numbers by comparison to the nitrate-respiring bacteria that they are unable to effect significant reduction of nitrite before it is absorbed into the blood. Bryant (1964) reported numbers of anaerobic and facultative anaerobic as high as $10^{10}$/ml and isolated many of the predominant species. Aerobes and denitrifiers were noticeably absent.

The quantitative implications of heterotrophic nitrification in alkaline aquatic environments seem to be minimal. Indeed, the highest concentrations so far reported amount to 28.0, 4.0, 114, and 10 mg N/liter for hydroxylamine, 1-nitrosoethanol, nitrite, and nitrate, respectively. However, some of these metabolites, such as hydroxylamine and the nitroso compounds as well as nitrite, are known to be toxic and even mutagenic or carcinogenic (Venulet and Van Etten, 1970). Hence, the presence and potential persistence in alkaline aquatic environments of such compounds, even at parts-per-million levels, may pose a hazard to humans and animals. The formation of such compounds in these environments, therefore, should be subject to careful and prompt scrutiny.

The accumulation of nitrite and amines in soils receiving heavy applications of animal waste has generated some concern about the potential formation of hazardous nitrosamines, which are potential carcinogens, although these is no evidence that nitrosamines are formed in soils receiving heavy applications of animal wastes. Nitrosamines can be produced chemically by condensation of nitrous acid with a secondary amine or microbiologically at neutral or alkaline pH with nitrite ion (Klubes and Jondorf, 1971; Ayanaba and Alexander, 1974; Mills and Alexander, 1976). Laboratory studies with lake water, sewage, and soil, each of which were amended with dimethylamine and nitrite at concentrations in excess of those found in nature, showed the bacterial formation of trace quantities of dimethylnitrosamine (Ayanaba and Alexander, 1974). Voets et al. (1975) did not detect nitrosamines in waste waters containing high nitrite concentrations and concluded that they presented no problems in waste disposal.

Mills and Alexander (1976) demonstrated that nitrosamine formation, though involving microorganisms, could not definitely be attributed to a biological process since boiled cells catalyzed the reaction. Sander (1968) showed that E. coli, Proteus vulgaris, and Serratia marcescens could bring about the formation of nitrosamines, although the only reaction that could clearly be shown to be biological was the dissimilatory reduction of nitrate. Similarly, Fong and Chan (1973) showed that the rate of nitrosamine formation in sterilized fish containing 40 $\mu$g nitrate and 5 $\mu$g nitrite/g was the same between the control and the one inoculated with Staphylococcus aureus, though a 1-day lag period was observed in the control. It is not clear precisely what role microorganisms have in the formation of nitrosamines, although their activities can certainly affect the formation of nitrite and other products that are necessary for nitrosamine formation. Since chemical nitrosation is concentration dependent (Friedman, 1972), it appears that the significant biotic factor is the formation of nitrite.

Nitrite is widely used as a preservative in meats to inhibit the outgrowth of spores of Clostridium botulinum, thereby greatly reducing the possibility of botulism food poisoning. The chemical formation of nitrosamines from nitrite and naturally-occurring secondary amines in meat with normal cooking procedures has been shown to occur (Lee, 1970). Though nitrosamine formation in the rat gut has been demonstrated (Klubes and Jondorf, 1971), it would appear to be chemical because of the low pH.

The microbial formation of nitrosamines in pickled cabbage has recently been offered as an explanation for the 1000-fold higher incidence of stomach and esophageal cancer in specific provinces of China, where this food is consumed (Bricklin, 1975). The isolation and identification of nitrosamines and the isolation of the causative (though unidentified) fungal agent suggest that nitrosamine formation, in this instance, occurs through heterotrophic nitrification.

# References

Abd-el-Malek, Y., Hosny, I., and Eman, N. F., 1975, Studies on some environmental factors affecting denitrification in soil, *Zentralbl. Bakteriol. Parasitenkd. Infektionskr. Hyg. Abt. 2.* **130**:644–653.

Adams, C. E., Krerkel, P. A., and Bingham, E. C., 1970, Investigations into the reduction of high nitrogen concentrations, in: *Advances in Water Pollution Research,* Vol. 1 (S. H. Jenkins, ed.), pp. 1–1311, Pergamon Press, New York.

Adriano, D. C., Pratt, P. F., and Bishop, S. E., 1971, Nitrate and salt in groundwaters from land disposal of dairy manure, *Soil Sci. Soc. Amer. Proc.* **35**:759.

Albrecht, B., Junge, C., and Zakosek, H., 1970, Der $N_2O$-Gehalt der Bodenluft in drei Bodenprofilen, *Z. Pflanzenernaehr. Bodenkd.* **125**:205.

Aleem, M. I. H., and Lees, H., 1963, Autotrophic enzyme systems. I. Electron transport systems concerned with hydroxylamine oxidation in *Nitrosomonas, Can. J. Biochem. Physiol.* **41**:763.

Aleem, M. I. H., and Nason, A. 1963. Metabolic pathways of bacterial nitrification, in: *Symposium on Marine Microbiology* (L. H. Oppenheimer, ed.), pp. 392–409, Charles C Thomas, Springfield, Ill.

Aleem, M. I. H., Lees, H., and Lyric, R., 1964, Ammonium oxidation by cell-free extracts of *Aspergillus wentii, Can. J. Biochem.* **42**:989.

Aleem, M. I. H., Hock, G. E., and Vanner, J. E., 1965, Water as the source of oxidant and reductant in bacterial chemosynthesis, *Proc. Natl. Acad. Sci. USA* **54**:869.

Alexander, M., 1965, Nitrification, in: *Soil Nitrogen* (W. V. Bartholomew and F. E. Clark, eds.), pp. 309–346, *Amer. Soc. Agron.,* No. 10, Madison, Wisc.

Alexander, M., Marshall, K. C., and Hirsch, P., 1960, Autotrophy and heterotrophy in nitrification, *Trans. Int. Congr. Soil Sci., 7th,* pp. 586–591, Madison, Wisc.

Allison, F. E., 1955, The enigma of soil nitrogen balance sheets, *Adv. Agron.* **7**:213.

Allison, F. E., 1966, The fate of nitrogen applied to soils, *Adv. Agron.* **18**:219.

Amarger, N., and Alexander, M., 1968, Nitrite formation from hydroxylamine and oximes by *Pseudomonas aeruginosa, J. Bacteriol.* **95**:1651.

Amer, F. M., and Bartholomew, W. V., 1951, Influence of oxygen concentration in soil air on nitrification, *Soil Sci.* **71**:215.

Anderson, J. H., 1964, The metabolism of hydroxylamine to nitrite by *Nitrosomonas, Biochem. J.* **91**:8.

Anderson, O. E., Boswell, F. C., and Harrison, R. M., 1971, Variations in low temperature adaptability of nitrifiers in acid soils, *Soil Sci. Soc. Amer. Proc.* **35**:68.

Anthonisen, A. C., Loehr, R. C., Prakasam, T. B. S., and Srinath, E. G., 1976, Inhibition of nitrification by ammonia and nitrous acid, *J. Water Pollut. Contr. Fed.* **48**:835.

Ardakani, M. S., Rehbock, J. T., and McLaren, A. D., 1973, Oxidation of nitrite to nitrate in a soil column, *Soil Sci. Soc. Amer. Proc.* **37**:53.

Ardakani, M. S., Rehbock, J. T., and McLaren, A. D., 1974a, Oxidation of ammonium to nitrate in a soil column, *Soil Sci. Soc. Amer. Proc.* **38**:96.

Ardakani, M. S., Schulz, R. K., and McLaren, A. D., 1974b, A kinetic study of ammonium and nitrite oxidation in a soil field plot, *Soil Sci. Soc. Amer. Proc.* **38**:273.

Ardakani, M. S., Belser, L. W., and McLaren, A. D., 1975, Reduction of nitrate in a soil column during continuous flow, *Soil Sci. Soc. Amer. Proc.* **39**:290.

Arima, K., Imanaka, H., Kousaka, M., Futuka, A., and Tamura, G., 1964, Pyrrolnitrin, a new antibiotic substance produced by *Pseudomonas*, *Agr. Biol. Chem.* **28**:575.

Arnold, P. W., 1954, Losses of nitrous oxide from soil, *J. Soil Sci.* **5**:116.

Asghar, M., and Kanehiro, Y., 1976, Effects of sugarcane trash and pineapple incorporation on soil nitrogen, pH, and redox potential, *Plant Soil* **44**:209.

Avnimelech, Y., 1971, Nitrate transformations in peat, *Soil Sci.* **111**:113.

Avnimelech, Y., and Raveh, A., 1974, The control of nitrate accumulation in soils by induced denitrification, *Water Res.* **8**:553.

Avnimelech, Y., and Raveh, J., 1976, Nitrate leakage from soils differing in texture and nitrogen load, *J. Environ. Qual.* **5**:79.

Ayanaba, A., and Alexander, M., 1973, Microbial formation of nitrosamines *in vitro*, *Appl. Microbiol.* **25**:862.

Ayanaba, A., and Alexander, M., 1974, Transformations of methylamines and formation of a hazardous product, dimethylnitrosamine, in samples of treated sewage and lake water, *J. Environ. Qual.* **3**:83.

Ayanaba, A., and Omayuili, A. P. O., 1975, Microbial ecology of acid tropical soils. A preliminary report, *Plant Soil* **43**:519.

Ayanaba, A., Verstraete, W., and Alexander, M., 1973, Possible microbial contribution to nitrosamine formation in sewage and soil, *J. Natl. Cancer Inst.* **50**:811.

Ayers, R. S., and Branson, R. L., 1973, Nitrates in the upper Santa Ana River Basin in relation to groundwater pollution, California Agricultural Experiment Station, Bulletin 861.

Bailey, L. D., and Beauchamp, E. G., 1973a, Effects of temperature on $NO_3^-$ and $NO_2^-$ reduction, nitrogenous gas production, and redox potential in a saturated soil, *Can. J. Soil Sci.* **53**:213.

Bailey, L. D., and Beauchamp, E. G., 1973b, Effects of moisture, added $NO_3^-$, and macerated roots on $NO_3^-$ transformation and redox potential in surface and subsurface soils, *Can. J. Soil Sci.* **53**:219.

Baker, J. R., Struempler, A., and Chaykin, S., 1963, A comparative study of trimethylamine-N-oxide biosynthesis, *Biochim. Biophys. Acta* **71**:58.

Balakrishnan, S., and Eckenfelder, W. W., 1969, Nitrogen relationships in biological treatment processes. III. Denitrification in the modified activated sludge process, *Water Res.* **3**:177.

Baldensperger, J., and Garcia, J. L., 1975, Reduction of oxidized inorganic nitrogen compounds by a new strain of *Thiobacillus denitrificans*, *Arch. Microbiol.* **103**:31.

Balderston, W. L., Sherr, B., and Payne, W. J., 1976, Blockage by acetylene of nitrous oxide reduction in *Pseudomonas perfectomarinus*, *Appl. Environ. Microbiol.* **31**:504.

Ballio, A., Bertholdt, H., Chain, E. B., and Di Vittorio, V., 1962, Structure of ferroverdin, *Nature (London)* **194**:769.

Bates, D. R., and Hayes, P. B., 1967, Atmospheric nitrous oxide, *Planet. Space Sci.* **15**:189.

Baxter, R. M., Wood, R. B., and Prosser, M. V., 1973, The probable occurrence of hydroxylamine in the water of an Ethiopian Lake, *Limnol. Oceanogr.* **18**:470.

Bellet, P., and Gerard, D., 1962, N-Oxydation microbiologique de la strychnine, *Ann. Pharm. Fr.* **20**:928.

Belser, L. W., 1974, The ecology of nitrifying bacteria, University of California, Berkeley. University Microfilms, Ann Arbor, Mich.

Bickel, M. H. 1969. The pharmacology and biochemistry of N-oxides, *Pharmacol. Rev.*, **21**:325.

Birch, H. F., 1958, The effect of soil drying on humus decomposition and nitrogen availability, *Plant Soil* **10**:9.

Birkinshaw, J. H., and Dryland, A. M. L., 1964, Studies in the biochemistry of microorganisms. 116, Biosynthesis of $\beta$-nitropropionic acid by the mould *Penicillium atrovenetum* G. Smith, *Biochem. J.* **93**:478.

Bisset, K. A., and Grace, Joyce B., 1954, The nature and relationships of autotrophic bacteria, in: *Autotrophic Micro-organisms* (B. A. Fry and J. L. Peel, eds.), pp. 28–53, Cambridge Univ. Press, London.

Bollag, J. M., and Tung, G., 1972, Nitrous oxide release by soil fungi, *Soil Biol. Biochem.* **4**:271.

Bollag, J. M., Drzymala, S., and Kardos, L. T., 1973, Biological vs. chemical nitrite decomposition in soil, *Soil Sci.* **116**:44.

Boon, B., and Laudelout, H., 1962, Kinetics of nitrite oxidation by *Nitrobacter winogradskyi*, *Biochem J.* **85**:440.

Bouwer, H., 1970, Ground water recharge design for renovating waste water, *J. Sanit. Eng. Div. Proc. Amer. Soc. Civil Eng.* **96**:59.

Bouwer, H., Lance, J. C., and Riggs, M. S., 1974, High-rate land treatment. II. Water quality and economic aspects of the Flushing Meadows Project, *J. Water Pollut. Control Fed.* **46**:844.

Bowman, R. A., and Focht, D. D., 1974, The influence of glucose and nitrate concentrations upon denitrification rates in sandy soils, *Soil Biol. Biochem.* **6**:297.

Boyland, E., and Manson, D., 1966, The biochemistry of aromatic amines. The metabolism of 2-naphthylamine and 2-naphthylhydroxylamine derivates, *Biochem. J.* **101**:84.

Brar, S. S., 1972, Influence of roots on denitrification, *Plant Soil* **36**:713.

Bray, H. G., and White, K., 1966, *Kinetics and Thermodynamics in Biochemistry*, Academic Press, New York.

Bremner, J. M., and Shaw, K., 1958, Denitrification in soil. II. Factors affecting denitrification, *J. Agric. Sci.* **51**:40.

Brian, P. W., Elson, G. W., Hemming, H. G., and Radley, M., 1965, An inhibitor of plant growth produced by *Aspergillus wentii* Whemer, *Nature (London)* **2078**:998.

Bricklin, M., 1975, What the Chinese have learned about cancer, *Prevention* **27**:33.

Broadbent, F. E., 1975, Field trials with isotopes, Davis site, in: *Nitrate in Effluents from Irrigated Lands* (P. F. Pratt, Principal Investigator), pp. 179–188, National Science Foundation Annual Report GI34733X.

Broadbent, F. E., and Clark, F. E., 1965, Denitrification, in: *Nitrogen* (W. V. Bartholomew and F. E. Clark, eds.), pp. 344–359, Amer. Soc. Agron., Madison, Wisc.

Broadbent, F. E., and Tusneem, M. E., 1971, Losses of nitrogen from some flooded soils in tracer experiments, *Soil Sci. Soc. Amer. Proc.* **35**:922.

Broadbent, F. E., Taylor, N. B., and Hill, G. N., 1957, Nitrification of ammoniacal fertilizers in some California soils, *Hilgardia* **27**:247.

Bryant, M. P., 1964, Some aspects of the bacteriology of the rumen, in: *Principles and Applications in Aquatic Microbiology* (H. Heukelekian and N. C. Dondero, eds.), pp. 366–393, John Wiley and Sons, New York.

Buchanan, R. E., and Gibbons, N. E., 1974, *Bergey's Manual of Determinative Bacteriology*, 8th Ed., Williams & Wilkins, Baltimore.

Burford, J. R., and Bremner, J. M., 1975, Relationships between the denitrification capacities of soils and total water-soluble and readily-decomposable soil organic matter, *Soil Biol. Biochem.* **7**:389.

Burford, J. R., and Millington, R. J., 1968, Nitrous oxide in the atmosphere of a red-brown earth, *Proc. 9th International Congr. Soil Sci. Transactions, Adelaide*, Vol. 2, pp. 505–511.

Burford, J. R., and Stefanson, R. C., 1973, Measurement of gaseous losses of nitrogen from soils, *Soil Biol. Biochem.* **5**:133.

Buswell, A. M., Shiota, T., Lawrence, N., and Meter, I. V., 1954, Laboratory studies on the kinetics of the growth of *Nitrosomonas* with relation to the nitrification phase of the BOD test, *Appl. Microbiol.* **2**:21.

Cady, F. B., and Bartholomew, W. V., 1960, Sequential products of anaerobic denitrification in Norfolk soil, *Soil Sci. Soc. Amer. Proc.* **24**:477.

Campbell, C. A., Biederbeck, V. O., and Warder, F. G., 1971, Influence of simulated fall and spring conditions in the soil system. II. Effect on soil nitrogen, *Soil Sci. Soc. Amer. Proc.* **35**:480.

Campbell, C. A., Stewart, D. W., Nicholaichuk, W., and Biederbeck, V. O., 1974, Effects of growing season, soil temperature, moisture and $NH_4^+$-N on soil nitrogen, *Can. J. Soil Sci.* **54**:403.

Campbell, N. E. R., and Aleem, M. I. H., 1965a, The effect of 2-chloro-6-(trichloromethyl)-pyridine on the chemoautotrophic metabolism of nitrifying bacteria. 1. Ammonia and hydroxylamine oxidation by *Nitrosomonas, Antonie van Leeuwenhoek J. Microbiol. Serol.* **31**:124.

Campbell, N. E. R., and Aleem, M. I. H., 1965b, The effect of 2-chloro-6-(trichloromethyl)-pyridine on the chemoautotrophic metabolism of nitrifying bacteria. 2. Nitrite oxidation by *Nitrobacter, Antonie van Leeuwenhoek J. Microbiol. Serol.* **31**:137.

Carlucci, A. F., and Schubert, H. R., 1969, Nitrate reduction in sea water of the deep nitrite maximum off Peru, *Limnol. Oceanogr.* **14**:187.

Castell, C. H., and Mapplebeck, E. G., 1956, A note on the production of nitrite from hydroxylamine by some heterotrophic bacteria, *J. Fish. Res. Board Can.* **13**:201.

Cawse, P. A., and Sheldon, D., 1972, Rapid reduction of nitrate in soil re-moistened after air-drying, *J. Agric. Sci. Camb.* **78**:405.

Chance, B., 1957, Cellular oxygen requirements, *Fed. Proc. Fed. Amer. Soc. Exp. Biol.* **16**:671.

Chatelain, R., 1969, Nitrate reduction by *Alcaligenes odorans* var. *viridans, Ann. Inst. Pasteur* **116**:498.

Chen, R. L., Keeney, D. R., Graetz, D. A., and Holding, A. J., 1972a, Denitrification and nitrate reduction in Wisconsin lake sediments, *J. Environ. Qual.* **1**:158.

Chen, R. L., Keeney, D. R., Konrad, J. G., Holding, A. J., and Graetz, D. A., 1972b, Gas production in sediments of Lake Mendota, Wisconsin, *J. Environ. Qual.* **1**:155.

Chinn, S. H. F., 1973, Effect of eight fungicides on microbial activities in soil as measured by a bioassay method, *Can. J. Microbiol.* **19**:771.

Clark, C., and Schmidt, E. L., 1966, Effect of mixed culture on *Nitrosomonas europaea* simulated by uptake and utilization of pyruvate, *J. Bacteriol.* **91**:367.

Clark, C., and Schmidt, E. L., 1967a, Growth response of *Nitrosomonas europaea* to amino acids, *J. Bacteriol.* **93**:1302.

Clark, C., and Schmidt, E. L., 1967b, Uptake and utilization of amino acids by resting cells of *Nitrosomonas europaea, J. Bacteriol.* **93**:1309.

Clemo, G. R., and McIlwain, H., 1938, The phenazine series. VII. The pigment of *Chromobacterium iodinum*; the phenazine di-*N*-oxides, *J. Chem. Soc.* 479–483.

Collins, F. M., 1955, Effect of aeration on the formation of nitrate-reducing enzymes by *P. aeruginosa, Nature (London)* **1975**:173.

Cooper, G. S., and Smith, R. L., 1963, Sequence of products formed during denitrification in some diverse western soils, *Soil Sci. Soc. Amer. Proc.* **27**:659.

Cooper, J. E., 1975, Nitrification in soils incubated with pig slurry, *Soil Biol. Biochem.* **7**:119.

Cornforth, J. W., and James, A. T., 1956, Structure of a naturally occurring antagonist of dihydrostreptomycin, *Biochem. J.* **63**:124.

Coutts, R. T., 1967, Hydroxamic acids, *Can. J. Pharm. Sci.* **2**:27.

Cox, C. D., and Payne, W. J., 1973, Separation of soluble denitrifying enzymes and cytochromes from *Pseudomonas perfectomarinus, Can. J. Microbiol.* **19**:861.

Cox, C. D., Jr., Payne, W. J., and Dervartanian, D. V., 1971, Electron paramagnetic resonance studies on the nature of hemoproteins in nitrite and nitric oxide reduction, *Biochim. Biophys. Acta* **253**:290.

Cramer, J. W., Miller, J. A., and Miller, E. C., 1960, *N*-Hydroxylation: A new metabolic reaction observed in the rat with the carcinogen 2-acetylaminofluorene, *J. Biol. Chem.* **235**:885.

Curtis, E. J. C., Jurrant, K., and Harman, M. M., 1975, Nitrification in rivers in the Trent Basin, *Water Res.* **9**:255.

Daniel, R. M., and Appleby, C. A., 1972, Anaerobic-nitrate, symbiotic, and aerobic growth of *Rhizobium japonicum*. Effects of cytochrome $P_{450}$, other hemoproteins, nitrate, and nitrite reductases, *Biochim. Biophys. Acta* **275**:347.

Das, M. L., and Ziegler, D. M., 1970, Rat liver oxidative *N*-dealkylase and *N*-oxidase activities as a function of animal age, *Arch. Biochem. Biophys.* **140**:300.

Daubner, I., and Ritter, R., 1973, Bakteriengehalt und Stoffumsatzaktivitat einiger physiolo-gischen Bakteriengruppen in zweikunstlichen Grundwasserseen (Baggerseen), *Arch. Hydrobiol.* **72**:440.

Dawson, R. W., and Murphy, K. L., 1972, The temperature dependency of biological denitrification, *Water Res.* **6**:71.

Dawson, R. W., and Murphy, K. L., 1973, Factors affecting biological denitrification of wastewater, in: *Advances in Water Pollution Research* (S. H. Jenkins, ed.), pp. 671–680, Pergamon Press, New York.

Deherain, P. P., 1897, La reduction des nitrates dans la terre arable, *C. R. Acad. Sci., Paris*, **124**:269.

Delwiche, C. C., 1970, The nitrogen cycle, *Scientific American* **223**:137.

Delwiche, C. C., and Finstein, M. S., 1965, Carbon and energy sources for the nitrifying autotroph, *Nitrobacter*, *J. Bacteriol.* **90**:102.

Deppe, K., and Engel, H., 1960, Untersuchungen über die Temperaturabhängigkeit der Nitratbildung durch *Nitrobacter winogradski* Buch. bei ungehemntem und gehemnten Wachstum, *Zentralbl. Bakteriol. Parasitenkd. Infektionskr. Hyg. Abt. II* **113**:561.

Devitt, D., Letey, J., Lund, L. J., and Blair, J. W., 1976, Nitrate–nitrogen movement through soil as affected by soil profile characteristics, *J. Environ. Qual.* **5**:283.

Doner, H. E., 1975, Disappearance of nitrate under transient conditions in columns of soil, *Soil Biol. Biochem.* **7**:257.

Doner, H. E., and McLaren, A. D., 1976, Soil nitrogen transformations: A modeling study, in: *Environmental Biogeochemistry*, Vol. I (J. O. Nriagu, ed.), pp. 245–258, Ann Arbor Science, Ann Arbor.

Doner, H. E., Volz, M. G., and McLaren, A. D., 1974, Column studies of denitrification in soil, *Soil Biol. Biochem.* **6**:341.

Doner, H. E., Volz, M. G., Belser, L. W., and Loken, J.-P., 1975, Short-term nitrate losses and associated microbial populations in soil columns, *Soil Biol. Biochem.* **7**:261.

Dowdell, R. J., and Smith, K. A., 1974, Field studies of the soil atmosphere. II. Occurrence of nitrous oxide, *J. Soil Sci.* **25**:231.

Downey, R. J., and Kiszkiss, D. F., 1969, Oxygen and nitrate-induced modification of the electron transfer system of *Bacillus stearothermophilus*, *Microbios* **2**:145.

Doxtader, K. G., 1965, Nitrification by heterotrophic microorganisms, Ph.D. Thesis, Cornell University, Ithaca.

Doxtader, K. G., and Alexander, M., 1966a, Nitrification by growing and replacement cultures of *Aspergillus*, *Can. J. Microbiol.* **12**:807.

Doxtader, K. G., and Alexander, M., 1966b, Nitrification by heterotrophic soil microorganisms, *Soil Sci. Soc. Amer. Proc.* **30**:351.

Doxtader, K. G., and Alexander, M., 1966c, Role of 3-nitropropanoic acid in nitrate formation by *Aspergillus flavus*, *J. Bacteriol.* **91**:1186.

Dulaney, E. L., 1963, Further studies on formation of N-formyl hydroxyaminoacetic acid by Penicillium, Mycologia 55:211.

Eckenfelder, W. W., 1967, A design procedure for biological nitrification and denitrification, Chem. Eng. Progr. Symp. 78(63):230.

Elliott, L. F., McCalla, T. M., Mielke, L. N., and Travis, T. A., 1972, Ammonium, nitrate, and total nitrogen in the soil water of feedlot and field soil profiles, Appl. Microbiol. 28:810.

Emery, T. F., 1963, Aspartase-catalyzed synthesis of N-hydroxyaspartic acid, Biochem. 2:1041.

Emery, T. F., 1967, Hadacidin, in: Antibiotics II. Biosynthesis (D. Gottlieb and P. D. Shaw, eds.), pp. 17–25, 439, Springer-Verlag, New York.

Engel, M. S., and Alexander, M., 1958, Growth and autotrophic metabolism of Nitrosomonas europaea, J. Bacteriol. 76:217.

Engel, M. S., and Alexander, M., 1960, Autotrophic oxidation of ammonium and hydroxylamine, Soil Sci. Soc. Amer. Proc. 24:48.

Etinger-Tulczynska, R., 1969, A comparative study of nitrification in soils from arid and semi-rid areas of Israel, J. Soil Sci. 20:473.

Eylar, O. R., and Schmidt, E. L., 1959, A survey of heterotrophic microorganisms from soil for ability to form nitrite and nitrate, J. Gen. Microbiol. 20:473.

Falcone, A. B., Shug, A. L., and Nicholas, D. J. D., 1962, Oxidation of hydroxylamine by particles from Nitrosomonas, Biochem. Biophys. Res. Comm. 9:126.

Faurie, G., Joseerand, A., and Bardin, R., 1975, Influence of clay minerals on ammonium retention and nitrification, Rev. Ecol. Biol. Sol. 12:201.

Feuer, H., 1969, The Chemistry of the Nitro and the Nitroso Groups, Wiley Interscience, New York.

Fewson, C. A., and Nicholas, D. J. D., 1961, Nitrate reductase from Pseudomonas aeruginosa, Biochim. Biophys. Acta 49:335.

Finsen, P. O., and Sampson, D., 1959, Denitrification of sewage effluents, Water Waste Treat. J. 7:298.

Fliermans, C. B., and Schmidt, E. L., 1975, Autoradiography and immunofluorescence combined for autoecological study of single cell activity with Nitrobacter as a model system, Appl. Microbiol. 30:676.

Fliermans, C. B., Bohlool, B. B., and Schmidt, E. L., 1974, Autoecological study of the chemoautotroph Nitrobacter by immunofluorescence, Appl. Microbiol. 27:124.

Flühler, J., 1973, Sauerstoffdiffusion in Boden, Mitt. Schweiz. Anst. Jorstl. Vers'wes. 49:125.

Focht, D. D., 1973, Isotope fractionation of $^{15}N$ and $^{14}N$ in microbiological nitrogen transformations: A theoretical model, J. Environ. Qual. 2:247.

Focht, D. D., 1974, The effect of temperature, pH, and aeration on the production of nitrous oxide and gaseous nitrogen—A zero-order kinetic model, Soil Sci. 118:173.

Focht, D. D., and Chang, A. C., 1975, Nitrification and denitrification processes related to waste water treatment, Adv. Appl. Microbiol. 19:153.

Focht, D. D., and Joseph, H., 1973, An improved method for the enumeration of denitrifying bacteria, Soil Sci. Soc. Amer. Proc. 37:698.

Focht, D. D., and Joseph, H., 1974, Degradation of 1,1-diphenylethylene by mixed cultures, Can. J. Microbiol. 20:631.

Focht, D. D., Fetter, N. R., Lonkerd, W., and Stolzy, L. H., 1975, Effects of moisture and manure upon gaseous concentrations of nitrous oxide in soils, Proc. 2nd Ann. NSF-RANN Trace Contam. Colnf., Asilomar, Calif. 1974.

Fong, Y. Y., and Chan, W. C., 1973, Bacterial production of dimethylnitrosamine in salted fish, Nature (London) 243:421.

Forget, P., and Dervartanian, D. V., 1972, Bacterial nitrate reductases. EPR studies on nitrate reductase from Micrococcus denitrificans, Biochim. Biophys. Acta 256:600.

Francis, C. W., and Callahan, M. W., 1975, Biological denitrification and its application in treatment of high-nitrate waste water, *J. Environ. Qual.* **4**:153.

Frederick, L. R., 1956, The formation of nitrate from ammonium nitrogen in soils. 1. Effect of temperature, *Soil Sci. Soc. Amer. Proc.* **20**:496.

Friedman, M. A., 1972, Nitrosation of sarcosine: Chemical kinetics and gastric assay, *Bull. Environ. Contam. Toxicol.* **8**:375.

Gambrell, R. P., Gilliam, J. W., and Weed, S. B., 1975, Denitrification in subsoils of the North Carolina coastal plain as affected by soil drainage, *J. Environ. Qual.* **4**:311.

Garcia, J. L., 1973, Sequence des produits formes au cours de la denitrification dans les sols de rizieres du Senegal, *Ann. Microbiol. Inst. Pasteur* **124B**:351.

Garcia, J. L., 1974, Reduction de l'oxyde nitreux dans les sols de rizieres du Senegal: Mesure de l'activite denitrificante, *Soil Biol. Biochem.* **6**:79.

Garribaldi, J. A., 1971, Influence of temperature on the iron metabolism of a fluorescent pseudomonad, *J. Bacteriol.* **105**:1036.

Gayon, F., and Dupetit, G., 1886, Reduction des nitrates par les infiniments petis, *Mem. Soc. Bordeaux, Ser. 3* **2**:201.

Germanier, R., and Wuhrmann, K., 1963, Über den aeroben mikrobeillen Abbau aromatischer Nitroverbindingen, *Pathol. Microbiol.* **26**:569.

Gibson, F., and Magrath, D. I., 1969, The isolation and characterization of a hydroxamic acid (aerobactin) formed by *Aerobacter aerogenes* 62–1, *Biochim. Biophys. Acta* **192**:175.

Gode, P., 1970, Untersuchungen über nitrifizierende Bakterien in einem geschichteten eutrophen See, Doktorats Dissertation, Universität Kiel.

Gode, P., and Overbeck, J., 1972, Untersuchungen zur heterotrophen Nitrification im See, *Z. Allg. Mikrobiol.* **12**:567.

Goering, J. J., 1968, Denitrification in the oxygen minimum layer of the eastern tropical Pacific Ocean, *Deep Sea Res.* **15**:157.

Goering, J. J., and Cline, J. D., 1970, A note on denitrification in seawater, *Limnol. Oceanogr.* **15**:306.

Goering, J. J., and Dugdale, V. A., 1966a, Estimates of the rates of denitrification in a subartic lake, *Limnol. Oceanogr.* **11**:113.

Goering, J. J., and Dugdale, V. A., 1966b, Denitrification rates in an island bay in the equatorial Pacific Ocean, *Science* **154**:505.

Gottlieb, D., 1967, Biosynthesis of chloramphenicol, in: *Antibiotics II* (D. Gottlieb and P. D. Shaw, eds.), pp. 32–43, Springer-Verlag, New York.

Gould, G. W., and Lees, H., 1960, The isolation and culture of the nitrifying organisms: Part 1, *Nitrobacter, Can. J. Microbiol.* **6**:299.

Grass, L. B., MacKenzie, A. J., Meek, B. D., and Spencer, W. F., 1973, Manganese and iron solubility changes as a factor in tile drain clogging. II. Observations during the growth of cotton, *Soil Sci. Soc. Amer. Proc.* **37**:17.

Greenland, D. J., 1962, Denitrification in some tropical soils, *J. Agric. Sci.* **58**:227.

Greenwood, D. J., 1961, The effect of oxygen concentration on the decomposition of organic materials in soil, *Plant Soil* **14**:360.

Gunner, H. B., 1963, Nitrification by *Arthrobacter globiformis*, *Nature (London)* **197**:1127.

Harms, H., Koops, H. P., and Wehrmann, H., 1976, An ammonia-oxidizing bacterium, *Nitrosovibrio tenuis*, nov. gen. nov. sp., *Arch. Microbiol.* **108**:105.

Hart, L. T., Larson, A. D., and McCleskey, C. S., 1965, Denitrification by *Corynebacterium nephridii*, *J. Bacteriol.* **89**:1104.

Hattori, A., and Wada, E., 1971, Nitrite distribution and its regulating processes in the equatorial Pacific Ocean, *Deep Sea Res.* **18**:557.

Hauck, R. D., and Melstead, S. W., 1956, Some aspects of the problem of evaluating denitrification in soils, *Soil Sci. Soc. Amer. Proc.* **20**:361.

Hauck, R. D., Bartholomew, W. V., Bremner, J. M., Broadbent, F. E., Cheng, H. H., Edwards, A. P., Keeney, D. R., Legg, J. O., Olsen, S. R., and Porter, L. K., 1972, Use of variations in natural isotope abundance for environmental studies: A questionable approach, *Science* **177**:454.

Haydon, A. H., Davis, W. B., Arceneaux, J. E. L., and Byers, B. R., 1973, Hydroxamate recognition during iron transport from hydroxamate-iron chelates, *J. Bacteriol.* **165**:912.

Herlichy, M., 1973, Distribution of nitrifying and heterotrophic microorganisms in cutover peats, *Soil Biol. Biochem.* **5**:621.

Hermann, H., 1961, Identifizierung eines Stoffwechselproduktes von *Clitocybe suaveolens* as 4-Methylnitrosaminobenzaldehyd, *Hoppe-Seyler's Z. Physiol. Chem.* **326**:13.

Herr, R. R., Jahnke, H. K., and Argaudelus, A. D., 1967, The structure of streptozotocin, *J. Amer. Chem. Soc.* **89**:4808.

Heukelekian, H., and Heller, A., 1940, Relation between food concentration and surface for bacterial growth, *J. Bacteriol.* **40**:547.

Hidy, P. H., Hodge, E. B., Young, V. V., Harmed, R. L., Brewer, G. A., Philips, W. F., Runge, W. F., Stavely, H. E., Rohland, A., Boaz, H., and Sullivan, H. R., 1955, Structure and reactions of cycloserine, *J. Amer. Chem. Soc.* **77**:2345.

Hirsch, P., Overrein, L., and Alexander, M., 1961, Formation of nitrite and nitrate by actinomycetes and fungi, *J. Bacteriol.* **82**:442.

Hofman, T., and Lees, H., 1952, The biochemistry of the nitrifying organisms. 2. The free-energy efficiency of *Nitrosomonas*, *Biochem. J.* **52**:140.

Hofman, T., and Lees, H., 1953, The biochemistry of the nitrifying organisms. 4. The respiration and intermediary metabolism of *Nitrosomonas*, *Biochem. J.* **54**:579.

Hollis, D. G., Wiggins, G. L., and Weaver, R. E., 1972, An unclassified gram-negative rod isolated from the pharynx on Thayer-Martin medium (selective agar), *Appl. Microbiol.* **24**:772.

Hora, T. S., and Ivengar, M. R. S., 1960, Nitrification by soil fungi, *Arch. Mikrobiol.* **35**:252.

Hylin, J. W., and Matsumoto, H., 1960, The biosynthesis of 3-nitropropanoic acid by *Penicillium atrovenetum*, *Arch. Biochem. Biophys.* **93**:542.

Ingraham, J. L., 1962, Temperature relationships, in: *The Bacteria* (I. C. Gunsalus and R. Y. Stanier, eds.), Vol. 4, pp. 265–296, Academic Press, New York.

Irving, C. C., 1964, Enzymatic *N*-hydroxylation of the carcinogen 2-acetylaminofluorene and the metabolism of *N*-hydroxy-2-acetylaminofluorene-9-$^{14}$C *in vitro*, *J. Biol. Chem.* **239**:1589.

Ishaque, M., and Aleem, M. I. H., 1973, Intermediates of denitrification in the chemoautotroph *Thiobacillus denitrificans*, *Arch. Microbiol.* **94**:269.

Ishaque, M., and Cornfield, A. N., 1972, Nitrogen mineralization and nitrification during incubation of East Pakistan "tea" soils in relation to pH, *Plant Soil* **37**:91.

Ishaque, M., and Cornfield, A. N., 1974, Nitrogen mineralization and nitrification in relation to incubation temperature in an acid Bangladesh soil lacking autotrophic nitrifying organisms, *Trop. Agric.* **51**:37.

Isono, M., 1954, Oxidative metabolism of phenylacetic acid by *Penicillium chryosogenum*, Q-176, *J. Agric. Chem. Soc. Jap.* **28**:566.

Jensen, H. L., and Lautrup-Larsen, G., 1967, Microorganisms that decompose nitro-aromatic compounds, with special reference to dinitro-*o*-cresol, *Acta Agr. Scand.* **17**:115.

John, P., and Whatley, F. R., 1970, Oxidative phosphorylation coupled to oxygen uptake and nitrate reduction in *Micrococcus denitrificans*, *Biochim. Biophys. Acta* **216**:241.

Johnson, P. W., and Sieburth, J. McN., 1976, *In situ* morphology of nitrifying-like bacteria in aquaculture systems, *Appl. Environ. Microbiol.* **31**:423.

Johnson, W. K., 1969, Removal of nitrogen by biological treatment, in: *Advances in Water Quality Improvement* (E. F. Gloyna and W. F. Eckenfelder, eds.), pp. 178–189, Univ. of Texas Press, Austin.

Johnson, W. K., and Schroepfer, G. J., 1964, Nitrogen removal by nitrification and denitrification, *J. Water Pollut. Control Fed.* **36**:1015.

Jones, E. J., 1951, Loss of elemental nitrogen from soils under anaerobic conditions, *Soil Sci.* **71**:193.

Justice, J. K., and Smith, R. L., 1962, Nitrification of ammonium sulfate in a calcareous soil as influenced by combinations of moisture, temperature, and levels of added nitrogen, *Soil Sci. Soc. Amer. Proc.* **26**:246.

Kaczka, E. A., Gitterman, C. O., Dulaney, E. L., and Folkers, K., 1962, Hadacidin, a new growth-inhibitory substance in human tumor systems, *Biochemistry* **1**:340.

Kao, I. C., Chiu, S. Y., Fan, L. T., and Erickson, L. E., 1973, ATP pools in pure and mixed cultures, *J. Water Pollut. Control Fed.* **45**:926.

Kawai, S., Kobayashi, F., Oshima, T., and Egami, F., 1965, Studies on the oxidation of *p*-aminobenzoate to *p*-nitrobenzoate by *Streptomyces thioluteus*, *Arch. Biochem. Biophys.* **112**:537.

Keeney, D. R., Herbert, R. A., and Hollding, A. J., 1971, Microbiological aspects of the pollution of fresh water with inorganic nutrients, in: *Microbial Aspects of Pollution* (G. Sykes and F. A. Skinner, eds.), pp. 181–200, Academic Press, London.

Kefauver, M., and Allison, F. E., 1957, Nitrite reduction by *Bacterium denitrificans* in relation to oxidation-reduction potential and oxygen tension, *J. Bacteriol.* **73**:8.

Kessel, J. F. van, 1976, Influence of denitrification in aquatic sediments on the nitrogen content of natural waters, *Agric. Res. Rep.* (Versl. Landbouwk. Onderz.) No. 858, Centre for Agricultural Publishing and Documentation, Wageningen.

Kessler, E., and Oesterfield, H., 1970, Nitrification and induction of nitrate reductase in nitrogen-deficient algae, *Nature* (*London*) **228**:287.

Kido, T., Yuamaoto, T., and Soda, K., 1975, Microbial assimilation of alkyl nitro compounds and formation of nitrite, *Arch. Microbiol.* **106**:165.

Kiese, M., and Rauscher, E., 1963, Isolation of phenylhydroxylamine produced from *N*-ethylaniline by microsomal enzymes, *Biochem. Z.* **338**:1.

Klubes, P., and Jondorf, W. R., 1971, Dimethynitrosamine formation from sodium nitrite and dimethylamine by bacterial flora of rat intestines, *Res. Commun. Chem. Pathol. Pharmacol.* **2**:24.

Knowles, G., Downing, A. L., and Barrett, M. J., 1965, Determination of kinetic constants for nitrifying bacteria in mixed culture, with the aid of an electronic computer, *J. Gen. Microbiol.* **38**:263.

Knuth, D. T., 1970, Nitrogen-cycle ecology of solid waste composting, *Compost Sci.* **11**:8.

Kohl, D. H., Shearer, G. B., and Commoner, B., 1971, Fertilizer nitrogen: Contribution to nitrate in surface water in a corn belt watershed, *Science* **174**:1331.

Kohl, D. H., Vitayathil, F., Whitlow, P., Shearer, G., and Chien, S. H., 1976, Denitrification kinetics in soil systems: The significance of good fits to mathematical forms, *Soil Sci. Soc. Amer. Proc.* **40**:249.

Koike, I., and Hattori, A., 1975a, Growth yield of a denitrifying bacterium, *Pseudomonas denitrificans*, under aerobic and denitrifying conditions, *J. Gen. Microbiol.* **88**:1.

Koike, I., and Hattori, A., 1975b, Energy yield of denitrification: An estimate from growth yield in continuous cultures of *Pseudomonas denitrificans* under nitrate-nitrite, and nitrous oxide-limited conditions, *J. Gen. Microbiol.* **88**:11.

Krul, J. M., 1976, Dissimilatory nitrate and nitrite reduction under aerobic conditions by an aerobically and anaerobically grown *Alcaligenes* sp. and by activated sludge, *J. Appl. Bacteriol.* **40**:245.

Lance, J. C., 1972, Nitrogen removal by soil mechanisms, *J. Water Pollut. Control Fed.* **44**:1352.

Lance, J. C., Whistler, F. D., and Rice, R. C., 1976, Maximizing denitrification during soil filtration of sewage water, *J. Environ. Qual.* **5**:102.

Latimer, W. M., 1952, *The Oxidation States of the Elements and Their Potentials in Aqueous Solutions*, 2nd Ed., Prentice-Hall, New York.

Laudelout, H., and van Tichelen, L., 1960, Kinetics of the nitrite oxidation by *Nitrobacter winogradski*, *J. Bacteriol.* **79**:39.

Laurent, M., 1971, La nitrification autotrophe et heterotrophe dans les ecosystems aquatiques, *Ann. Inst. Pasteur* **121**:795.

Lee, D. H. K., 1970, Nitrates, nitrites, and methemoglobinemia, *Environ. Res.* **3**:484.

Lees, H., 1954, The biochemistry of the nitrifying bacteria, in: *Autotrophic Micro-organisms* (B. A. Fry and J. L. Peel, eds.), pp. 84–98, Cambridge Univ. Press, London.

Lees, H., and Simpson, J. R., 1957, The biochemistry of the nitrifying organisms. 5. Nitrite oxidation by *Nitrobacter*, *Biochem. J.* **65**:297.

Lees, H., and Quastel, J. H., 1946a, Biochemistry of nitrification in soil. I, *Biochem. J.* **40**:803.

Lees, H., and Quastel, J. H., 1946b, Biochemistry of nitrification in soil. II, *Biochem. J.* **40**:815.

Lemee, G., 1967, Investigations sur la mineralisation de l'azote et son evolution anjuelle dans les humus forestiers *in situ*, *Oecol. Plant.* **2**:285.

Lemee, G., 1975, Recherches sur les ecosystemes des reserves biologiques de la Foret de Fontainebleau. III. Influence du peuplement gramineen sur les caracteres et l'activite biologique du mull acide, *Rev. Ecol. Biol. Sol* **12**:157.

Lemoigne, M., Monguillon, P., and Desveaux, R., 1936, Recherches sur le role biologique de l'hydroxylamine, *Bull. Soc. Biol.* **18**:1291.

Lipman, J. G., 1908, *Bacteria in Relation to Country Life*, Macmillan, New York.

Little, H. N., 1951, Oxidation of nitroethane by extracts from Neurospora, *J. Biol. Chem.* **193**:347.

Loveless, J. E., and Painter, H. A., 1968, The influence of metal ion concentration and pH value on the growth of a *Nitrosomonas* strain isolated from activated sludge, *J. Gen. Microbiol.* **52**:1.

Lund, L. J., Adriano, D. C., and Pratt, P. F., 1974, Nitrate concentration in deep soil cores as related to soil profile characteristics, *J. Environ. Qual.* **3**:78.

MacDonald, J. C., 1961, Biosynthesis of aspergillic acid, *J. Biol. Chem.* **236**:512.

MacDonald, J. C., 1962, Biosynthesis of hydroxyaspergillic acid, *J. Biol. Chem.* **237**:1977.

MacDonald, J. C., 1965, Biosynthesis of pulcherriminic acid, *Biochem. J.* **96**:533.

MacGregor, A. N., 1972, Gaseous losses of nitrogen from freshly wetted desert soils, *Soil Sci. Soc. Amer. Proc.* **36**:594.

Mahendrappa, M. K., and Smith, R. L., 1967, Some effects of moisture on denitrification in acid and alkaline soils, *Soil Sci. Soc. Amer. Proc.* **31**:212.

Mahendrappa, M. K., Smith, R. L., and Christianson, A. T., 1966, Nitrifying organisms affected by climatic region in western United States, *Soil Sci. Soc. Amer. Proc.* **30**:60.

Mainwright, M., and Pugh, G. J. F., 1973, The effect of three fungicides on nitrification and ammonification in soil, *Soil Biol. Biochem.* **5**:577.

Malavolta, E., de Camargo, R., and Haag, H. P., 1955, Nota sobra a nitrificacao por fungos do solo, *Bol. Inst. Zimotec. (Sao Paulo)*, No. 13.

Mann, L. D., Focht, D. D., Joseph, H. A., and Stolzy, L. H., 1972, Increased denitrification in soils by additions of sulfur as an energy source, *J. Environ. Qual.* **1**:329.

Marshall, K. C., and Alexander, M., 1962, Nitrification by *Aspergillus flavus*, *J. Bacteriol.* **83**:572.

Martin, J. P., and Ervin, J. O., 1953, Nitrogen losses during oxidation of sulfur in soils, *Calif. Citrogr.* **39**(38):54.

Matsubara, T., and Iwasaki, H., 1971, Enzymatic steps of dissimilatory nitrate reduction in *Alcaligenes faecalis*, *J. Biochem.* **69**:859.

Matsubara, T., and Mori, T., 1968, Studies on denitrification. IX. Nitrous oxide, its production and reduction to nitrogen, *J. Biochem.* **64**:863.

McCarty, P. L., 1972, Energetics of organic matter degradation, in: *Water Pollution Microbiology* (R. Mitchell, ed.), pp. 91–118, Wiley Interscience, New York.

McCarty, P. L., and Haug, R. T., 1971, Nitrogen removal from waste waters by biological nitrification and denitrification, in: *Microbial Aspects of Pollution* (G. Sykes and F. A. Skinner, eds.), pp. 215–232, Academic Press, New York.

McCarty, P. L., Beck, L., and St. Amant, P., 1969, Biological denitrification of agricultural wastewaters by addition of organic materials, *Purdue Univ. Eng. Ext. Serv.* **135**:1271.

McElroy, M. B., Elkins, J. W., Wofsky, S. C., and Yung, Y. L., 1976, Sources and sinks for atmospheric $N_2O$, *Rev. Geophy. Space Phys.* **14**:143.

McGarity, J. W., and Meyers, R. J. K., 1968, Denitrifying activity in solodized solonetz soils of eastern Australia, *Soil Sci. Soc. Amer. Proc.* **32**:812.

McHarness, D. D., and McCarty, P. L., 1973, Field study of nitrification with submerged filters, *Environ. Protection Agency Tech. Ser.* EPA-R2-73-158.

McLaren, A. D., 1971, Kinetics of nitrification in soil: Growth of the nitrifiers, *Soil Sci. Soc. Amer. Proc.* **35**:91.

McLaren, A. D., and Skujins, J. J., 1963, Nitrification by *Nitrobacter agilis* on surfaces and in soils with respect to hydrogen ion concentration, *Can. J. Microbiol.* **9**:729.

Mechsner, K., and Wuhrmann, K., 1963, Beitrag zur Kenntnis der mikrobiellen Denitrifikation, *Pathol. Microbiol.* **26**:579.

Meek, B. D., Grass, L. B., and MacKenzie, A. J., 1969, Applied nitrogen losses in relation to oxygen status of soils, *Soil Sci. Soc. Amer. Proc.* **33**:575.

Meek, B. D., Grass, L. B., Willardson, L. S., and MacKenzie, A. J., 1970, Nitrate transformations in a column with a controlled water table, *Soil Sci. Soc. Amer. Proc.* **34**:235.

Meek, B. D., MacKenzie, A. J., Donovan, T. J., and Spencer, W. F., 1973, The effect of large applications of manure on movement of nitrate and carbon in an irrigated desert soil, *J. Environ. Qual.* **3**:253.

Meiklejohn, J., 1954, Some aspects of the physiology of the nitrifying bacteria, in: *Autotrophic Micro-organisms* (B. A. Fry and J. C. Peel, eds.), pp. 68–83, Cambridge Univ. Press, London.

Meyerhof, O., 1916a, Untersuchungen über den Atmungsvorgang nitrifizieren der Bakterien. I. Die Atmung des Nitratbildners, *Arch. Ges. Physiol.* **164**:353.

Meyerhof, O., 1916b, Untersuchungen über den Atmungsvorgang nitrifizierender Bakterien. II. Beeinflussungen der Atmung des Nitratbildners durch chemische Substanterien, *Arch. Ges. Physiol.* **164**:229.

Micetich, R. G., and MacDonald, J. C., 1965, Biosynthesis of neoaspergillic and neohydroxyaspergillic acids, *J. Biol. Chem.* **240**:1692.

Mills, A. L., and Alexander, M., 1976, *N*-Nitrosamine formation by cultures of several microorganisms, *Appl. Environ. Microbiol.* **31**:892.

Mishustin, E., 1926, Zur Frage von der Nitrit-Bildung durch metatrophe Bakterien, *Ber. Bakteriol.-Agron. Sat.* **42**:28, in Russian.

Misra, C., Nielsen, D. R., and Biggar, J. W., 1974, Nitrogen transformations in soil during leaching. II. Steady state nitrification and nitrate reduction, *Soil Sci. Soc. Amer. Proc.* **38**:294.

Miyazaki, T., Wada, E., and Hattori, A., 1973, Capacities of shallow waters of Sagani Bay for oxidation of inorganic nitrogen, *Deep-Sea Res.* **20**:571.

Moore, S. F., and Schroeder, E. D., 1971, The effect of nitrate feed on denitrification, *Water Res.* **5**:445.

Morrill, L. G., and Dawson, J. E., 1967, Patterns observed for the oxidation of ammonium to nitrate by soil organisms, *Soil Sci. Soc. Amer. Proc.* **31**:757.

Mulbarger, M. C., 1971, Nitrification and denitrification in activated sludge systems, *J. Water Pollut. Control Fed.* **43**:2059.

Murray, E. D., and Sanwal, B. D., 1963, An immunological enquiry into the identity of assimilatory and dissimilatory nitrate reductase for *Escherichia coli*, *Can. J. Microbiol.* **9**:781.

Murray, I., Parsons, J. W., and Robinson, K., 1975, Inter-relationships between nitrogen balance, pH and dissolved oxygen in an oxidation ditch treating farm animal waste, *Water Res.* **9**:25.

Murthy, Y. K. S., Thiemann, J. E., Colronelli, C., and Sensi, P., 1966, Alanosine, a new antiviral and antitumor agent isolated from a *Streptomyces*, *Nature (London)* **211**:1198.

Myers, R. J. K., 1975, Temperature effects on ammonification and nitrification in a tropical soil, *Soil Biol. Biochem.* **7**:83.

Myers, R. J. K., and McGarity, J. W., 1972, Denitrification in undisturbed cores from a solodized solenetz B horizon, *Plant Soil* **37**:81.

National Academy of Sciences, 1971, *Accumulation of Nitrate*, Washington, D.C.

Neilands, J. B., 1967, Hydroxamic acids in nature, *Science* **156**:1443.

Nelson, D. W., and Bremner, J. M., 1970a, Role of soil minerals and metallic cations in nitrite decomposition and chemodenitrification in soils, *Soil Biol. Biochem.* **2**:1.

Nelson, D. W., and Bremner, J. M., 1970b, Gaseous products of nitrite decomposition in soils, *Soil Biol. Biochem.* **2**:203.

Nelson, L. B., 1975, Fertilizers for all-out food production, in: *Strategy and Resource Implications* (W. P. Martin, ed.), pp. 15–20, Amer. Soc. Agron., Madison, Wisc.

Nicholas, D. J. D., and Jones, O. T. G., 1960, Oxidation of hydroxylamine in cell-free extracts of *Nitrosomonas europaea*, *Nature (London)* **165**:512.

Nommik, H., 1956, Investigations on denitrification in soil, *Acta Agr. Scand.* **6**:195.

Novak, J. T., 1974, Temperature-substrate interactions in biological treatment, *J. Water Pollut. Control Fed.* **46**:1984.

Obaton, M., Amarger, N., and Alexander, M., 1968, Heterotrophic nitrification by *Pseudomonas aeruginosa*, *Arch. Mikrobiol.* **63**:122.

Odu, C. T. I., and Adeoye, K. B., 1970, Heterotrophic nitrification in soils—A preliminary investigation, *Soil Biol. Biochem.* **2**:41.

Overrein, L. N., 1971, Isotope studies in nitrogen in forest soil. I. Relative losses of nitrogen through leaching during period of forty months, *Medd. Nor. Skogsforskningsinst.* **29**:261.

Paerl, H. W., Richards, R. C., Leonard, R. L., and Goldman, C. R., 1975, Seasonal nitrate cycling as evidence for complete vertical mixing in Lake Tahoe, California-Nevada, *Limnol. Oceanogr.* **20**:1.

Painter, H. A., 1970, A review of literature of inorganic nitrogen metabolism in microorganisms, *Water Res.* **4**:393.

Pang, P. C., Cho, C. M., and Hedlin, R. A., 1975, Effects of pH and nitrifier population on nitrification of band-applied and homogeneously mixed urea nitrogen in soils, *Can. J. Soil Sci.* **55**:15.

Pasveer, A., 1971, Verdere ontwikkeling. Het oxydenitroproces, $H_2O$ *(Netherlands)* **4**:499.

Patrick, W. H., 1960, Nitrate reduction rates in submerged soil as affected by redox potential, *7th Int. Congr. Soil Sci.*, Madison, Wisc., pp. 494–500.

Patrick, W. H., Jr., and Gotoh, S., 1974, The role of oxygen in nitrogen loss from flooded soils, *Soil Sci.* **118**:78.

Paul, E. A., and Myers, R. J. K., 1971, Effect of soil moisture stress on uptake and recovery of tagged nitrogen by wheat, *Can. J. Soil Sci.* **51**:37.

Payne, W. J., 1973, Reduction of nitrogenous oxides by microorganisms, *Bacteriol. Rev.* **37**:409.

Pearsall, W. H., and Mortimer, G. H., 1939, Oxidation-reduction potentials in water-logged soils, natural waters and muds, *J. Ecol.* **27**:483.

Pilot, L., and Patrick, W. H., Jr., 1972, Nitrate reduction in soils: Effect of soil moisture tension, *Soil Sci.* **114**:312.

Postgate, J. R., 1974, Evolution within nitrogen-fixing systems, in: *Evolution in the Microbial World* (M. J. Carlile and J. J. Skehel, eds.), pp. 263–292, Cambridge Univ. Press, London.

Potter, L. F., 1964, Planktonic and benthic bacteria of lakes and ponds, in: *Principles and Application in Aquatic Microbiology* (H. Heukelekian and N. C. Dondero, eds.), pp. 148–166, John Wiley and Sons, New York.

Pourbaix, M., 1966, *Atlas of Electrochemical Equilibria in Aqueous Solutions*, Pergamon Press, Oxford.

Prakasam, T. B. S., and Loehr, R. C., 1972, Microbial nitrification and denitrification in concentrated wastes, *Water Res.* **6**:859.

Prakash, O., and Sadana, J. C., 1973, Metabolism of nitrate in *Achromobacter fischeri*, *Can. J. Microbiol.* **19**:15.

Pratt, P. F., Jones, W. W., and Hunsaker, V. E., 1972, Nitrate in deep soil profiles in relation to fertilizer rates and leaching volume, *J. Environ. Qual.* **1**:97.

Raymond, D. G. M., 1970, Metabolism and cometabolism of nitrophenols by a soil microorganism, M.S. Thesis, Cornell Univ., Ithaca.

Reddy, K. R., and Patrick, W. H., Jr., 1975, Effect of alternate aerobic and anaerobic conditions of redox potential, organic matter decomposition and nitrogen loss in a flooded soil, *Soil Biol. Biochem.* **7**:87.

Rees, M., and Nason, A., 1965, Incorporation of atmospheric oxygen into nitrite formed during ammonia oxidation by *Nitrosomonas europaea*, *Biochim. Biophys. Acta* **113**:398.

Remacle, J., and Froment, A., 1972, Mineral nitrogen contents and microbial counts in calcareous soils under oak at Virelles (Belgium), *Oecol. Plant.* **7**:69.

Renner, E. D., and Becker, G. L., 1970, Production of nitric oxide and nitrous oxide during denitrification by *Corynebacterium nephridii*, *J. Bacteriol.* **101**:821.

Richards, F. A., and Broenkow, W. W., 1971, Chemical changes including nitrate reduction in Darwin Bay, Galapagos Archipelago over a 2-month period, 1969, *Limnol. Oceanogr.* **16**:758.

Rimer, A. E., and Woodward, R. L., 1972, Two-stage activated sludge pilot-plant operations at Fitchburg, Massachusetts, *J. Water Pollut. Control Fed.* **44**:101.

Rittenberg, S. C., 1963, Marine bacteriology and the problem of mineralization, in: *Symposium on Marine Microbiology* (C. H. Oppenheimer, ed.), pp. 48–60, Charles C Thomas, Springfield, Ill.

Rolston, D. E., Fried, M., and Goldhalmer, D. A., 1976, Denitrification measured directly from nitrogen and nitrous oxide gas fluxes, *Soil Sci. Soc. Amer. Proc.* **40**:259.

Runge, M., 1974, Die Stikstoff-Mineralisation im Boden eines Sauerhumus-Buchenwaldes. II: Die Nitratproduktion, *Oecol. Plant.* **9**:219.

Sabey, B. R., Bartholomew, W. V., Shaw, R., and Pesek, J., 1956, Influence of temperature on nitrification in soils, *Soil Sci. Soc. Amer. Proc.* **20**:357.

Sacks, L. E., and Barker, H. A., 1949, The influence of oxygen on nitrate and nitrite reduction, *J. Bacteriol.* **58**:11.

Sander, J., 1968, Nitrosaminsythese durch Bakterien, *Hoppe-Seyler's Z. Physiol. Chem.* **349**:429.

Sapshead, L. M., and Wimpenny, J. M. T., 1972, The influence of oxygen and nitrite on the formation of the cytochrome pigments of the aerobic and anaerobic respiratory chain of *Micrococcus denitrificans*, *Biochim. Biophys. Acta* **267**:388.

Saris, N. E., and Virtanen, A. I., 1957, On hydroxylamine compounds in *Azotobacter* cultures, *Acta Chem. Scand.* **11**:1438.

Satoh, T., Hoshino, Y., and Kitamura, H., 1974, Isolation of denitrifying photosynthetic bacteria, *Agr. Biol. Chem.* **38**:1749.

Schloesing, J. J. T., and Muntz, A., 1877, Sur la nitrification par les ferments organises, *C. R. Acad. Sci., Paris* **84**:301.

Schmidt, E. L., 1974, Quantitative autecological study of microorganisms in soil by immunofluorescence, *Soil Sci.* **118**:141.

Schoenbein, C. F., 1868, Über die Umwandlung der Nitrate in Nitrite durch Conserven und andere organische Gerbilde, *J. Prakt. Chem.* **105**:208.

Scholberl, P., and Engel, H., 1964, Das Verhalten der nitrifizierenden Bakterien gegenüber gelöstem Sauerstoff, *Arch. Mikrobiol.* **48**:393.

Schuman, G. E., McCalla, T. M., Saxton, K. E., and Knox, H. T., 1975, Nitrate movement and its distribution in the soil profile of differentially fertilized corn watersheds, *Soil Sci. Soc. Amer. Proc.* **39**:1192.

Schütz, K., Junge, C., Beck, R., and Albrecht, B., 1970, Studies of atmospheric $N_2O$, *J. Geophys. Res.* **75**:2230.

Seaman, G. R., 1957, The metabolism of pyruvic oxime by extracts of *Tetrahymena pyriformis* S., *Biochim. Biophys. Acta* **26**:313.

Seeler, G., and Engel, J., 1959, Die Inaktivierung des Oxydations vermogens von *Nitrobacter winogradskyi*, Buch. *Arch. Mikrobiol.* **33**:387.

Shaw, P. D., and Gottlieb, D., 1965, Origin of 3-nitropropionic acid in fungi, in: *Biogenesis of Antibiotic Substances* (Z. Vanek and Z. Hostalek, eds.), pp. 261–269, Academic Press, New York.

Shih, C. N., McCoy, E., and Marth, E. H., 1974, Nitrification by aflatoxigenic strains of *Aspergillus flavus* and *Aspergillus parasitus*, *J. Gen. Microbiol.* **84**:357.

Shirata, K., Takashi, D., Hayashi, T., Matsubara, I., and Suzuki, T., 1970, The structure of fluopsins C and F, *J. Antibiot.* **23**:546.

Silver, W. S., 1961, Studies on nitrite oxidizing microorganisms. I. Nitrite oxidation by *Nitrobacter*, *Soil Sci. Soc. Amer. Proc.* **25**:197.

Sing, R. B., Green, R., Kelly, B. K., Miller, G. A., and Gordon, J. J., 1966, *Actinonin. Chem. Abst.* **65**:2968g.

Skerman, V. B. D., and MacRae, I. C., 1957, The influence of oxygen on the reduction of nitrate by adapted cells of *Pseudomonas denitrificans*, *Can. J. Microbiol.* **3**:215.

Skinner, F. A., 1972, The removal of nitrate from solution by floc-forming bacteria on decomposing cellulose particles, *J. Appl. Bacteriol.* **35**:453.

Skinner, F. A., and Walker, N., 1961, Growth of *Nitrosomonas europaea* in batch and continuous culture, *Arch. Mikrobiol.* **38**:339.

Smith, A. J., and Hoare, D. S., 1968, Acetate assimilation by *Nitrobacter agilis* in relation to its "obligate autotrophy," *J. Bacteriol.* **95**:844.

Smith, P. A. S., 1966, *The Chemistry of Open Chain Organic Nitrogen Compounds*, Vol. II, W. A. Benjamin, New York.

Snow, G. A., 1970, Mycobactins: Iron-chelating growth factors from mycobacteria, *Bacteriol. Rev.* **34**:99.

Sperl, G. T., and Hoare, D. S., 1971, Denitrification with methanol: A selective enrichment for *Hyphomicrobium* species, *J. Bacteriol.* **108**:733.

Stanford, G., Dzienia, S., and Vander Pol, R. A., 1975a, Effect of temperature on denitrification rate in soils, *Soil Sci. Soc. Amer. Proc.* **39**:867.

Stanford, G., Legg, J. O., Dzienia, S., and Simpson, E. C., Jr., 1975b, Denitrification and associated nitrogen transformations in soils, *Soil Sci.* **170**:147.

Stanford, G., Vander Pol, R. A., and Dzienia, S., 1975c, Potential denitrification rates in relation to total and extractable soil carbon, *Soil Sci. Soc. Amer. Proc.* **39**:284.

Starr, J. L., and Parlange, J. Y., 1975, Nonlinear denitrification kinetics with continuous flow in soil columns, *Soil Sci. Soc. Amer. Proc.* **39**:875.

Starr, J. L., Broadbent, F. E., and Nielsen, D. R., 1974, Nitrogen transformations during continuous leaching, *Soil Sci. Soc. Amer. Proc.* **38**:283.

Stefanson, R. C., 1972a, Soil denitrification in sealed soil-plant systems. I. Effect of plants, soil water content and soil organic matter content, *Plant Soil* **37**:113.

Stefanson, R. C., 1972b, Soil denitrification in sealed soil-plant systems. II. Effect of soil water content and form of applied nitrogen, *Plant Soil* **37**:129.

Stefanson, R. C., 1972c, Soil denitrification in sealed soil-plant systems. III. Effect of disturbed and undisturbed soil samples, *Plant Soil* **27**:141.

Stefanson, R. C., 1973, Evolution patterns of nitrous oxide and nitrogen in sealed soil-plant systems, *Soil Biol. Biochem.* **5**:167.

Steinberg, R. A., 1939, Effects of nitrogen compounds and trace elements on growth of *Aspergillus niger*, *J. Agr. Res.* **59**:731.

Stensel, H. D., 1973, Process kinetics for denitrification, *J. Sanit. Eng. Div. Amer. Soc. Civil Eng.* **99**:EE3, 388.

Stevens, R. L., and Emery, T. F., 1966, The biosynthesis of hadacidin, *Biochemistry* **5**:74.

Takita, T., Naganawa, H., Maeda, K., and Umezawa, H., 1962, The structures of ilamycin and ilamycin B, *J. Antibiot.* **17**:129.

Tamura, S., Murayama, A., and Hata, K., 1967, The isolation and structural elucidation of fragin, a new plant growth inhibitor, *Agr. Biol. Chem.* **31**:758.

Tateson, J. E., 1970, Early steps in the biosynthesis of mycobactins P and S, *Biochem. J.* **118**:747.

Taylor, B. F., and Heeb, M. J., 1972, The anaerobic degradation of aromatic compounds by a denitrifying bacterium, *Arch. Mikrobiol.* **83**:165.

Terai, H., and Mori, T., 1975, Studies on phosphorylation coupled with denitrification and aerobic respiration in *Pseudomonas denitrificans*, *Bot. Mag. (Tokyo)* **88**:231.

Terry, R. E., and Nelson, D. W., 1975, Factors influencing nitrate transformations in sediments, *J. Environ. Qual.* **4**:549.

Thaer, R., Ahlers, R., and Grabbe, K., 1975, Behandlung von Rinderflussigmist. I. Behandlung in aeroben Verfahren mit erhohlen Temperaturen, *Sonder. Ber. Landwirtsch.* **192**:836.

Tuffey, T. J., and Matulewich, V. A., 1974, Zones of nitrification, *Water Res. Bull.* **10**:555.

Ulken, A., 1963, Die Herkunft die Nitrites in der Elbe, *Arch. Hydrobiol.* **59**:486.

Vagnai, S., and Klein, D. A., 1974, A study of nitrite-dependent dissimilatory micro-organisms isolated from Oregon soils, *Soil Biol. Biochem.* **6**:335.

Valera, C. L., and Alexander, M., 1961, Nutrition and physiology of denitrifying bacteria, *Plant Soil* **15**:268.

Van Cleemput, O., and Patrick, W. H., Jr., 1974, Nitrate and nitrite reduction in flooded gamma-irradiated soil under controlled pH and redox potential conditions, *Soil Biol. Biochem.* **6**:85.

Van der Staay, R. L., and Focht, D. D., 1977, Effects of surface area upon bacterial denitrification rates, *Soil Sci.* **123**:18.

Van't Riet, J., Knook, D. L., and Planta, R. J., 1972, The role of cytochrome b in nitrate assimilation and nitrate respiration in *Klebsiella aerogenes*, *Fed. Europ. Biochem. Soc. Lett.* **23**:44.

Venulet, J., and Van Etten, R. L., 1970, Biochemistry and pharmacology of the nitro and nitroso groups, in: *Chemistry of the Nitro and Nitroso Groups, Part II* (H. Feuer, ed.), p. 201, Wiley Interscience, New York.

Verhoeven, W., 1950, On a spore-forming bacterium causing the swelling of cans containing cured ham, *Antonie Van Leeuwenhoek J. Microbiol. Serol.* **16**:269.

Verstraete, W., 1971, Heterotrophic nitrification by *Arthrobacter* sp., Ph.D. Thesis, Cornell Univ., Ithaca, N.Y.

Verstraete, W., and Alexander, M., 1972a, Heterotrophic nitrification in samples from natural environments, *Naturwissenschaften* **59**:79.

Verstraete, W., and Alexander, M., 1972b, Heterotrophic nitrification by *Arthrobacter* sp., *J. Bacteriol.* **110**:955.

Verstraete, W., and Alexander, M., 1972c, Mechanisms of nitrification by *Arthrobacter* sp., *J. Bacteriol.* **110**:962.

Verstraete, W., and Alexander, M., 1972d, Formation of hydroxylamine from ammonium by N-oxygenation, *Biochim. Biophys. Acta* **261**:59.

Verstraete, W., and Alexander, M., 1973, Heterotrophic nitrification in samples of natural ecosystems, *Environ. Sci. Technol.* **7**:39.

Verstraete, W., and Voets, J. P., 1976, Unpublished results.

Viets, F. G., Jr., 1971, Water quality in relation to farm use of fertilizer, *Bioscience* **21**:460.

Vives, J., and Parés, R., 1975, Enumeracion y caracterizacion de la flora desnitricante quimioorganotrofa en una pradera experimental, *Microbiol. Espan.* **28**:43.

Voets, J. P., Van Staen, H., and Verstraete, W., 1975, Removal of nitrogen from highly nitrogenous wastewaters, *J. Water Poll. Control Fed.* **47**:394.

Volz, M. G., Belser, L. W., Ardakani, M. S., and McLaren, A. D., 1975a, Nitrate reduction and associated microbial populations in a ponded Hanford sandy loam, *J. Environ. Qual.* **4**:99.

Volz, M. G., Belser, L. W., Ardakani, M. S., and McLaren, A. D., 1975b, Nitrate reduction and nitrite utilization by nitrifiers in an unsaturated Hanford sandy loam, *J. Environ. Qual.* **4**:179.

Wada, E., and Hattori, A., 1971, Nitrite metabolism in the euphotic layer of the central North Pacific Ocean, *Limnol. Oceanogr.* **16**:766.

Wada, E., Kadonaga, T., and Matsuo, S., 1975, $^{15}N$ abundance in nitrogen of naturally occurring substances and global assessment of denitrification from isotopic viewpoint, *Geochem. J.* **9**:139.

Waelsch, H., 1952, Certain aspects of intermediary metabolism of glutamine, asparagine and glutathione, *Advan. Enzymol.* **13**:237.

Wagner, G. H., and Smith, G. E., 1960, Recovery of fertilizer nitrogen from soils, *Missouri Agric. Exp. Sta. Res. Bull.* **738**:1.

Wagner, P., 1895, Die geringe Ausnutzung des Stallmiststickstoffs und ihre Ursachen, *Deut. Landw. Presse.* **92**.

Waid, J. S., 1975, Hydroxamic acids in soil systems, in: *Soil Biochemistry*, Vol. 4 (E. A. Paul and A. D. McLaren, eds.), pp. 655–701, Marcel Dekker, New York.

Waksman, S. A., 1927, *Principles of Soil Microbiology*, Williams & Wilkins, Baltimore.

Waksman, S. A., and Lechevalier, H. A., 1962, *The Actinomycetales, Vol. III, Antibiotics of Actinomycetes*, Williams & Wilkins, Baltimore.

Walker, N., 1975, Nitrification and nitrifying bacteria, in: *Soil Microbiology* (N. Walker, ed.), pp. 133–146, John Wiley and Sons, New York.

Warington, R., 1884, On nitrification. Part III, *J. Chem. Soc.* **45**:637.

Watson, S. W., 1963, Autotrophic nitrification in the ocean, in: *Symposium on Marine Microbiology* (C. H. Oppenheimer, ed.), pp. 73–84, Charles C Thomas, Springfield, Ill.

Watson, S. W., 1971, Reisolation of *Nitrosospira briensis*, *Arch. Mikrobiol.* **75**:179.

Watson, S. W., and Waterbury, J. B., 1971, Characteristics of two marine nitrite-oxidizing bacteria, *Nitrospira gracilis* nov. gen. nov. sp. and *Nitrococcus mobilis* nov. gen. nov. sp., *Arch. Mikrobiol.* **77**:203.

Watson, S. W., Graham, L. B., Remsen, C. C., and Valois, F. W., 1971, A lobular ammonia-oxidizing bacterium, *Nitrosolobus multiformis* nov. sp., *Arch. Mikrobiol.* **76**:183.

Webb, K. L., and Wiebe, W. J., 1975, Nitrification on a coral reef, *Can. J. Microbiol.* **21**:1427.

Weber, D. F., and Gainey, P. L., 1962, Relative sensitivity of nitrifying organisms to hydrogen ions in soils and in solutions, *Soil Sci.* **94**:138.

Weissenberg, H., 1902, Über die Denitrifikation, *Zentralbl. Bakteriol. Parasitenk II, Abt.* **8**:166.

Wesseling, J., and von Wijk, W. R., 1957, Land drainage in relation to soils and crops. I. Soil physical conditions in relation to drain depth, in: *Drainage of Agricultural Lands* (L. N. Luthin, ed.), pp. 461–504, Amer. Soc. Agron., Madison, Wisc.

Wheatland, A. B., Barrett, M. J., and Bruce, A. M., 1959, Some observations on denitrification in rivers and estuaries, *J. Proc. Inst. Sew. Purif.* 149–159.

White, D. C., and Sinclair, P. R., 1971, Branched electron-transport systems in bacteria, *Adv. Microbiol. Physiol.* **5**:173.

Wiley, P. F., Herr, R. R., MacKellar, F. A., and Agroundelis, A. D., 1965, Three chemically related metabolites of *Streptomyces*. II. Structural studies, *J. Org. Chem.* **30**:2330.

Wiljer, J., and Delwiche, C. C., 1954, Investigations on the denitrifying process in soil, *Plant Soil* **5**:155.

Wimpenny, J. W., 1969, Oxygen and carbon dioxide as regulators of microbial growth and metabolism, in: *Microbial Growth* (P. M. Meadow and S. J. Pirt, eds.), pp. 161–198, Cambridge Univ. Press, Cambridge.

Winogradsky, S., 1890, Sur les organismes de la nitrification, *C. R. Acad. Sci., Paris* **110**:1013.

Winogradsky, S., and Winogradsky, H., 1933, Nouvelles recherches sur les organismes de la nitrification, *Ann. Inst. Pasteur (Paris)* **50**:350.

Woldendorp, J. W., 1962, The quantitative influence of the rhizosphere on denitrification, *Plant Soil* **17**:267.

Woldendorp, J. W., 1968, Losses of soil nitrogen, *Stikstof* **12**:32.

Wong-Chong, G. M., and Loehr, R. C., 1975, The kinetics of microbial nitrification, *Water Res.* **9**:1099.

Wuhrmann, K., 1963, Effects of oxygen tension on biochemical reactions in sewage purification plants, in: *Biological Waste Treatment Processes, 3rd Conf. Biol. Waste Treat., 1960* (W. W. Eckenfelder and J. McCabe, eds.), pp. 27–38, Pergamon Press, New York.

Wuhrmann, K., 1964, Microbial aspects of water pollution control, *Adv. Appl. Microbiol.* **6**:119.

Wuhrmann, K., and Mechsner, K., 1973, Discussion by K. Wuhrmann, p. 682, Fig. 1, on paper by Dawson and Murphy (1973). Factors affecting biological denitrification of wastewater, in: *Advances in Water Pollution Research* (S. H. Jenkins, ed.), pp. 671–680, Pergamon Press, New York.

Yamane, I., and Sato, K., 1968, Initial rapid drop of oxidation–reduction potential in submerged air-dried soils, *Soil Sci. Pl. Nutr.* **14**:68.

Yoshida, T., and Alexander, M., 1970, Nitrous oxide formation by *Nitrosomonas europaea* and heterotrophic microorganisms, *Soil Soc. Amer. Proc.* **34**:880.

Yoshinari, T., and Knowles, R., 1976, Acetylene inhibition of nitrous oxide reduction by denitrifying bacteria, *Biochem. Biophys. Res. Comm.* **69**:705.

Youatt, J. B., 1954, Denitrification of nitrite by a species of *Achromobacter*, *Nature (London)* **173**:826.

ZoBell, C. E., 1935, Oxidation reduction potential and the activity of marine nitrifiers, *J. Bacteriol.* **29**:78.

ZoBell, C. E., 1943, The effect of solid surface upon bacterial activity, *J. Bacteriol.* **40**:39.

ZoBell, C. E., and Anderson, D. O., 1936, Observation of the multiplication of bacteria in different volumes of stored sea water and the influences of oxygen tension and solid surfaces, *Biol. Bull.* **71**:324.

**5**

# Microbial Ecology of Liquid Waste Treatment

## J. W. M. LA RIVIÈRE

## 1. Introduction

Biological waste treatment is now changing from a technology for defensive disposal of waste materials towards a technology for utilizing waste. This change will not happen overnight, so we shall be faced with a period in which the traditional methods coexist with new methods, many of which will also be based on biotechnology. As a result, knowledge of the ecology of the traditional systems is important for improving the existing methods as well as for guiding their change into more profitable waste utilization.

Waste problems arise from the fact that natural resources only very seldom possess the composition that is consistent with total utilization. In processing them, we selectively concentrate some components for one particular use, discarding as much as possible components relevant to minor uses. Thus, more than anything else, waste technology has to focus on procedures for sorting out components and redistributing them in space and time over various different use targets. In short, it is a fight against the stoichiometry of nature, constituted by the reshuffling of impractically composed packages of, in themselves, useful materials.

The value of organic waste resides mainly in its mineral components and the energy the sun directly or indirectly appended to them; only in some cases, a small structural value is also present, like in straw that can be used for fiber. The traditional methods of biological waste treatment

**J. W. M. LA RIVIÈRE** • International Institute for Hydraulic and Environmental Engineering, Delft University of Technology, Delft, The Netherlands.

aim at destruction of pathogens and at conversion of organic waste into minerals and of necessity also into microbial cell material. The cells ultimately are also mineralized. Usually the processes require an input of energy for aeration, and only when anaerobic digestion is applied is part of the energy harvested. Because water is generally used as a vehicle for waste transportation, the resulting effluent solutions of minerals are both too concentrated, often causing eutrophication, and too dilute, precluding economical harvesting and transport of the minerals. The inadequacy of the traditional methods manifests itself most strongly in big cities and feedlots for cattle, as the minerals in the food are easily transported to a high concentration of consumers but cannot normally be returned economically to the agricultural land where this was grown. Until recently, the rationale of waste treatment was restricted to preserving an acceptable $O_2$ balance in the receiving water, i.e., of discharging an effluent of an acceptable, low BOD.* Now that eutrophication and the rising costs of fertilizer and of energy have become problems, the stage has been set for development of methods aiming, in addition, at recovery of energy and minerals. From this point of view, waste treatment becomes an operation at the intersection of the hydrological cycle and cycles of carbon, nitrogen, phosphorus, and sulfur in which both water and pollutants are separated and, as far as possible, recycled (Fig. 1).

Biological waste treatment started its development at the end of the last century, but until recently, application and innovation of these methods were in the hands of civil engineers. It is only during the last 30 years that microbiologists started to become interested in the microbiology of what meanwhile had become the largest microbiological industry. Most of this work was inspired by the wish for better understanding and improving the process, but in retrospect a great deal of the results must be considered to be of restricted, academic value and not immediately relevant to the process itself. In addition, a lot of work of dubious quality emerged as many workers lacked microbiological knowledge (engineers) or engineering knowledge (microbiologists), and good teamwork seldom occurred.

The present paper will not discuss the often tantalizing ecological curiosities that happen to be encountered in treatment plants but that are isolated in themselves and only circumstantial to the functioning of the process. Instead, it will be confined to ecological events vital to the most important treatment processes and the extent to which technology has exploited the microbiological potential of these processes. Hence, it will be necessary to precede these considerations with an analysis of these treatment methods in ecological terms.

---

* BOD, Biochemical oxygen demand, the amount of oxygen consumed by full biodegradation of the organic matter in 1 liter of sample (mg $O_2$/liter); $BOD_5^{20}$ = BOD found over a period of 5 days at 20°C, amounting to 60–65% of total BOD.

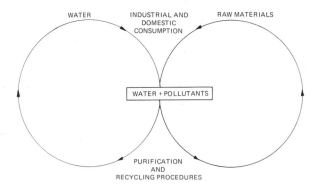

**Figure 1.** Waste treatment as an operation at the intersection of the hydrologic cycle and the biogeochemical cycles of C, N, P, S, etc.

The ecological stage on which biological waste treatment is performed takes an intermediate position between that of the fully controlled systems of ordinary biotechnology and the mineralization process encountered in nature, of which it is an intensification. The lack of asepsis provides for an unlimited cast of microbes. However, the flow and composition of the medium to be treated, although usually beyond control, possess as a rule a sufficient degree of constancy for imposing some selection of organisms and some reproducibility of the system's performance, while in addition the design and operation of the treatment plant allow for some measure of process manipulation.

In microbiological terms, we are dealing with Beijerinckian enrichment cultures in more or less continuous flow systems. These can be characterized as some form of continuous culture fed with a multiple substrate and inhabited by a flexible, mixed microbial population selected by the medium and the process conditions. It is generally acknowledged that such systems do not lend themselves easily to rigorous and meaningful analysis, as aptly stated by Jannasch and Mateles (1974): "Compromises are necessary, and the art of microbial ecology appears to reside in operating in that narrow area where those two seemingly incongruous aims accord: the scientific soundness of data and their ecological relevance."

Books and review articles demonstrating developments in the present subject over the past 15 years have been prepared by Hawkes (1963), Heukelekian and Doudero (1964), Gaudy and Gaudy (1966), Pipes (1966), Sykes and Skinner (1971), Mitchell (1972), and Curds and Hawkes (1975). In addition, the *Journal of the Water Pollution Control Federation* includes in its annual literature surveys a section on the microbiology of waste treatment.

## 2. Composition of Waste

The composition of the waste is probably the most important determinant in selecting and maintaining the mixed microbial population that has to degrade it, and while doing so, utilizes it as growth medium.

The water content of wastes can range from almost zero for dry solids to 95 % for sludges and almost 100 % for dilute waste streams, varying with the amount of water intentionally used for transportation or fortuitously added from, for example, storm sewers. In most instances wastes contain dissolved as well as suspended matter, part of which is settleable.

Their chemical composition reflects the nature of the consumption or production process that has rejected them and hence shows a great diversity (Sierp, 1959; Painter, 1971; Koziorowski and Kucharski, 1972; Bond et al., 1974). One important overall characteristic is their C : N : P ratio, as this determines the nature of the limiting substrate. As a rule, the carbon and energy source is limiting and, if not, is made to be so by addition of nitrogen and/or phosphorus compounds or by mixing with domestic waste. This is, of course, based on the desire to discharge an effluent with as little oxygen demand as possible.

Table I gives a view of the suitability of some wastes as a growth medium for microbes, with particular reference to the C : N : P ratio. These data illustrate that serious imbalances may occur and must be rectified to render microbial treatment effective. It should be realized that the possibilities for influencing medium composition are very few as compared to the practice in the fermentation industry, where the economic constraints allow for medium optimization with respect to the most profitable production. Similarly, strain selection is economically impossible because of the high costs sterilization of wastes would require. Hence, one has to use the population that emerges by enrichment.

In contrast to laboratory enrichment media, most waste flows have such a diverse composition that their selective power is very weak. In the common waste product urine, for instance, more than 200 different compounds have been identified (Bond et al., 1974). Besides organic compounds, also inorganic compounds and ions like sulfides, ammonia, nitrites, cyanides, and thiocyanates are amenable to biological oxidation, and these are more capable of selecting specific microbial populations. The same holds for toxic degradable organic compounds like phenol and its derivatives, which select through substrate specificity as well as suppression of sensitive organisms.

Since biological waste treatment, by its very nature, allows undegradable substances to leave the plant with the effluent, a cumulative increase in the concentration of such substances occurs when river water is used successively by different towns, as is the case along the river Rhine. This is

**Table I. C : N : P Ratios of Various Wastes Compared to That of Bacterial Cells and of a Common Laboratory Medium**

| | $BOD_5^{20}$ (g/liter) | Glucose equivalents[a] (g/liter) | C : N : P weight basis | Reference |
|---|---|---|---|---|
| Escherichia coli | — | — | 50 : 14 : 3 | Stanier et al., 1971 |
| Theoretical medium for "total" consumption[b] | — | — | 100 : 14 : 3 | |
| Routine laboratory medium[c] | 7.4 | 10 | 133 : 8.6 : 3 | |
| Optimum ratio in waste treatment practice | — | — | 146 : 16 : 3 | Koot, 1974 |
| Settled domestic sewage | 0.37 | 0.5 | 86 : 29 : 3 | la Rivière, 1972 |
| Brewery waste | 1.6 | 2.2 | 263 : 8.7 : 3 | Sierp, 1959 |
| Slaughterhouse waste | 0.8 | 1.1 | 165 : 53 : 3 | Sierp, 1959 |
| Potato-starch factory waste | 3.2 | 4.3 | 96 : 7.5 : 3 | Koot, 1974 |
| Sugar beet washing water | 4.6 | 6.2 | 573 : 16 : 3 | Koot, 1974 |
| Retting waste | 2.5 | 3.4 | 156 : 4.6 : 3 | Koot, 1974 |

[a] Equivalents calculated from the theoretical $BOD_5^{20}$ of glucose.
[b] Assuming 50% of carbon source being assimilated.
[c] 10 g glucose, 1 g $NH_4Cl$, and 0.5 g $K_2HPO_4$ per liter and minerals.

of special importance for drugs with high potency at low concentrations. This consideration prompted Stumm-Zollinger and Fair (1965) to examine and establish the biodegradability of steroid hormones as used in contraceptive preparations.

Although some of the undegradable substances can be oxidized by chlorination during the preparation of drinking water, this operation may also produce potentially harmful and persistent haloforms, as shown by Rook (1974) for chlorination of humic substances.

Every waste stream is further characterized by temporal fluctuations in dilution and composition as a consequence of the characteristics of the waste-producing process involved. These fluctuations may possess some regularity over periods of 24 hr, upon which also, for example, weekly variations may be superimposed, as is the case with domestic waste where production during the night is low and often once-a-week peaks occur on washing day. Food industries show, in addition, seasonal fluctuations, and chemical industries may produce peak amounts of waste by clean-up operations at irregular intervals. Thus, combined flows of different wastes may show considerable quality variations of a stochastic nature, often empirically known to the treatment plant operator who has come to recognize "bad" and "easy" periods.

Some of the fluctuations in composition may involve changes in temperature, pH, and salt content, while others may deal with heavy metals, pesticides, detergents, or other potentially toxic substances. Since almost every treatment plant operates within a legislative framework preventing excessive disturbance of its operation, it seems safe to conclude that in many cases the treatment process is being taxed just to the limit of effective operation by inputs of potentially deleterious substances. This borderline is empirically marked by "accidents" rather than by rational understanding.

The effect of quality change almost always leads to a shift in the composition of the microbial population, which as a rule will adapt to the new situation through selection and induction. Since the shift in composition takes time, almost any change of quality is accompanied by a period of low-grade performance. Thus, it is good practice to avoid peak loads as much as possible. We must, however, take into account the interesting possibility that regular quality fluctuations will have a special selective effect of their own by weeding out the slowly adapting organisms or by "shaving" away at intervals those segments of the population that are highly sensitive to substances introduced periodically for a short time. It is, therefore, necessary to consider not only waste composition by itself, but also the fluctuations in the waste flow as a determinant of the composition of the mixed microbial populations in the treatment plant, as they must bear the imprint of the previous history of waste inputs.

Finally, we should not forget that the microbial cells carried along with the sewage as well as those produced during waste treatment constitute themselves a substrate for enrichment, in the Beijerinckian sense, of other organisms that can in turn degrade them. The large masses of cells, for example, in aeration units are a constant challenge to predatory protozoa and bacteria, including *Bdellovibrio*, and, of course, the bacteriophages so much feared in the bioengineering industry. The aims of waste treatment include also the disposal of these large cell masses the treatment unavoidably generates, and usually this is carried out in the process of anaerobic digestion.

## 3. Main Characteristics of Ecosystems Encountered in Biological Waste Treatment

### 3.1. Transport Systems

Sewage is usually transported by gravity through sewers, i.e., partially filled concrete pipes or channels. Microbial decomposition starts, of course, already during transport, leading successively to consumption of dissolved oxygen, nitrate reduction, fermentation, sulfate reduction, and finally methane production (Stumm and Morgan, 1970). Hence, most sewage flows are anaerobic and give off $H_2S$, which may dissolve in the water film covering the sewer wall exposed to air. There chemical (Chen and Morris, 1972) and biological oxidation takes place, leading to corrosion by $H_2SO_4$. In addition, $CH_4$ can be generated, and instances are known of formation of explosive mixtures which, when accidentally ignited, have led to blowing off of manhole covers. Perhaps the most interesting feature of the sewer ecosystem is its capacity for rapid $O_2$ uptake, which, in cases of long transport lines and small liquid depth, may lead to considerable reduction in BOD. This makes the system an example of many naturally occurring systems in which $O_2$ is consumed under virtually anaerobic conditions, characterized by anaerobic compartments in which $O_2$ is more rapidly consumed than it is imported through the interface with air. It is likely that $O_2$ uptake by such systems is more rapid than the Monod relationship between growth rate constant and limiting $O_2$ concentration would predict, indicating a possible intermediate role for inorganic carrier molecules like reduced iron, sulfur, or nitrogen compounds. It should be mentioned in passing that the opportunity for exploiting sewers for transport as well as aeration has not been applied in practice.

In addition, important changes in the composition of the microbial population must be occurring: The increase in the number of degraders is accompanied by a decrease in those segments of the "natural" population of the waste that can no longer thrive under the sewer conditions. An example is, of course, the intestinal microbes of man as they occur in domestic sewage.

Finally we must realize that the temperature of sewage is always a few degrees centigrade higher than the raw water before it was used; hence, water temperatures in sewage treatment plants are higher than in neighboring surface waters.

## 3.2. Treatment Systems

The various types of biological waste treatment plants are listed in Fig. 2. Treatment can be preceded by separation into settleable matter and supernatant fluid, each to be treated in the most appropriate way. Alternatively, the waste can be comprehensively treated without initial separation. The treatment methods themselves can be oxidative conversion into minerals and cells, anaerobic transformation into minerals, methane, and cells, or photosynthetic transformation into algal cells.

The treatment actually carried out in practice consists as a rule of one of these methods or a combination of several of them. A simple procedure is just sedimentation followed by anaerobic digestion of the settled material only. A very frequently applied combination consists of sedimentation

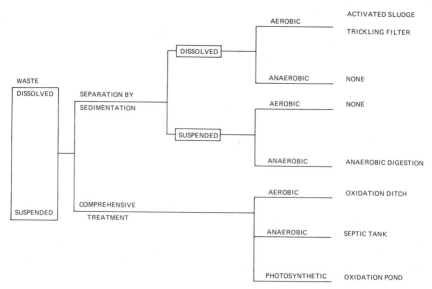

**Figure 2.** Main types of biological treatment methods for organic wastes.

followed by aerobic treatment and a second sedimentation, the sludges of both sedimentations being anaerobically digested (Fig. 3).

In general, the sewage flows first through a comminutor (i.e., a grinder to cut down the size of larger objects) and a rough screening device, followed by a sand-catching channel (flow velocity 0.3 m/sec) in which only heavy particles can settle. The flow then enters the primary sedimentation tank, which is gently stirred to prevent short circuiting. It also serves to remove floating matter like fats and oils. Detention times range from 1 to 2 hr. In these basins, the water and sediment become anaerobic,

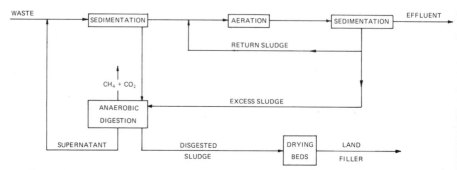

**Figure 3.** A frequently used combination of oxidative treatment and anaerobic digestion. With trickling filters no sludge is returned.

and fermentation and anaerobic respiration may already take place, some-times leading to the formation of gas bubbles which may upset the sedimentation process as they provide buoyancy to particles to which they adhere. The organic matter settling in this tank, called primary sludge, may constitute up to 40% of the BOD of the raw sewage in the case of domestic waste.

The supernatant fluid from the primary sedimentation tank then enters the aeration installation.

### 3.2.1. The Activated Sludge Method

The activated sludge process is in general comparable to an open homogeneous culture system of the fully mixed reactor type with partial feedback of cells (Herbert, 1961). There exists a great diversity of aeration equipment on the market designed to provide oxygenation and stirring capacity at minimum energy expenditure. In order to promote oxygen transfer, the prevailing $O_2$ concentration in the culture liquid is kept as low as possible, but not lower than 1–2 mg/liter, so as not to make $O_2$ the limiting factor. The special feature that lends the process its name resides in the settleability of the microbial cell mass which makes feedback of cell material possible; this feedback reduces the growth rate of the cells which, in turn, promotes the formation of settleable flocs. The process can be described approximately by means of the ordinary mathematical model for this type of reactor, based on the following equations:

$$D = \frac{f}{V} \tag{1}$$

$$\frac{dx}{dt} = \mu x - \frac{Dx}{R} \tag{2}$$

$$\frac{ds}{dt} = Ds_R - Ds - \frac{\mu x}{Y} \tag{3}$$

$$\mu = \mu_{max} \frac{s}{K_s + s} \tag{4}$$

in which $f$ = liquid flow in liters/hr; $V$ = liquid volume of the aeration basin in liters; $D$ = dilution rate in $hr^{-1}$; $x$ = microbial cell mass in g dry wt./liter; $t$ = time in hr; $\mu$ = specific growth rate constant in $hr^{-1}$; $R$ = recycle ratio = $(1 - a)/(1 - b)$, in which $a$ = fraction of liquid volume returned and $b$ = fraction of cell mass returned, both per liter of aeration basin; $s$ = concentration of limiting nutrient in aeration basin and effluent in g/liter; $s_R$ = concentration of limiting substrate in reservoir or inflowing liquid in g/liter; $Y$ = yield constant, i.e. grams of cell mass (dry weight) produced per gram of limiting substrate consumed; $\mu_{max}$ = maximum specific growth rate constant in $hr^{-1}$; $K_s$ = saturation constant in g/liter, i.e. the concentration of limiting substrate at which $\mu = 1/2\mu_{max}$.

When a steady state is achieved both $dx/dt$ and $ds/dt$ become 0 and equations (2) and (3) lead to

$$\mu = D/R \qquad (5)$$

$$x = RY(s_R - s) \qquad (6)$$

If there is no recycling, $R = 1$ and equations (5) and (6) are reduced to

$$\mu = D \qquad (7)$$

and

$$x = Y(s_R - s) \qquad (8)$$

respectively. With recycling, $R$ must have values greater than 1, which leads to smaller values of $\mu$ and larger values of $x$. By introducing appropriate terms and/or factors this mathematical model can be further refined, e.g., to make allowance for substrate consumption for cell maintainance, for dying off of cells, or for concentration-dependent toxic action of the limiting substrate.

The validity of the model hinges upon the crucial question: To what extent can we extrapolate the continuous culture theory valid for single substrates and pure cultures to multisubstrate, mixed-culture systems? In other words, can we find values for $\mu_{max}$, $Y$, $K_s$ and factors for decay, maintainance, and toxicity that are sufficiently "constant" to make the equations meaningful?

The answer to this question is further complicated by the fact that the ideal of complete mixing is seldom reached in practice. The same holds true for the so-called "tapered aeration" system, in which the piston flow principle is used, which includes locally adjusted aeration intensity along the flow through the aeration basin. Finally, the set of parameters used by civil engineers in designing and further developing the process is not immediately transparent to the microbiologist, who tries to describe the process in his own terms and is inclined to use a somewhat different approach.

Nevertheless, some progress has been made in finding conceptual frameworks that are both scientifically and technologically acceptable (Lawrence and McCarty, 1970). After all, it is highly unlikely that mixed populations would behave in a completely irreproducible manner and very much unlike pure cultures. Hence, it appears justified to study at least qualitatively the mixed populations of the sludge in the framework of that developed for pure cultures, and in several instances this approach has already borne fruit also in a practical, quantitative manner. Cases in point are nitrification and the function of protozoa, which will be discussed in Section 4.

Table II presents the most important parameters of three types of activated sludge systems together with the microbiological parameters one would use when dealing with a pure culture in a simple, defined medium. The numerical values present ranges or orders of magnitude with the limited aim of circumscribing the system in which the ecological events to be discussed are taking place. The differences between low- and high-rate activated sludge systems follow from the figures in the table. The oxidation ditch, however, differs from these conventional systems in that it treats both the suspended and the dissolved organic matter in one basin, which often also serves as secondary sedimentation tank. In this case there is no primary sedimentation, and partial feedback of cells is performed by removing the sludge more slowly than the clear effluent.

**Table II. Process Parameters of Some Activated Sludge Processes for Domestic Sewage[a]**

| Engineering parameter | Range of practical values | | | Comparable microbiological parameter |
| --- | --- | --- | --- | --- |
| | Oxidation ditch | Low rate plant | High rate plant | |
| $BOD_5^{20}$ of settled sewage, $g/m^3$ | 200–400 | 200–400 | 200–400 | $s_R$ |
| $BOD_5^{20}$ of effluent, $g/m^3$ | 20–40 | 20–40 | 20–40 | $s$ |
| Treatment efficiency, % | >95 | >90 | <90 | $\dfrac{s_R - s}{s_R} \times 100$ |
| Hydraulic load, $m^3$ sewage/$m^3$ aeration basin/day, $day^{-1}$ | 0.2–0.4 | 3–6 | 20–50 | $D$ |
| BOD load, kg BOD/$m^3$ aeration basin/day, $kg/m^3$/day | 0.1–0.2 | 0.4–1 | >3 | $D \times s_R$ |
| Aeration (detention) time, hr | 60–70 | 5–8 | <2 | $\dfrac{1}{D}$ |
| Mixed liquor suspended solids (MLSS), $kg/m^3$ | 4 | 0.9–4.6 | 0.2–2.5 | $x$ |
| Sludge loading, BOD load/kg MLSS, $day^{-1}$ | 0.05 | 0.5 | 1–4 | $\dfrac{D \times s_R}{x}$ |
| Specific sludge wastage rate = net amount of sludge produced or removed/day/$m^3$/kg MLSS, $day^{-1}$ | 0.04 | 0.03–0.3 | 1–2.5 | $\mu$ |
| Sludge age, reciprocal value of sludge wastage rate, days | 25 | 3–30 | 0.4–1 | $\dfrac{1}{\mu}$ |

[a] Imhoff et al., 1971; Koot, 1974; Fair et al., 1968.

In the table, values for $\mu_{max}$, $K_s$, and $Y$ are missing because they are not normally explicitly used in design and operation procedures. The recirculation ratio is often expressed in hydraulic terms rather than in terms of weight percent of sludge returned; it amounts in practice to 30 to 70 % of the sludge settling in the secondary sedimentation tank.

The most striking datum in Table II is the extremely low specific growth rate of the activated sludge, which places the process outside the common experience of most microbiologists and biotechnologists. An important consequence of this low growth rate is the increased impact of the maintenance energy to be discussed in Section 4.

### 3.2.2. Trickling Filters

The "trickling filter" aeration method is an example of a closed heterogeneous continuous culture system. It consists of a cylindrical column with a diameter of up to 30–40 m and a height of 1.5–4 m. The column is filled with pieces of lava, rock, or plastic with dimensions of 5–9 cm, providing a surface of 100–200 $m^2/m^3$. The settled sewage is evenly distributed over the top surface by a sprinkling device and flows downwards over the stone surfaces on which a microbial film maintains itself and mineralizes the dissolved organic matter as it passes by. Oxygen supply is ensured by an upward (winter) or downward (summer) draft of air through the column, which has ventilation holes in the bottom. As a rule, no forced ventilation is required. This type of film reactor is heterogeneous because the microbial composition of the film differs with depth as a result of change of medium composition with progressing mineralization. It is "closed" because the film cannot keep growing indefinitely without bringing the process to a standstill. In practice, however, the balance of shearing forces and cohesion of the film is such that filters can be kept in operation for indefinite periods of time as sloughing off of pieces of film ("humus") keeps pace with its growth. This three-phase system (air, water, microbial film) is then followed by a secondary sedimentation tank, as in the activated sludge process. Table III presents some process parameters.

Different filters can be used in parallel or in series, and recirculation is frequently applied as the hydraulic load is instrumental in keeping the thickness of the biological film thin enough to prevent clogging. Recirculation is also an instrument to dilute peak loads of toxic substances. Filter performance is more sensitive to seasonal drops in temperature than that of activated sludge units, and their purification efficiency on the average is somewhat lower. The filters may further give rise to fly and odor nuisances, and their top layers may require occasional cleaning to remove algal growth that can cause local clogging (ponding). A newly built filter requires 1–3 months for "ripening," i.e., developing an adequate biological film. Their

**Table III. Process Parameters for Trickling Filters for Domestic Sewage Treatment**[a]

| Parameter | Value for low rate performance | Value for high rate performance |
|---|---|---|
| Hydraulic load, $m^3/m^2/hr$ | 0.05–0.3 | 0.6–2 |
| BOD load, $kg/m^3/day$ | 0.1–0.4 | 0.6–1.8 |
| Biological film material, $kg/m^3$ | 5–7 | 3–6 |
| BOD load/kg film material, $day^{-1}$ | 0.01–0.08 | 0.1–0.6 |
| Contact time | 1–3 hr | 10–30 min |
| BOD removal efficiency, % | 80–90 | 65–85 |

[a] Imhoff et al., 1971; Koot, 1974; Fair et al., 1968.

resistance to toxic materials and shock loads does not differ much from that of the activated sludge system. The aeration costs spent in the latter process are used in the trickling filters for elevating the settled sewage to the top of the column and for recirculation pumping. Trickling filters are simpler and more easy to run than activated sludge units, and this constitutes one of their main assets.

### 3.2.3. Anaerobic Digestion

Anaerobic digestion is a far from ideal version of the fully mixed homogeneous continuous culture without feedback of cells. The substrate consists of the settled material of the first sedimentation tank to which, in the case of complete treatment, the surplus flocs or the humus from the second sedimentation are added.

Because of the high costs that continuous good mixing of the substrate would require, the mixing that occurs in the digestion tanks is mainly restricted to that caused by the gas bubbles arising from fermentation. Thus, in most cases, the digestion process is carried out in tanks provided with inlets and outlets and a covering device that permits collection of the gas and at the same time prevents bad odors from escaping and $O_2$ from the air from interfering with the obligately anaerobic process. Essentially, it is not much more than a contained and covered manure heap left to anaerobic decay. Especially in milder climates, heat exchangers are incorporated into the systems in order to provide for higher temperatures (usually 35 or 55°C), i.e., faster digestion, which means smaller tank volume. Often two tanks are used in series, the last one serving as a separator for the digested sludge. In view of the lack of mixing and of feedback of cells, it is not surprising that detention times are long, ranging from 10 to 50 days. Normally the digestion process manages to transform the primary sludge, which is malodorous, hygienically hazardous, and untransportable because of its bad dewatering properties, into a sludge that has lost about 40% of the original solids and possesses only 40–50% of the original volume. This material is much less hazardous, can be dried more easily, and does not produce offensive odors during drying. During the process also many plant seeds are killed. In addition, a gas mixture of $CH_4$ and $CO_2$ (5:8 to 8:5) with traces of $NH_3$, $H_2S$, and $H_2$ is obtained, about 30 liters being produced from the waste of one person per day.

The technology of the process can obviously be improved by better mixing, working at high cell density and using continuous instead of discontinuous input and output.

A classic, small-scale forerunner of the anaerobic digester is the well-known septic tank, which is still used today. It is a compartmented holding tank which provides a much longer detention time for the settleable matter than for the supernatant fluid. Its efficiency is therefore largely restricted to the extent of anaerobic digestion occurring in the sediment, leading to an overall BOD reduction of 25–60%.

Since anaerobic digestion is usually the end of the line in sewage treatment, it is worthwhile realizing that the digested sludge and its supernatant fluid are the ultimate products requiring disposal. The latter is usually fed back to the flow of incoming sewage, and the sludge is dried, if possible, in the open air on drying beds, so as to make it suitable for soil improvement. Its beneficial components (nutrients and organic matter) make this a logical and desirable solution, but such use can be jeopardized by the presence of pathogens, heavy metals, and other objectionable materials.

### 3.2.4. Oxidation Ponds

Oxidation ponds (or waste-stabilization ponds or sewage lagoons) consist of large tanks exposed to sunlight. The sewage is mineralized by bacteria in the bottom layers, and the resulting minerals are converted to algal cells, which are usually discharged with the effluent into the receiving water. In the receiving water, they are less harmful than the original untreated sewage would have been. Mineralization and photosynthesis are linked to one another by the $O_2$ produced by the algae, which enables the bacteria to set free the $CO_2$ and minerals on which these algae grow in turn. Since there are losses of both $O_2$ produced by the algae and $CH_4$ produced by the bacteria, the conversion reached in practice does not fully conform to the theoretical stoichiometry of the process. The process requires more land (20–100 $m^2$/kg BOD/day) than the methods discussed above, and it only works well when there is enough sunlight. Hence, it is mostly applied in tropical regions where land is cheap. Low investment and operation costs are an important advantage. The depth of the ponds usually ranges between 0.9 and 1.5 m with detention times of 3–12 days; oxygen contents range from 0 to more than 300% of saturation, depending on depth and time of day (Arceivala *et al.*, 1970).

Also for this process, a number of variations occur, including provisions for mixing, for preventing short circuiting, and for multistage operation. In addition, the resulting algal suspension may be used for irrigation or for feeding fish ponds. Sometimes the algae are harvested and used as fodder.

The oxidation pond is one of the most interesting systems from the point of view of waste recycling: If control of the process makes cultivation of a desired alga possible, it offers an excellent opportunity for reuse of the minerals contained in the waste and at the same time solves the problem of eutrophication of the receiving water. The efficiency could be further enhanced by first removing and using the energy in the waste by an anaerobic digestion step preceding the pond system. Since the production of algae is now also being studied from the bioengineering viewpoint, using pilot plants and synthetic media, merging of these two developments holds very interesting prospects as envisaged in an early stage by Oswald and Golueke (1960, 1968).

## 4. Microbial Ecology of Activated Sludge Systems

In analyzing the microbial ecology of a given system, the microbiologist instinctively starts by making a qualitative and quantitative inventory of the organisms found, an inventory which then should serve as a basis for attempts to understand the system and its dynamics as a whole.

The opposite approach, which is complementary, focuses first on the system and its behavior under different conditions. This approach then leads to delineation of the characteristics of the microbial population, first physiologically and then only finally, as a last priority, taxonomically.

Both approaches have been followed with regard to the activated sludge processes, but a satisfactory coherent picture has not yet emerged. Although not always easily distinguishable, an attempt has been made in the following to deal with these two approaches separately.

### 4.1. Microbial Populations of Activated Sludge

In an activated sludge aeration basin, we must distinguish between the large population of mostly bacterial cells clustered together in flocs and a smaller but significant population of freely suspended cells, including bacteria as well as protozoa. Since only the flocs are recycled via the sedimentation tank, the conditions under which these two populations maintain themselves are fundamentally different.

## 4.1.1. The Activated Sludge Floc

Table IV presents some of the most important features of the floc. It is a heterogeneous particle in that it contains inert sewage material, microbially produced matrix material, and dead cells along with living ones. Its size, providing different conditions for cells on the outside and inside, is regulated by the balance between its cohesion and the shearing forces caused by the aeration/agitation process. Furthermore, it is likely that larger particles would be subject to internal anaerobiosis and hence to disruptive fermentation.

The activity must be small as the cells are growing at very low rates under a starvation diet as far as carbon and energy source is concerned, and cells inside are under even worse conditions than those on the outside. It is, therefore, not surprising that their activity is 100 to 1000 times smaller than that of a pure bacterial suspension under nonlimiting conditions, measured in terms of $O_2$ consumption rate per unit of dry weight.

This state of affairs is confirmed by the results of Pike and Carrington (1972) who, working with refined techniques, found that the viable plate count of organisms in the floc was between 1 and 2% of the total microscopic count. Even when one assumes that improved techniques will extend the viable count to include the most fastidious and debilitated

### Table IV. Characteristics of Activated Sludge Floc

| Enzymatically active material: | Inert material: |
|---|---|
| Growing cells | Dead cells, partially decomposed |
| "Resting" cells | Matrix material (biopolymers) |
| Nonviable cells, some enzymes still operating | Nonsettleable organic and inorganic matter from sewage |

Shape: irregular, filamentous protrusions occur
Size: 0.02–0.2 mm (Mueller et al., 1967)
Specific gravity: 1.005 g dry wt./ml (Fair et al., 1968)
Usual suspension density: 0.2–5 g dry wt./liter (Koot, 1974)
Cells by microscopic count: $10^{12}$/g dry wt. (Pike, 1975)
Cells by plate count: $10^{10}$–$10^{11}$/g dry wt. (Pike, 1975)
Oxygen uptake rate: 10–60 mg $O_2$/g dry wt./hr (Weddle and Jenkins, 1971)
Usual specific growth rate: smaller than 0.3/day (Weddle and Jenkins, 1971)

Bacterial genera reported as floc constituents by at least two different authors (Pike, 1975):

| | |
|---|---|
| Pseudomonas | Flavobacterium |
| Zoogloea | Bacillus |
| Sphaerotilus | Arthrobacter |
| Achromobacter | Bdellovibrio |
| Nitrosomonas | Coliform bacteria |

organisms, it is not likely that the two will ever be found to coincide. Meanwhile, we are still in need of quantitative data on the relative amounts of the various categories of active and inert floc material in relation to system parameters.

The conditions that provoke microbial growth in the form of flocs rather than as freely suspended cells are as yet insufficiently understood. For the moment, we have to accept the empirical fact that populations exist that do grow reproducibly in settleable flocs in systems designed to select for them, like the activated sludge system. Since these are characterized by low imposed growth rates and low nutrient concentrations, it is logical to suppose that these conditions are operative in floc formation.

It appears most likely that the flocs behave as self-propagating entities—comparable, for example, to mycelium pellets of a mold—which consist of a more or less specific population, part of which takes care of the formation of the floc's matrix material. In this way the floc becomes capable of maintaining itself in the face of inputs of all sorts of other organisms which do not form flocs and are not recycled. As the flocs grow slowly, as imposed by their size, segments are broken off by the shearing forces in the liquid that in turn serve as nuclei for outgrowth into new flocs. If the population is to some extent specific, one may further expect some interdependence and division of labor to have evolved among its members by the pressure of optimal economics, leading, for example, to loss of superfluous enzymes.

An extreme example of such an "organism-like" conglomerate is the kefir grain, which consists of a mixture of yeasts and lactic acid bacteria cemented together by the polysaccharide kefiran produced by the latter, the cells within the matrix thus acquiring an "immobilized" status. These grains grow slowly in milk, although none of the constituent organisms can do so by themselves. Such grains can be propagated indefinitely by transferring them every 24–48 hr to fresh milk by simple mechanical separation. No aseptic techniques are required. Reconstitution of these grains from the isolated constituents has so far not been possible (la Rivière et al., 1967).

The mechanism of floc formation has attracted a great deal of attention because feedback of cells by means of sedimentation, though cheap, is vulnerable to disturbances and incapable of concentrating the sludge beyond the limits dictated by its variable settling characteristics. Thus various authors (cf. Harris and Mitchell, 1973) have sought to devise means by which freely suspended cells can be reproducibly precipitated as well-settling flocs. In some of these, essentially physico-chemical, studies it has been implied that the floc formation as it normally occurs in aeration tanks also is a physico-chemical phenomenon, bringing about aggregation of freely suspended proliferating cells. This in turn implies a two-stage time sequence of growth and floc formation which is difficult to reconcile with the characteristic time independence of the continuous culture. Hence, it appears logical to prefer the model of the floc as a proliferating entity, outlined above, a model which stresses the special significance of the biological formation of matrix material in the absence of excess carbon.

Extraction of activated sludge flocs has yielded several polymeric materials that could be instrumental in holding the cells together, but it is difficult to establish this with certainty because internal cell polymers also can be leached out by the extraction procedure. Cellulosic fibrils, found by Deinema and Zevenhuizen (1971) with the electron microscope as a linking material between the cells, may well qualify as an important matrix material, being present in amounts of 1–4% of the dry weight of the floc. The same holds for a heteropolysaccharide extracted with hot water by Wallen and Davis (1972); this heteropolysaccharide amounted to 2.7% of the dry weight of the floc. Scanning electron microscopy of flocs showed that hot water treatment did indeed remove a substantial part of the interlinking slime material. Furthermore, a poly-$\beta$-hydroxy-alkanoate, composed of $\beta$-hydroxyvaleric and $\beta$-hydroxybutyric acids, was found to occur in flocs from two different sewage treatment plants in amounts of 1.3% dry wt. (Wallen and Rohwedder, 1974). It is not yet certain whether or not this substance is an internal reserve material like the poly-$\beta$-hydroxybutyrate found to occur in floc cells by several authors. The work in this area is further stimulated by the possibility of extracting valuable biopolymers from activated sludge, which would improve the economy of the process. Meanwhile, we must hope that further work along these lines will bring irrefutable proof, fulfilling Koch's postulates, for microbial matrix production by organisms found systematically in sufficient numbers as an integral part of the floc population. *Zoogloea* still remains a persistent candidate in this respect (Unz and Farrah, 1976).

Although one can isolate almost any microbe from activated sludge by applying the proper selective enrichment method, it is generally agreed that heterotrophic aerobic bacteria play the most important role. The exhaustive review by Pike (1975) confirms this, as shown in Table IV. Since the workers concerned have used different media and methods as well as different types of activated sludge, it is at the moment impossible to come to a conclusion as to the taxonomic definition of the essential organisms of activated sludge. Also, one cannot definitely exclude the fact that important constituents of the floc so far have escaped isolation.

The problem of characterizing the population of the floc has also been approached from the angle of operational needs (Weddle and Jenkins, 1971). Active biomass has been assessed by measurements of organic nitrogen, DNA, and ATP instead of dry organic matter. A higher resolution is offered by combining methods which determine specific enzyme activities of the floc, of which the determination of oxidative capacity by means of reducible dyes or Warburg manometry is an example. Finally, the best resolution would be obtained by quantitative listing of the relative numbers of a series of physiological types, chosen on criteria of practical relevance. Automated procedures combined with numerical analysis by computer (Hedén and Illéni, 1975) appear to bring such a characterization within reach, and the work of Pike and Carrington (1972) demonstrates the feasibility of this approach. Nevertheless, it would seem that a concerted effort towards identifying the matrix-forming organisms in the floc would afford a shortcut to distinguishing the essential core of the floc population from the variable population segment of cooperative "fellow travellers."

Algae do not occur in significant numbers in activated sludge as light penetration is minimal. Although various fungi can be isolated from them (Cooke, 1963), flocs do not normally contain appreciable amounts of fungal material except under special conditions like temporarily low pH or in the case of specific industrial wastes.

Protozoa, on the other hand, are almost certainly an essential element of the population in the aeration basin. They occur attached to the floc as well as freely suspended in the water phase and will be more conveniently discussed in later sections.

### 4.1.2. Freely Suspended Population of the Activated Sludge Liquor

The nonsettleable part of the microbes in the aeration tank merits separate attention because these organisms are not being recycled. The size and composition of this population is determined by (1) nonsettling cells entering with the sewage, (2) exchange with the floc population, and (3) processes of predation and growth.

The total numbers of bacteria in settled sewage entering the aeration basin amount to about $10^8$/ml, of which $1\%$ can be accounted for as viable cells. Effluents contain about $10^7$ cells/ml, of which $2\%$ is viable (Pike, 1975). Such a bacterial density in an effluent is in accordance with its usual clarity, as it is well known from laboratory experience that suspensions of most bacteria begin to show visible turbidity in layers of a few centimeters at a little over $10^7$ cells/ml.

Protozoa occur only in small numbers in sewage but may have densities of 50,000 cells/ml in the aeration tank. Their identification is much easier than that of bacteria: Although much harder to isolate and cultivate, protozoa can be recognized microscopically by their morphological characteristics, a situation just opposite to that with bacteria. For the class Phytomastigophorea, Curds (1975) listed 3 genera with 4 species; for the Zoomastigophorea, 3 genera with 6 species; for the Rhizopodea, 5 genera with 8 species; and for the Ciliatea, 13 genera with 24 species. These 42 species were most frequently encountered out of the 228 totally reported in activated sludge.

By themselves, these identifications do not carry us much further in understanding the role of the protozoa. It is evident from Curds' (1975) review that only dynamic studies provide an insight, and these will be discussed below. Meanwhile we should point to the low bacterial density of the freely suspended population as one of the most prominent features of this part of the ecosystem. Like the observation that "the dog did not bark during the night," this negative characteristic provides the clue to the understanding of a most interesting constellation of dynamic events.

## 4.2. Dynamic Relationships in the Activated Sludge Process

### 4.2.1. Activity of the Population as a Whole

The first question arising in this context is to what extent the full potential offered by the activated sludge ecosystem has been exploited by the existing types of treatment plants. For an answer one has to define the borderlines of the elasticity of the system as seen from the point of view of the microbial ecologist.

Depending on prevailing socioeconomic constraints, technological development has attempted to respond to desires for maximization of hydraulic and BOD load and minimization of aeration costs, excess sludge production, effluent BOD, and operational risks, singly and in combination. As we shall see, pushing one of these objectives to its extreme often leads to conflict with one of the others.

*Increasing the hydraulic load* for a given flow of sewage is equivalent to decreasing the volume of the tank required for aeration and hence also to increasing $D$, $\mu$, and $s$. If one cannot accept a higher BOD in the effluent, $R$ must be pushed to its maximum value so as to keep $\mu = D/R$ from rising. This leads to a demand for devices for floc separation better than sedimentation. High-rate activated sludge plants with their lower quality effluents are the practical outcome of this trend, in which sedimentation, however, is still retained. If, in addition, the sewage has a high BOD, another borderline appears, which is the technical possibility of introducing enough $O_2$ per unit volume per unit time to cope with the higher specific $O_2$ requirements in the tank. Replacing air by $O_2$ or $O_2$-enriched air (Scherb, 1973) and so-called "deep well" aeration respond to this by providing greater $O_2$ transfer rates from the gas to the liquid phase. In the latter system, the aeration tank consists of a vertical tubular vessel about 120 m deep in which the air is introduced near the bottom, where the $O_2$ saturation value is some 12 times higher than at the surface. In this case, the ultimate borderline appears to be technological rather than microbiological: It is provided by the thickest suspension of settleable flocs that can still be effectively fed with waste and $O_2$.

Another borderline is met when one allows $\mu$ to rise and accepts a considerable decline in effluent quality. With rising $\mu$, not only the amount of excess sludge increases if $R$ is kept constant but also the point may be reached where flocs will no longer settle or be formed, as the stability of the floc appears to depend on low $\mu$ and $s$ values. It is our lack of knowledge of the conditions under which stable growth in flocs is assured that makes this truly microbiological borderline rather vague. Indications that it can be overstepped are given by the well-known phenomenon of bulking sludge, a condition in which the sludge settles poorly, an occurrence provoked by a variety of incompletely known factors.

*Maximizing the BOD load* can be important when agricultural or industrial waste is available in a concentrated form and a minimum of dilution is desired. We then run into the technical borderline of aeration, as seen above. If one gives up the condition that settleable flocs must be produced, however, and it is desired to oxidize a thick waste of high BOD, the interesting possibility of working at higher temperatures (40–60°C) presents itself through the heat production of the bacterial population itself, as shown in the pilot experiments of Pöpel and Ohnmacht (1972). The advantage of higher growth rates and more effective killing of pathogens has to be balanced against the high energy requirements for stirring and aerating the thick suspension. Nevertheless, this possibility points to the advantages that can be gained from working at higher temperatures, which may be feasible where heated cooling waters can be combined with waste streams.

Minimization of aeration costs can be obtained by producing as much excess sludge as possible by maximizing $\mu x$, a common practice in the fermentation industry but in our case one limited by the borderlines mentioned above for increasing $\mu$. The higher sludge production shifts part of the purification burden from the aeration tank to the digester, which produces energy rather than consumes it. Again, the high-rate activated sludge plant is characterized by high sludge production and higher specific growth rate which makes this type of plant the compromise reached in practice on several counts.

*Minimization of sludge production* is generally preferred and practiced by taking $\mu$ as low as possible and feeding back as much of the flocs as the system permits. Theoretically, zero sludge production is possible thanks to the fact that the yield constant $Y$ is not constant but decreases at lower $\mu$ values, at which the maintenance requirements of the cells are no longer negligible, and in fact can use up all of the energy produced by substrate oxidation, leading to zero net growth (Stouthamer and Bettenhausen, 1973;

Veldkamp, 1975). Minimizing $\mu$ $(=D/R)$ is done by reducing $D$ and increasing $R$. The system producing the least amount of excess sludge is the oxidation ditch with a detention time of 3 days. With careful handling, the sludge production approaches zero, but this is in practice seldom realized. The sludge that is produced is highly "mineralized," i.e., contains very little endogenous reserve materials and, in contrast to sludges of high- and low-rate plants, can be dried on drying beds without risk of nuisance.

When, like in domestic sewage, nitrogen and phosphorus are in excess, it is not possible to uncouple growth and dissimilation except by the above means. When nitrogen and phosphorus are limiting factors, restricted dosing of them can assist in keeping the growth limited and make the cells behave as "resting cells," thus decreasing excess sludge production.

Various authors (Servizi and Bogan, 1963; McCarty, 1972) have calculated maximum possible cell yields on the basis of the energy characteristics of the substrate. Their results are valuable because they indicate maximum theoretical limits. So far, however, it is not possible to extrapolate our present meager knowledge on cell yield in pure cultures metabolizing simple substrates toward practical data for waste treatment systems. In their design, empirical experience from practice or pilot experiments is still the main basis for obtaining yield data.

*Minimization of effluent BOD* happily coincides with measures for minimization of $\mu$ and maximization of $R$ and hence is subject to the same considerations as those given for minimum sludge production.

*Minimization of risk* is practiced by using an overcapacity for tanks and pumps and keeping standby storage facilities in reserve against exceptional hazards, like peak loads of toxic materials or accidental oil discharges. Low-rate plants provide a comfortable reserve capacity in themselves, while increased recirculation can be used for diluting waste streams when necessary. In some industrial treatment plants, small batches of flocs are kept actively growing as potential seed material for rapidly starting up after a calamity.

It would appear from the above that technical developments have struck out in various directions and, in general, have come close to what is microbiologically feasible, although the exact definition of the borderlines in some instances leaves much to be desired. This holds in particular for the definition of the conditions delineating growth in flocs, although a great deal of work has been done to explore the causes of bulking sludge (Pipes, 1967). This defect has been associated in general with the predominance of filamentous bacteria or molds. The reason why the filaments cause the sludge to settle less well has not been clearly defined; it could be increased friction or reduced specific density, such as might be caused by gas vacuoles (Walsby, 1972; 1974). Filamentous flocs also may more easily retain gas bubbles, thereby providing buoyancy. The following organisms have been implicated: *Sphaerotilus natans, Geotrichum candidum, Bacillus* sp., *Zoophagus, Beggiatoa, Thiothrix, Nocardia*, members of the Vitreoscillaceae, and some types of coliform bacteria. Given the enormous diversity of the nutritional conditions these organisms require for optimum growth, it is not surprising that the factors leading to bulking sludge have been ascribed to overloading as well as underloading of the plant, presence of $H_2S$, low pH, and low phosphate contents.

A valuable consequence of work in this area is a much improved knowledge of filamentous bacteria (Mulder, 1964; Eikelboom, 1975; Dondero, 1975), but meanwhile the question remains as to whether the attention paid to bulking sludge and filamentous organisms has not been disproportionate in comparison with more fundamental studies of the ecological conditions required for growth in settleable flocs.

A further conclusion from the above inquiry is that the low growth rates generally applied point to a poor exploitation of the microbe's unique capacity for rapid growth and transformation. This is unavoidable as long as sedimentation has to be used for feedback of cells unless we learn to grow settleable flocs at high specific growth rates or, alternatively, develop means for cell separation other than sedimentation.

As far as temperature is concerned, we can safely assume that activated sludge treatment is in principle possible at any temperature between, for example, 5 and 50°C, provided the system has been designed for the desired temperature range. Clearly, transformation rates will be higher at elevated temperatures, and this is reflected in the shorter detention times applied in installations in the tropics as compared with those of the same installations in temperate climates. Oxidation ditches are designed in the Netherlands for detention times of about 3 days, while in India detention times of 0.5 to 1 day can be used. Although a rise in temperature does reduce the solubility of $O_2$, it also increases the rate of $O_2$ transfer from air to water and thus enhances the capacity of the aeration equipment. Temperature-associated changes of the floc population have not yet been clearly described.

Similarly, we may expect that plants can be operated at any pH within the range of 5 to 8 or even beyond that, in both directions. The appearance of molds and yeasts at lower pH values is well known. Also, not surprisingly, salinities ranging from that of fresh water to that of sea water permit activated sludge operation, making it possible to use sea water for waste transport where fresh water is scarce. In contrast, one-time peak loads of heat, acid or salts will throw the system out of balance and lead to periods of lowered performance until the system has recovered, and as a result, such shock loads, though not always avoidable, are a constant source of worry to the plant operator faced as he is with having to produce an inferior effluent.

The effects of toxic substances not added in peak loads but present as more or less constant constituents of the waste stream are in general much less severe than expected on the basis of laboratory data obtained for pure cultures in simple media. Some undegradable toxic substances like metal ions are easily rendered less harmful by partial precipitation, depending on pH, or by chelation, for which the great diversity of organic compounds present offers ample opportunity. Undegradable organic substances are less easily inactivated, but the effect of those that are biodegradable under the prevailing conditions should be judged on the basis of their steady-state concentration in the aeration tank and not on that in the incoming waste. The latter can be high enough so as to be in the toxic range, whereas the former may easily be harmless. It goes without saying that temperature, pH, and synergism all play a role that is superimposed on any inactivation that may be operative. Obviously, nonsevere toxic effects may be masked and go unnoticed if plant performance does not suffer when a selectively killed segment of the population is replaced by an equally effective insensitive subpopulation. A survey of maximum allowable concentrations in waste streams, used in practice for the protection of treatment plants, is offered by Liebmann (1960).

Apart from direct effects upon the oxidative capacity or the settleability of sludge, some substances like detergents may upset treatment through other mechanisms. Excessive foam formation in the aeration and second sedimentation basins marked the initial phase of the use of detergents, which at that time had not yet been designed to be easily biodegradable. The detergent molecules are easily absorbed at solid/liquid and gas/liquid interfaces. Foaming occurs more severely after aeration than before, because in the later stages of treatment fewer solids are available. Clogging of the air/water interface impairs aeration efficiency even when the foaming itself does not constitute a nuisance.

### 4.2.2. Activity of Population Segments

*4.2.2a. Nitrification.* Even though for obscure reasons the $O_2$ required for biological oxidation of ammonia is not conceptually included in the BOD, reduced nitrogen compounds do exert an $O_2$ demand in surface waters, just like organic compounds do. Hence, nitrification is usually included in the objectives of sewage treatment, transformation to nitrate offering the additional advantage of providing a good start for nitrogen removal by denitrification, should this be desired.

The realization that nitrification does not always automatically take place along with organic matter oxidation stimulated further study, in spite of the somewhat laborious cultivation technique the autotrophic nature of nitrifiers imposes. However, in view of their strict substrate specificity, nitrifying organisms can be singled out and studied as a separate group, for which substrate and product concentrations can be easily determined, and the problems of multisubstrate-mixed population cultures are much less severe. In addition, nitrification can be stopped by selective inhibitors like thiourea. Taking advantage of this situation, Downing and co-workers (1964) have been able to analyze nitrification dynamics to the extent that the ecological conditions required for the process to occur could be clearly defined.

Maximum specific growth rate, yield, and saturation constants of *Nitrosomonas* and *Nitrobacter* as affected by temperature, $O_2$ concentration, and pH were used to formulate the ranges of values within which process parameters had to be chosen to make nitrification possible. The low growth rate of the organisms and their requirement for relatively high $O_2$ concentrations of at least 0.5 mg $O_2$/liter were found to be the main causes for the precarious position of nitrification as a process component. The inhibitory effect formerly ascribed to the presence, of organic matter could be explained by taking into account that heterotrophic $O_2$ consumption can be so rapid as to create local $O_2$ deficiency for nitrification. From growth rate determinations, minimum aeration times could be calculated above which no washout of the nitrifiers would occur.

The harmony found to exist between theory and practice establishes nitrification as a classical example of an area in which microbial ecology has had a practical impact on the activated sludge process. A detailed review is offered by Focht and Chang (1975).

*4.2.2b. Role of Protozoa.* The recognizability, feeding habits, and growth characteristics of protozoa make them a population segment amenable to separate study. By means of extensive laboratory experiments including pilot-activated sludge plants with and without protozoa, Curds (1975) showed that the low number of freely suspended bacteria in the aeration tank is a result of predation by protozoa. Although various workers were inclined to ascribe to protozoa also a role in floc formation, it has now become clear that predation is their main function.

In view of the importance that must be attached to maintaining an effective protozoal population, an ecological delineation of the conditions for their persistence had become desirable. This need led Curds to develop a mathematical model to test the mechanisms he postulated for the predator–prey relationship on the basis of field observations and laboratory experiments. In a first approach, he divided the protozoa into those that are attached to the flocs and those that are freely suspended, both groups feeding on the freely suspended bacteria. Both these bacteria and the freely suspended protozoa must have a growth rate at least as high as $D$ in order to maintain themselves without recirculation, while the attached protozoa, like the flocs themselves, possess the lower growth rate $D/R$ as they are partially recycled. Assuming classical Monod relationships for the growth rate–substrate relationships for the bacteria as well as the protozoa and using realistic values for his constants, Curds showed with his model that the numbers of protozoa and freely suspended bacteria were in balance under standard plant conditions, while bad plant performance leading to turbid effluents could be correlated to imbalance between protozoa and bacteria in favor of the latter. Since attached protozoa were much less easily washed out, their role showed up as the larger one. Counting of cells leads to the estimate that protozoa constitute about 5% of the suspended material in the activated sludge culture liquid. If the steady-state mass of the freely suspended bacteria they feed on is of the same order of magnitude and one takes into account that the growth rate and hence also their specific metabolic turnover rate must be considerably higher than that of the floc, it appears likely that the contribution to the overall plant performance of the microbial population that forms no part of the flocs, in combination with the attached protozoa, amounts to more than 10%.

### 4.2.3. Interrelationships within the Population

Competition in the activated sludge aeration basin is much more complex than in a batch or continuous culture inoculated once with a mixed population. This is caused by the fact that the incoming settled waste continually carries in large numbers of organisms of varying identity that, depending on their predilections and susceptibility to inhibition and predation, will grow or die under the conditions they encounter. But even when they die, they may establish themselves as a permanent segment of the steady-state population when their input rate is balanced by the sum of death rate and discharge into the effluent. For freely suspended organisms, the equation reads

$$\frac{dx}{dt} = Dx_R - kx - Dx$$

in which $x$ represents the density of the species in question in the aeration tank and effluent and $k$ its specific death rate. For $dx/dt = 0$, a steady state is reached at which the fraction of remaining viable cells, $x/x_R$, amounts to $D/(k + D)$. Hence 50% remains if $k$ equals $D$.

A case in point is the pathogens and indicator organisms like coliform bacteria which occur in domestic sewage at levels of about $10^7$ cells/ml (Painter, 1971) and are reduced to 3–20% of the original value (Geldreich, 1972; Pike, 1975) by activated sludge treatment. Such results should not be viewed with too much optimism since even 1% of $10^7$ is still a high number, illustrating the need for chlorinating the effluent if the use of the receiving stream demands high hygienic standards (Edel et al., 1972). The specific death rate of such organisms is, of course, a composite of many different mechanisms, including the action of protozoa, phages, *Bdellovibrio*, and other aggressive organisms besides normal dying off by adverse conditions. Curds (1975) showed that about half of the 95% reduction in the numbers of *Escherichia coli* was due to the action of protozoa. The relative contribution of each of the other mechanisms is less clear. In the case of viruses, adsorption to the floc is a further means of removal (Grabow, 1968).

Although the competitive situation in the aeration tank will be largely determined by the $K_s$ and $\mu_{max}$ characteristics of the individual strains, this does not mean that all organisms will have the same growth rate. We have already discussed the different competitive situation for organisms in the floc and those freely suspended. In addition, growth rates even higher than those imposed by $D$ and $D/R$ have to occur when organisms are at the same time growing and subject to predation.

The fact that most waste streams constitute multiple substrates has led to studies trying to connect the events in aeration tanks with known behavior of pure cultures towards mixed substrates of known composition. The latter may show simultaneous or sequential substrate utilization, depending upon the organism and the substrates involved. In some laboratory experiments (Gaudy et al., 1963; Stumm-Zollinger, 1966), activated sludge and other mixed populations also could be made to show sequential substrate consumption of the classical diauxic type, indicating that some substrates inhibit the utilization of others. Such experiments are only possible in batch cultures where the decline of substrates can be followed in time and thus the initial concentrations have to be substantial. Since the actual process proceeds in a continuous culture system in which the substrate concentrations are as low as those in the effluent, the element of time is eliminated, and the concept of sequential utilization becomes meaningless. If the composition of the medium would offer the possibility of one substrate inhibiting the consumption of another, this would either show up in a peak concentration of the latter substrate in the effluent or, more likely, it would be circumvented by the selection of an organism not susceptible to this particular inhibition. Moreover, the selective pressure of the process gives an advantage to multivorous organisms, i.e., organisms capable of simultaneous dissimilation of various carbon sources in such a way that their cumulative concentrations constitute the limiting concentration operative in the Monod equation: If each of the many carbon sources present would be degraded by its own special organism, the concentration of each of the sources would be no further reduced than to the value demanded by the Monod equation for the prevailing $\mu$ and the $K_s$ and $\mu_{max}$ values of the organism in question. As a result, the total concentration of carbon sources in the effluent would be rather high. Such selection for multivorous organisms was postulated in 1969 by la Rivière (1971), and experimental evidence supporting this was provided by Silver and Mateles (1969) for *E. coli* growing on lactose–glucose mixtures. Mutants constitutive for β-galactosidase overgrew the parent strain in which the enzyme was inducible. These mutants could simultaneously utilize lactose and glucose at concentrations at which the parent strain was not capable of doing so. For further elucidation of the dynamics of selection in the activated sludge process, detailed analysis of the organic components of the effluents would yield important clues, since the concentration of each of them reflects singly, or in combination with others, a growth-limiting factor for a segment of the operative population.

Simultaneous oxidation of more than one substrate by one organism can be performed also by the mechanism of cooxidation, in which case the secondary substrate has no energetic feedback to the cell that degrades it. This has been well documented in many instances (Horvath, 1972), but actual proof that cooxidation occurs in sewage treatment has not yet emerged.

While several substrates thus may be degraded by one and the same organism, single substrates may also require concerted action of combinations of two or more organisms in order to be degraded. It is, of course, not implausible that a chain of consecutive enzymes required for the degradation of a compound foreign to nature is not easily found in one single organism and that it is more likely that segments of the chain can be found in various different organisms which then may finish the job together. The relationships between the team partners may range from completely loose to fully interdependent, but selective pressure is likely to favor the latter since a close association of two organisms must possess many enzyme systems in duplicate, part of which becomes redundant as the association persists. Loss of a nonessential enzyme may lead to better cell economy and, as shown by Zamenhof and Zamenhof (1971), to higher growth rates at one and the same limiting substrate concentration, thus giving an advantage in competition.

The degradation of dimethylformamide might well be indicative of such interaction: A biotin and thiamin deficient isolate from an industrial waste treatment plant was found to be capable of degrading this compound completely, its cleavage into dimethylamine and formic acid being the first step. In a vitamin-free medium, degradation and growth only took place when to the isolate was added a *Protaminobacter* sp. which could grow on dimethylamine but not on dimethylformamide, the *Protaminobacter* serving as a vitamin source to the isolate and its presence permitting both to grow in a medium that supported growth of neither one separately (la Rivière, unpublished).

Even when degradable, synthetic or unusual chemicals may be expected to persist when their concentration is too low for effective induction or for selection of potential degraders. In general, however, we should recognize that the evident success of the activated sludge treatment is based on the ecological law that every niche should be filled, and it appears that study of effluents, as mentioned earlier, could help us find out what niches are left unused and why this came about.

## 5. Microbial Ecology of Trickling Filters

Much of the knowledge assembled about the activated sludge process also holds for the trickling filter. The latter has been studied less intensively, and this is understandable in view of its heterogeneity both as to medium composition and microbial composition of the film, which change with depth. This heterogeneity also renders mathematical modeling more difficult.

When the filter is not covered, which is only sometimes done to prevent freezing in winter, algae grow in the top layer and may there contribute to local clogging without assisting the overall purification effect of the system. According to Benson-Evans and Williams (1975), the following genera have been encountered most frequently: *Chlorella*, *Chlorococcum*, *Phormidium*, *Oscillatoria*, *Stigeoclonium*, and *Ulothrix* as well as diatoms.

The bacteria found in trickling filters are mainly aerobic heterotrophs, and their species distribution shows a great similarity to that of the activated sludge population.

The same holds for the protozoa, although here a somewhat higher diversity at the class level is found than in activated sludge. A detailed review has been presented by Curds (1975).

Fungi are generally a nuisance, as their loose growth tends to increase the risk of clogging of the filter. Tomlinson and Williams (1975) exhaustively reviewed the factors that selectively promote fungal growth in the filter, pH and waste composition being the most important ones. Since the oxidative powers of fungi are certainly no less than those of bacteria, one wonders if filters designed for fungal films would not be most suitable for industrial wastes that by their composition and pH would tend to be enriching for them.

Higher organisms like nematodes, rotifers, and annelids play a greater role in the trickling filter process than in the activated sludge process because they are important in the sloughing off of the film, having only a minor impact with regard to predation and concomitant oxidative activity. Detailed reviews have been presented by Schiemer (1975), Doohan (1975), and Solbé (1975).

The heterogeneity of the microbial film demonstrates itself vertically downwards in the reactor for the film as a whole, but also for each film element by itself in a direction perpendicular to the surface of the solid medium to which the film adheres. The thickness of the film can amount to 2 mm, and hence a gradient manifests itself in which concentrations of $O_2$, minerals, and carbon and energy sources diminish from outside toward the substratum surface. Corresponding changes in population are therefore to be expected. Hawkes (1963) concluded from practical experience that the optimum film thickness is 0.1–0.2 mm, which coincides with the thickness calculated by Wuhrmann (1964) to allow adequate $O_2$ supply to all bacteria in the matrix. Mack and co-workers (1975) studied the development of the film during the first 5 days of the ripening process in the top 30 cm of an operating trickling filter. It is evident from their scanning electron micrographs that the film has a rough surface marked by various types of colony-like structures of bacteria. Near the surface, many of these consisted of hollow cylinders with a slit pointing all in the same direction, presumably formed by water draining out from the hollow core as a consequence of the intermittent dosing. At the 30 cm depth, more conventionally shaped colonies were found.

Their pictures of thin sections also tend to confirm that the bacteria grow *in situ* within a matrix produced by themselves, making it most unlikely that the film would be a result of physico-chemical aggregation of initially freely suspended bacteria. Since ripening of the film takes a few months, it is to be hoped that extension of these studies will ultimately include fully ripened films as well as films throughout the entire depth of the reactor.

Theoretical considerations indicate that at a steady state of the filter at each depth, more or less constant conditions prevail in the medium that in turn have determined a more or less constant population in the film by selective pressure. This pressure tends to select, of course, for organisms that have the best survival capability under the given conditions, i.e., the lowest $K_s$ and/or highest $\mu_{max}$ for the prevailing limiting nutrient which also in this case, at least for domestic sewage, is the carbon and energy source. Another selective pressure is the hydraulic flow, which rigorously flushes away organisms incapable of embedding themselves firmly in the film. Thus, the selection for matrix-forming organisms holds for both trickling filters and the activated sludge process. As the organic matter concentration decreases with depth, the growth rate of film organisms must also decrease; hence, "ripening" of the filter must take more time for the deeper layers. Other constituents like $NH_3$ may increase in concentration with depth as they become liberated through hydrolysis in the upper layers. At least in theory, similar considerations could be applied to the gradient within the film itself, but in view of the much smaller space dimension and the fact that each segment in the film has once been on the outside of the film, the building up of the composition gradient in the film itself is necessarily a much more complex and intractable process.

The overriding question from the ecological point of view appears to be what relationship exists between specific degrading capabilities and determinants of the outcome of competition, i.e., $K_s$ and $\mu_{max}$. Very little work has been done on this question, but again the nitrifiers offer a clear-cut example. As expected, nitrification occurs more strongly in the lower layers of the filter (Pike, 1975), the low growth rate of the nitrifiers causing them to be outgrown in the upper layers. It would seem that a detailed "finger printing" analysis of the sewage at successive points during its passage through the filter would yield interesting data as to the sequence by which groups of organic compounds are being dissimilated. From these data, perhaps conclusions could be drawn about the successive location of the operative organisms. In contrast to the activated sludge, in filters ample scope is available for sequential, polyauxic degradation.

As in the activated sludge process, also here the entire spectrum from actively growing cells to "resting cells" to dead cells must be present leading to "immobilized" cell and enzyme activity superimposed upon that of the normal, active cells. Since the organic matter/minerals ratio for "humus" is smaller than for activated sludge, it must be concluded that exhaustion of endogenous organic matter occurs more intensely in the filters.

The sloughing off of film is very important for the practical operation of the reactor since it is this process that turns the theoretically closed nature of the system into an open one in practice, permitting long-term operation without cleaning. The rate of humus production should equal the net rate of cell growth. Accordingly, film growth rate and humus production are higher in high-rate than in low-rate trickling filters. Various mechanisms for the sloughing off process can be envisaged: (1) mechanical scouring by the flow of water, which begins to become effective when the film gets so thick that internal coherence no longer stands up to hydraulic shearing forces, which in turn are affected by free space in the filter; (2) weakening of coherence by gas-producing fermentation processes in the film which start when the film is thick enough to create anaerobic conditions permitting their initiation; (3) action by predators ranging from protozoa to larger grazers. By mere mechanical disturbance, these predators may cause film fragments to become loose. Although it is at present not clear which of these mechanisms plays the dominant role, it is quite likely that all three of them contribute. Since the performance of the filter suffers when removal of the film is either too fast or too slow, the art of designing a good trickling filter appears to hinge upon achieving a proper balance between film growth and disposal of excess film. It is likely that cohesion in the film plays the role of an important self-regulatory mechanism.

Furthermore, a considerable bacterial population must be present in the flowing liquid phase itself, a population arising by growth as well as release from the film, and these bacteria obviously cannot but contribute to the purification process. In this respect, it is of interest to realize that the time for a single input of chloride to pass through a filter is of the order of a few hours (Hawkes, 1963), indicating the existence of a considerable quantity of adhering water with which exchange takes place. Also here protozoa play an important role in keeping the freely suspended population within reasonable bounds so as to make delivery of a clear effluent possible. In contrast to the activated sludge process, the humus, as a result of its greater density never leads to sedimentation problems.

Although the film mass is much more compact than activated sludge, which contains about 20 times more water, the biomass per unit of BOD treated is about the same for each process (Pike, 1975).

Manipulation of the process is less easy than in the activated sludge system because neither aeration intensity nor feedback of cells can be changed. For this reason, often so-called alternating double filtration is used, in which two filters operate in series with periodical change of sequence. Obviously, the first filter receiving the most concentrated medium has the highest film growth; when this filter is switched to second position, the well-developed film is exposed to much lower nutrient concentrations and is then more effective than a film that would have been created and maintained by that low concentration. Also hydraulic recirculation is an important operational instrument for regulating the BOD load as well as maintaining the hydraulic load required for containing film thickness.

Changes of temperature, loading, and other process parameters would be expected to change the entire successive populations in the film, but the sessile organisms are not immediately washed out like in the activated sludge system. Thus the trickling filter is more resistant to minor disturbances in pH, temperature, nutrient concentration, and presence of toxic materials than the activated sludge process, but severe shocks that wipe out important segments of the film population can only be overcome by renewed "ripening," which takes longer than the recovery of an activated sludge plant subjected to the same shock.

While our insight into the processes going on in trickling filters is undoubtedly more limited than in the case of activated sludge systems, it would appear that the exploitation of the ecosystem offered by the trickling filter film reactor has reached the limits inherent in this system and that only in minor respects are improvements possible, such as utilization of waste heat and application of new filter materials. Also, the prospects for useful innovation by mathematical modeling appear small. We should realize that the considerations given above are qualitative and offer little basis for quantitative treatment, if only for the blurring effect that intermittent dosing as well as fluctuating loads must have on the distribution of different organisms over the film at various depths.

# 6. Microbial Ecology of Anaerobic Digestion

Like aerobic treatment, anaerobic digestion is no more than technological intensification of a natural process. Marshes and the bottoms of streams and lakes often produce copious amounts of $CH_4$ when organic matter has accumulated there and $O_2$, nitrate, and sulfate have been exhausted as electron acceptors. The study of the process has been retarded by the experimental difficulties presented by obligate anaerobes and by the fact that the ecological niche for $CH_4$ formation is almost entirely restricted to heterogeneous solid/liquid mixtures rather than more tractable solutions, a state of affairs caused by nature's most important force for concentrating organic matter, i.e., gravity, which in the aqueous environment leads to sedimentation. Thus, solid and not immediately dissolving fragments of organic matter become the main substrate for anaerobic digestion in nature.

Although we have had ample evidence showing that like any other microbial transformation, $CH_4$ production can proceed equally well or even better in properly constituted solutions, so far anaerobic digestion has been mainly applied in the treatment of sludge. This has also had a restrictive effect on the scope of studies in this area, which makes it likely that the full potential and flexibility of the process is at the moment only partially understood.

## 6.1. Microbial Populations Involved in Anaerobic Digestion

The initial substrate for anaerobic digestion consists mainly of macromolecular substances like polysaccharides, proteins, and fats, fully in agreement with their particulate and insoluble status which led to their concentration. The main ultimate products—$CH_4$ and $CO_2$—are formed by the true methanogenic bacteria which, as far as we now know (Wolfe, 1971), can only use one-carbon compounds, acetate, and $H_2$ as substrates. Hence we can separate the operative population into methanogenic and nonmethanogenic segments (Toerien et al., 1969), the latter taking care of hydrolysis and degrading the resulting compounds to acetate, one-carbon compounds, and $H_2$. Since hydrolysis by itself does not furnish enough energy for growth, it appears impracticable to distinguish a special hydrolytic population segment, as is done by some authors. We can best subdivide the nonmethanogenic population into two parts: (1) those transforming the initial polymers to fermentation products that are not substrates for methanogenesis, like butyrate, propionate, lactate, succinate, and ethanol; and (2) those further transforming the products of group 1 to methanogenic substrates. Of course, part of the products of group 1, like acetate, may already be a suitable methanogenic substrate.

Thus, in contrast to aerobic treatment, anaerobic digestion heavily depends on a successive, coordinated interaction of different specific populations and hence on the efficiency of internal regulation mechanisms. It is, therefore, not surprising that one of the main disturbances in practical performance is a disruption between the relative conversion rates of the two groups, manifesting itself in excessive acid production to lower the pH to less than 7; this slows down the methanogenic bacteria and that in turn promotes further acid formation, bringing the process to a standstill unless prevented by timely application of lime.

The multiple-stage nature of the process is further illustrated by the separate occurrence of the nonmethanogenic part in the digestive tract of herbivores, where 90% of the end-products are channeled into animal metabolism and 10% to methane production. This provides about 200 liters of methane per cow per day (Wolfe, 1971).

Counting the viable bacteria in anaerobic digesters is even more difficult than in aerobic systems because of the strict anaerobiosis that must be maintained and the variety of media required to represent the stages in the degradative chain. Fortunately, the Hungate (1950) technique, invented for rumen studies, has helped to surmount many of the problems. Another obstacle was encountered when attempts were made to identify the wealth of obligate anaerobes obtained, and existing keys were found to be inadequate. This led Toerien and Kotzé (1970) to develop special computer-based techniques for ordering some hundred of their isolates in physiological groups to which ecological significance could be attached. In this way, it should become possible to follow changes in the population in relation to modification of system parameters and also to compare populations of different digesters. To what extent this can be usefully done depends on the ease with which the different physiological groups can be selectively cultivated and counted.

From the reviews by Toerien and co-workers (1969) and by Crowther and Harkness (1975), the following picture of the population emerges: Total viable counts of nonmethanogenic obligate anaerobes range from $10^8$–$10^{10}$/ml. In terms of orthodox taxonomy, the genera *Clostridium*, *Bacteroides*, and *Bifidobacterium* take a prominent position among many unidentified organisms. Facultative anaerobes, though consistently present and for a long time considered the major group of nonmethanogenic bacteria, must be assigned a minor role, possibly as scavengers of traces of $O_2$. While many of the isolates were shown to produce the usual array of fermentation products from sugars, including $H_2$, it must be concluded that at least one physiological group remains to be found and defined because the very narrow substrate specificity of the methanogenic bacteria demands transformation of the nonmethanogenic substrates into the methanogenic ones.

Total viable counts of methanogenic bacteria range from $10^5$–$10^{10}$/ml, but as in the case of the nonmethanogenic bacteria, there is no certainty that all viable cells present were also successfully cultivated to give visible colonies.

According to Wolfe (1971), the following methanogenic bacteria have so far been isolated in pure culture from enrichments inoculated with digester sludge: *Methanobacterium ruminantium*, *Methanobacterium* strain M.o.H., *Methanobacterium formicicum*, *Methanosarcina barkeri*, *Methanospirillum* sp., and *Methanococcus* sp. They are all capable of utilizing $H_2$ and carbon dioxide, and some can use in addition formate, methanol, or acetate. Zeikus and Wolfe (1972) further reported the isolation from a digester of *Methanobacterium thermoautotrophicus*, which has an optimum temperature of 65–70°C. All of these bacteria can use sulfide and ammonia as sources of sulfur and nitrogen, respectively, and some but not all require growth factors.

*Methanobacterium* strain M.o.H was found by Bryant *et al.* (1967) to be associated with a nonmethanogenic short rod, called S organism, which formed acetate and $H_2$ from ethanol. The association had been considered a pure culture under the name *Methanobacillus omelianskii* since its isolation in 1940. The symbiosis proved to be based on the inhibitory effect of $H_2$ produced by the S organism and the exclusive dependence of *Methanobacterium* strain M.o.H on $H_2$ for methane formation from carbon dioxide. This clearly illustrates the difficulties besetting the isolation and study of both the methanogenic bacteria and the special physiological group responsible for preparing their limited diet. Which of the methanogenic bacteria predominate(s) in digestion is not yet known, nor can we exclude the possibility that a part of the responsible population so far has escaped isolation.

For obvious reasons, fungi and algae do not play any significant role in anaerobic digesters. In his review, Curds (1975) lists 63 species of protozoa that are found in digesters, including some obligate anaerobes. Their role, if any, has not been determined.

For a full understanding of the coordinated interactions of the three groups of bacteria involved, it would be necessary to know the constants $Y$, $K_s$, and $\mu_{max}$ for the operative organisms as well as the steady-state concentrations of the intermediate substrates functioning as limiting factors, inhibitors, or stimulators of growth.

Overall cell yields are, of course, much lower than for aerobic organisms. For the methanogenic bacteria, they amount to 0.04–0.05 while $\mu_{max}$ is also ranging from 0.01 to 0.02 $hr^{-1}$ (Lawrence and McCarty, 1969). But $\mu_{max}$ is as high as 0.14 $hr^{-1}$ for the thermophilic organism (Zeikus and Wolfe, 1972). For the nonmethanogenic bacteria, one would also expect low cell yields but high $\mu_{max}$ values, in the range of 0.1–0.5 $hr^{-1}$. These differences in potential growth rates can only be reconciled in an integrated, steady-state process by appropriately balanced cell masses and limiting substrate concentrations. The hydrolysis of insoluble initial substrates also may play a role. Since it is known that the system can be disturbed by rising acid concentrations, one wonders why this is not the rule rather than the exception. A possible answer to this question is suggested by Wolfe (1971): If the withdrawal of $H_2$ by the methanogenic bacteria would in general promote the transformation of nonmethanogenic substrates into acetate and $H_2$, as shown for the $M.$ $omelianskii$ association, the $H_2$ concentration itself would constitute an important regulatory factor. In this respect, it would be important to know how much $CH_4$ is generated by $CO_2$ reduction by $H_2$ and from other substrates. Various authors have estimated that 70% of the $CH_4$ originates from acetate and as much as 30% through reductive action by $H_2$ (cf. Toerien et al., 1969); furthermore, selective inhibition experiments with $CH_4$ analogs like chloroform have shown that, in the absence of $CH_4$ formation, much more $H_2$ is produced.

In addition to the interactions discussed above, most of the interactive mechanisms that also occur in aerobic systems must play a role in anaerobic digestion, but in view of our restricted knowledge, it is too early to analyze them.

### 6.2. Flexibility of the Process

For a long time, sanitary engineers (Imhoff et al., 1971) recognized two temperature optima for anaerobic digestion, around 35°C and around 55°C, but this dogma was shaken by the isolation of $M.$ $thermoautotrophicus$ with its optimum between 65–70°C. The pH range for the process is usually considered to be 6.4–8, but the formation of $CH_4$ in peat bogs at pH 4 still invites attempts to isolate acid-tolerant species (Barker, 1956). To ensure stability of the process, a buffering capacity equivalent to more than 1 g of bicarbonate ion per liter is recommended (Toerien et al., 1969).

Salt-tolerant populations are operative in $CH_4$ formation in marine sediments and can function in laboratory enrichments at NaCl concentrations of at least 5% (la Rivière, unpublished).

Anaerobic digestion is sensitive to various groups of toxic substances including heavy metals, detergents, chlorinated hydrocarbons, and also hydrogen sulfide. The pH and the presence of complexing agents and of ions capable of precipitating the toxic entity strongly affect the admissible loads of metals and, of course, also of sulfate, which very quickly turns into sulfide.

In comparison to the aerobic processes, the potential for improving the exploitation of anaerobic digestion appears very large indeed as the ecological boundaries of the anaerobic system have been far from reached in the existing technology. The renewed interest in energy economy has shifted a great deal of applied research capacity toward innovation of anaerobic digestion, which almost seems to have been rediscovered. Unfortunately, much of the results obtained remain commercial property not immediately available to the scientific literature.

One of the main avenues of improvement is lowering detention times from the current 10–50 days to just a few days, which even the low growth rates of the methanogenic bacteria would permit. Obvious measures to help achieve this are elevated digestion temperatures, continuous feeding, and discharge of the reactor and full mixing. Lawrence and McCarty (1969) have shown in pilot experiments at 35°C that close to 100% substrate removal can be obtained for dissolved fatty acids at detention times of less than 5 days in continuously fed, fully mixed reactors without feedback of cells.

Another promising approach is afforded by increasing cell detention time with respect to hydraulic detention time, that is, using anaerobic film reactors or anaerobic activated sludge systems. Pilot experiments (El-Shafie and Bloodgood, 1973; Lettinga and van Velsen, 1974) have shown that it is possible to obtain films as well as settleable flocs. Work in this area is discouraged by the long time it takes for building up such populations, but the proposition is particularly interesting because anaerobic processes have low cell yields and low nutrient requirements and also can be expected to operate at much higher cell densities than aerobic ones, as there is no need to step up aeration intensity accordingly. Cell retention or recirculation is, of course, only feasible for dissolved organic matter but would be most useful in treating relatively dilute streams like settled domestic sewage. One could even envisage a gradual shift from aerobic to anaerobic treatment as the main step followed by an aerobic treatment step as a polishing operation.

Two-stage operation has been often considered a promising means of optimizing the nonmethanogenic and the methanogenic steps separately to a level not possible for the traditional combined system. Because of our lack of knowledge of the regulatory mechanisms between the two, it is only possible to recommend experiments in this direction, if for no other reason than learning more about the regulation.

The advantages of anaerobic treatment are now leading to a further diversification as to the kind of waste treated, domestic sludges having been for a long time the traditional substrate. Other than hydrocarbons, very few organic compounds are refractory to digestion, and as a result several agricultural and industrial wastes are now being digested. Wastes of fermentation industries, slaughter houses, and especially feedlots are examples, and many others are being tested in pilot plants, as the economic climate not only favors energy production but also welcomes the fertilizer constituted by the digested sludge. Furthermore, anaerobic digestion plays an important role in studies of biological capture of solar energy as one of the means to transform plant or algal photosynthate into a transportable form of energy, a transformation which has the advantage of returning at the place of production of the photosynthate (rather than of its use) the minerals required for its formation, making nutrient recycling possible (Oswald, 1973).

In this respect, the several thousands of gobar-gas plants (gobar refers to cow dung) set up in the rural areas of India and other Asian countries constitute a most interesting development. These small $CH_4$ generators, the smallest type of which is designed for dealing with the waste of one family and five cows, are now being tested as an alternative to rural electric energy supply. Cow dung has already served for a long time as a fuel to rural populations, but the practice was highly wasteful in energy efficiency and conservation of fertilizer. Transformation into $CH_4$ with concomitant return of the fermented sludge to the land has proved attractive enough for generating large government schemes that promote the installation of these simple digesters and include the provision of loans to the farmers. The economical analysis by Prasad et al. (1974) of this enterprise and its potential include the possibility of energy farming by way of the water hyacinth.

## 7. Microbial Ecology of Oxidation Ponds

The oxidation pond system is more complex and less amenable to control than the previously discussed systems. It combines anaerobic digestion and oxidative mineralization with photosynthesis, without being optimized for any one of these processes. The oxidation pond thus rather closely resembles a natural, heavily eutrophic, aqueous system. It is strongly dependent on climate and/or weather fluctuation, causing variation in temperature, insolation, and also barometric pressure, which affects $O_2$ release from supersaturated surface layers. Superimposed upon these are diurnal fluctuations of $O_2$ concentration and pH, which are inherent to photosynthesis in aquatic ecosystems. Selective pressure is weak and sluggish and is mainly confined to that of waste composition and concentration in combination with the loading parameters of the pond. Whatever control measures the design provides are geared to minimize effluent BOD, mainly through maximizing the transformation of waste organic matter into algal cells.

Obviously, there are very few bacteria that cannot be found in oxidation ponds, and because of the varying heterogeneity of the system, there is little merit in attempting a descriptive analysis of the bacterial population. It can be noted, however, that special wastes and operational conditions may narrow and strengthen selective pressures so as to give rise to consistent enrichment of sometimes conspicuous groups of bacteria such as Thiorhodaceae; this has been reported by various authors. These bacteria appear below the algal layer in the top of the anaerobic zone if sufficient red light can penetrate there and $H_2S$ is available (Pike, 1975).

Very little work has been done on the occurrence and possible role of fungi, protozoa, and higher organisms, which undoubtedly are all part of the population. Understandably, most work has been concentrated on algae. In their review, Benson-Evans and Williams (1975) listed 138 algal species representing 56 genera in oxidation ponds. They singled out *Chlorella, Scenedesmus, Chlamydomonas*, and *Euglena* as the most constant and cosmopolitan genera.

The work so far done has helped to identify some of the major problems but has not yet produced conclusive solutions. The most important problems are: (1) the determinants of the composition of the algal population; (2) the relative contributions of heterotrophic algal metabolism as compared with photosynthetic growth; (3) the role of growth-promoting and inhibitory substances excreted by algae; (4) the mechanisms for (auto)degradation of algal populations (Shilo, 1975); and (5) the mechanisms determining vertical distribution, such as phototaxis, possesion of gas vacuoles, and chemotaxis.

The insufficient knowledge in these areas does not greatly hamper the effective use of simple oxidation ponds operated empirically. This lack of knowledge is painfully felt, however, when the objectives of the oxidation pond are narrowed down to higher efficiency, when available space is a problem, or when the objectives are envisaged to extend towards the production of an easily harvestable alga that is suitable as fodder, a food component or a substrate for $CH_4$ production. Although we can say that existing technology has far from reached the potential the ecosystem offers, the boundaries to which technology can be stretched cannot now be assessed on the basis of existing knowledge. This is in clear contrast to anaerobic digestion, where technology lags behind our information; in the case of oxidation ponds, technology pushes ahead in advance of basic ecological knowledge.

This is illustrated by the production of 1 ton dry wt. of *Spirulina* per day by the firm Sosa Texcoco in Mexico (personal communication). In this operation, the alga is continuously filtered off from part of a natural, alkaline salt lake, which is then replenished with nutrient minerals. Stanton (1977) applied for a patent for production of *Chlorella* grown on rubber wastes. Oswald (1973) and co-workers conducted successful experiments on $CH_4$ generation from algae grown on sewage; they obtained 4–5% light conversion efficiency in the photosynthetic step and 65% conversion efficiency in the transformation of this energy to combustible gas.

It would appear that the greatest single obstacle to the success and expansion of such studies resides in the simple fact that, with few exceptions, we are not able to select for a given desirable algal population. Preoccupation with the dazzling algal morphology and subcellular physiology should not make algologists neglect their responsibility for expanding the ecological knowledge of algal populations, particularly with respect to competition. Studies of the physiological basis for persistent, specific algal blooms and of the physiology of desirable algae in batch and continuous culture, in pure state as well as mixed with competitors are obvious approaches; in addition, it would be interesting to see what harvest the classical enrichment method would yield in illuminated cultures on specific media, including various kinds of common wastes.

Susceptibility to selective pressure applied by cheap methods will, of course, not automatically coincide with desirable properties like rapid growth, high yield, high protein content, good digestibility, and a morphology conducive to cheap harvesting methods. Present experience, however, is as yet so limited that experiments in this area appear more than justified.

## 8. Conclusions and Perspectives

Biological waste treatment plants are man-made microbial ecosystems in which the transformation flux is much higher than in any natural or agricultural ecosystem. An ordinary activated sludge treatment plant, for instance, converts daily 1–2 kg of dry organic matter/m$^2$ of its aeration basin; these figures are a few hundred times higher than the mineralization occurring in a highly productive forest, which produces and mineralizes about the same amount in 1 yr. Although from this point of view, a successful, concentrated mineralization has been achieved, the methods are wasteful with respect to mineral, energy, and water economy. Notwithstanding their empirical origin, the oxidative systems appear to have reached a degree of efficiency that is only subject to marginal improvement within the limited scope of mere BOD removal. In contrast, the anaerobic and photosynthetic methods have far from attained their full potential and at the same time hold much promise for energy and mineral recovery.

As far as the *aerobic methods* are concerned, the activated sludge process poses the unresolved riddle of the floc: Is it only at the expense of very low growth rates that bacteria are willing to perform the trick of growing in flocs in the absence of excess carbon? It appears that the answer to this question is more important to further deployment of the method than developing kinetic models, which after all do not offer much more than an alternative, albeit more rational, description of the observed performance of existing systems.

Changes, both in waste concentration and in objectives, are calling for adaptation of the processes: When water will be less luxuriously used for transport, higher BOD loads will provide the opportunity for working at higher temperatures generated by the process itself, leading to variations almost approaching composting, like "solid-state fermentation" of feedlot wastes (Hesseltine, 1972; Hrubant, 1975); when anaerobic methods will find greater application for main treatment, special oxidative polishing systems will be required for which the trickling filter is a reliable candidate; the rising value of sludge as fodder and fertilizer may call for optimization of cell yield; decentralization of waste treatment under the pressure of rising standards for rural sanitation is creating the need for small-scale household units of refrigerator size, now tentatively approaching the markets.

The *anaerobic digestion* method is in even greater need of improvement and innovation, but this need is fortunately matched by an ample margin of unexploited potential. As stressed already, good mixing, continuous feeding, feedback of cells, working at elevated temperatures, adaptation to dissolved waste flows and, possibly, multiple-step operation are means for increasing scope and efficiency.

The small-scale gobar-gas plants already seem to fulfill a need for decentralized energy generation as well as fertilizer production, but the incorporation of anaerobic digestion in schemes for producing energy from renewable resources offers even wider prospects. This cannot but emphasize the need for a better understanding of the ecological balance between the three interacting bacterial populations, of which the one responsible for formation of the methanogenic substrate is still very poorly known.

While least developed of all, the *photosynthetic method* offers the greatest scope for novel application, particularly in sunny climates, where the production of algae can be used at various scales to help alleviate shortages of fertilizer, fodder, energy, and food. Especially in Asia, there exists already a traditional infrastructure in which sewage farming, on land as well as in fish ponds, and collection of night soil (that is, undiluted human waste) by bucket instead of by water are accepted practices (Soemarwoto, 1975). A further evolution consisting of some innovation combined with rationalization of existing methods could lead here to integrated rural recycling systems of great efficiency. The oxidation pond would be an essential element in such recycling. Methods for sustained, selective growth of economically important algae may, however, constitute an important bottleneck.

Since the existing treatment systems were not originally designed on the basis of microbiological engineering principles, the *microbial ecologist* has almost been forced to approach them in his inquiry as if they were natural ecosystems. It now seems likely that the traditional systems are ready for a burst of rapid evolution and diversification, even before the microbial ecologist has fully caught up with them. At the same time, it is clear that such evolution, in order to be successful, must be based on sound knowledge of the various microbial ecosystems to be used as instruments that can be plugged in at strategic positions in various infrastructures of environmental conservation, geared also to a maximum utilization of renewable resources by recycling. This provides microbial ecology in a felicitous manner with a wealth of tasks, not only scientifically interesting, but also directly corresponding to urgent societal needs.

# References

Arceivala, S. J., Lakshminarayana, J. S. S., Alagarsamy, S. R., and Sastry, C. A., 1970, *Waste Stabilisation Ponds*, Central Public Health Engineering Research Institute, Nagpur.

Barker, H. A., 1956, *Bacterial Fermentations*, John Wiley and Sons, New York.

Benson-Evans, K., and Williams, P. F., 1975, Algae and bryophytes, in: *Ecological Aspects of Used-Water Treatment*, Vol. I (C. R. Curds and H. A. Hawkes, eds.), pp. 153–202, Academic Press, London.

Bond, R. G., Straub, C. P., and Prober, R. (eds.), 1974, *Handbook of Environmental Control*, Vol. IV, CRC Press, Cleveland.

Bryant, M. P., Wolin, E. A., Wolin, M. J., and Wolfe, R. S., 1967, *Methanobacillus omelianskii*, a symbiotic association of two species of bacteria, *Arch. Mikrobiol.* **59**:20.

Chen, K. Y., and Morris, J. C., 1972, Kinetics of oxidation of aqueous sulfide by $O_2$, *Environ. Sci. Technol.* **6**:529.

Cooke, W. B., 1963, A laboratory guide to fungi in polluted waters, sewage and sewage treatment systems, Public Health Service Publication No 999-WP-1, U.S. Department of Health, Education and Welfare, Cincinnatti.

Crowther, R. F., and Harkness, N., 1975, Anaerobic bacteria, in: *Ecological Aspects of Used-Water Treatment*, Vol. I (C. R. Curds and H. A. Hawkes, eds.), pp. 65–91, Academic Press, London.

Curds, C. R., 1975, Protozoa, in: *Ecological Aspects of Used-Water Treatment*, Vol. I (C. R. Curds and H. A. Hawkes, eds.), pp. 203–268, Academic Press, London.

Curds, C. R., and Hawkes, H. A. (eds.), 1975, *Ecological Aspects of Used-Water Treatment*, Vol. I, *The Organisms and Their Ecology*, Academic Press, London.

Deinema, M. H., and Zevenhuizen, L. P. T. M., 1971, Formation of cellulose fibrils by gram-negative bacteria and their role in bacterial flocculation, *Arch. Mikrobiol.* **78**:42.

Dondero, N. C., 1975, The *Sphaerotilus-Leptothrix* group, *Annu. Rev. Microbiol.* **29**:407.

Doohan, M., 1975, Rotifera, in: *Ecological Aspects of Used-Water Treatment*, Vol. I (C. R. Curds and H. A. Hawkes, eds.), pp. 289–304, Academic Press, London.

Downing, A. L., Painter, H. A., and Knowles, G., 1964, Nitrification in the activated sludge process, *Proc. Inst. Sewage Purif.* Part 2:130.

Edel, W., Guinée, P. A. M., van Schothorst, M., and Kampelmacher, E. H., 1972, The role of effluents in the spread of Salmonellae, *Zentralbl. Bakteriol. Parasitenkd. Infektionskr. Hyg.*, Abt I, Orig. Reihe A, **221**:547.

Eikelboom, D. H., 1975, Filamentous organisms observed in activated sludge, *Water Res.* **9**:365.

El-Shafie, A. T., and Bloodgood, D. E., 1973, Anaerobic treatment in a multiple upflow filter system, *J. Water Pollut. Control Fed.* **45**:2345.

Fair, G. M., Geyer, J. C., and Okun, D. A., 1968, *Water and Wastewater Engineering*, Vol. 2, *Water Purification and Wastewater Treatment and Disposal*, John Wiley and Sons, New York.

Focht, D. D., and Chang A. C., 1975, Nitrification and denitrification processes related to waste water treatment, *Adv. Appl. Microbiol.* **19**:153.

Gaudy, A. F., and Gaudy, E. T., 1966, Microbiology of waste waters, *Annu. Rev. Microbiol.* **20**:319.

Gaudy, A. F., Gaudy, E. T., and Komolrit, K., 1963, Multicomponent substrate utilization by natural populations and a pure culture of *Escherichia coli*, *Appl. Microbiol.* **11**:157.

Geldreich, E. E., 1972, Water-borne pathogens, in: *Water Pollution Microbiology* (R. Mitchell, ed.), pp. 207–241, Wiley Interscience, New York.

Grabow, W. O. K., 1968, The virology of waste water treatment, *Water Res.* **2**:675.

Harris, R. H., and Mitchell, R., 1973, The role of polymers in microbial aggregation, *Annu. Rev. Microbiol.* **27**:27.

Hawkes, H. A., 1963, *The Ecology of Waste Water Treatment*, Pergamon Press, Oxford.

Hedén, C-G., and Illéni, T. (eds.), 1975, *New Approaches to the Identification of Microorganisms*, John Wiley and Sons, New York.

Herbert, D., 1961, A theoretical analysis of continuous culture systems, in: *Continuous Culture of Microorganisms*, pp. 21–53, Society of Chemical Industry, Monograph No. 12, London.

Hesseltine, C. W., 1972, Solid state fermentations, *Biotechnol. Bioeng.* **14**:517.

Heukelekian, H., and Dondero, N. C. (eds.), 1964, *Principles and Applications in Aquatic Microbiology*, John Wiley and Sons, New York.

Horvath, R. S., 1972, Microbial co-metabolism and the degradation of organic compounds in nature, *Bacteriol. Rev.* **36**:146.

Hrubant, G. R., 1975, Changes in microbial population during fermentation of feedlot waste with corn, *Appl. Microbiol.* **30**:113.

Hungate, R. E., 1950, The anaerobic mesophilic cellulolytic bacteria, *Bacteriol. Rev.* **14**:1.

Imhoff, K., Müller, W. J., and Thistlethwayte, D. K. B., 1971, *Disposal of Sewage and Other Water-Borne Wastes*, Butterworths, London.

Jannasch, H. W., and Mateles, R. I., 1974, Experimental bacterial ecology, studied in continuous culture, *Adv. Microbial Physiol.* **11**:165.

Koot, A. C. J., 1974, *Behandeling van afvalwater*, Uitgeverij Waltman, Delft.

Koziorowski, B., and Kucharski, J., 1972, *Industrial Waste Disposal*, Pergamon Press, Oxford.

Lawrence, A. W., and McCarty, P. L., 1969, Kinetics of methane fermentation in anaerobic treatment, *J. Water Pollut. Control Fed.* **41**:R1.

Lawrence, A. W., and McCarty, P. L., 1970, Unified basis for biological treatment design and operation, *J. Sanit. Eng. Div. Amer. Soc. Civ. Eng.* **96**:757.

Lettinga, G., and van Velsen, A. F. M., 1974, Toepassing van methaangisting voor behandeling van minder geconcentreerd afvalwater, $H_2O$ **7**:281.

Liebmann, H., 1960, *Handbuch der Frischwasser-und Abwasserbiologie*, Band II, pp. 679–974, Oldenbourg, Munich.

Mack, W. N., Mack, J. P., and Ackerson, A. O., 1975, Microbial film development in a trickling filter, *Microb. Ecol.* **2**:215.

McCarty, P. L., 1972, Energetics of organic matter degradation, in: *Water Pollution Microbiology* (R. Mitchell, ed.), pp. 91–118, Wiley Interscience, New York.

Mitchell, R. (ed.), 1972, *Water Pollution Microbiology*, Wiley Interscience, New York.

Mueller, J. A., Morand, J., and Boyle, W. C., 1967, Floc sizing techniques, *Appl. Microbiol.* **15**:125.

Mulder, E. G., 1964, Iron bacteria, particularly those of the Sphaerotilus-Leptothrix group, and industrial problems, *J. Appl. Bacteriol.* **27**:151.

Oswald, W. J., 1973, Solar energy fixation with algal-bacterial systems, Proceedings of the Workshop on Bio-solar Conversion (NSF/RANN), pp. 38–41, Indiana University.

Oswald, W. J., and Golueke, C. G., 1960, Biological transformation of solar energy, *Adv. Appl. Microbiol.* **2**:223.

Oswald, W. J., and Golueke, C. G., 1968, Large-scale production of algae, in: *Single Cell Protein* (R. I. Mateles and S. R. Tannenbaum, eds.), pp. 271–305, MIT Press, Cambridge, Mass.

Painter, H. A., 1971, Chemical, physical and biological characteristics of wastes and waste effluents, in: *Water and Water Pollution Handbook*, Vol. I (L. L. Ciaccio, ed.), pp. 329–364, Marcel Dekker, New York.

Pike, E. B., 1975, Aerobic bacteria, in: *Ecological Aspects of Used-Water Treatment*, Vol. I (C. R. Curds and H. A. Hawkes, eds.), pp. 1–63, Academic Press, London.

Pike, E. B., and Carrington, E. G., 1972, Recent developments in the study of bacteria in the activated-sludge process, *Water Poll. Control* **71**:583.

Pipes, W. O., 1966, The ecological approach to the study of activated sludge, *Adv. Appl. Microbiol.* **8**:77.

Pipes, W. O., 1967, Bulking of activated sludge, *Adv. Appl. Microbiol.* **9**:185.

Pöpel, F., and Ohnmacht, Ch., 1972, Thermophilic bacterial oxidation of highly concentrated substrates, *Water Res.* **6**:807.

Prasad, C. R., Krishna Prasad, K., and Reddy, A. K. N., 1974, Bio-gas plants: Prospects, problems and tasks, *Economic and Political Weekly*, Vol. IX, Nos. 32–34, Special No. August 1974, 1347–1364.

la Rivière, J. W. M., 1971, New approaches to waste treatment and water recovery, in: *Proceedings of the Third International Conference on Global Impacts of Applied Microbiology* (Y. M. Freitas and F. Fernandes, eds.), pp. 148–167, University of Bombay.

la Rivière, J. W. M., 1972, A critical view of waste treatment, in: *Water Pollution Microbiology* (R. Mitchell, ed.), pp. 365–388, John Wiley and Sons, New York.

la Rivière, J. W. M., Kooiman, P., and Schmidt, K., 1967, Kefiran, a novel polysaccharide produced in the kefir grain by *Lactobacillus brevis*, *Arch. Mikrobiol.* **59**:269.

Rook, J. J., 1974, Formation of haloforms during chlorination of natural waters, *Water Treat. Exam.* **23**:234.

Schiemer, F., 1975, Nematoda, in: *Ecological Aspects of Used-Water Treatment*, Vol. I (C. R. Curds and H. A. Hawkes, eds.), pp. 269–288, Academic Press, London.

Scherb, K., 1973, Der Einsatz von reinem Sauerstoff oder sauerstoff-angereicherter Luft beim Belebungsverfahren, *Wasser Abwasser Forsch.* **1973(5)**:160.

Servizi, J. A., and Bogan, R. H., 1963, Free energy as a parameter in biological treatment, *J. Sanit. Eng. Div. Amer. Soc. Civ. Eng.* **89**, No SA3:17.

Shilo, M., 1975, Factors involved in dynamics of algal blooms in nature, in: *Unifying Concepts in Ecology* (W. H. van Dobben and R. H. Lowe-McConnell, eds.), pp. 127–132, Junk, the Hague.

Sierp, Fr., 1959, *Die gewerblichen und industriellen Abwässer*, Springer, Berlin.

Silver, R. S., and Mateles, R. I., 1969, Control of mixed-substrate utilization in continuous cultures of *Escherichia coli*, *J. Bacteriol.* **97**:535.

Soemarwoto, O., 1975, Rural ecology and development in Java, in: *Unifying Concepts in Ecology* (W. H. van Dobben and R. H. Lowe-McConnell, eds.), pp. 275–281, Junk, the Hague.

Solbé, J. F. de L. G., 1975, Annelida, in: *Ecological Aspects of Used-Water Treatment*, Vol. I (C. R. Curds and H. A. Hawkes, eds.), pp. 305–335, Academic Press, London.

Stanier, R. Y., Doudoroff, M., and Adelberg, E. A., 1971, *General Microbiology*, Macmillan, London.

Stanton, W. R., 1977, Algae in waste recovery, in: *Proceedings IV International Conference on Global Impacts of Applied Microbiology*, Sao Paulo.

Stouthamer, A. H., and Bettenhausen, C., 1973, Utilization of energy for growth and maintenance in continuous and batch cultures of microorganisms, *Biochim. Biophys. Acta* **301**:53.

Stumm, W., and Morgan, J. J., 1970, *Aquatic Chemistry*, Wiley Interscience, New York.

Stumm-Zollinger, E., 1966, Effects of inhibition and repression on the utilization of substrates by heterogeneous bacterial communities, *Appl. Microbiol.* **14**:654.

Stumm-Zollinger, E., and Fair, G. M., 1965, Biodegradation of steroid hormones, *J. Water Pollut. Control Fed.* **37**:1506.

Sykes, G., and Skinner, F. A. (eds.), 1971, *Microbial Aspects of Pollution*, Academic Press, London.

Toerien, D. F., and Kotzé, J. P., 1970, Population description of the non-methanogenic phase of anaerobic digestion, *Water Res.* **4**:129 (I); 285 (II); 305 (III); 315 (IV).

Toerien, D. F., Hattingh, W. H. J., Kotzé, J. P., Thiel, P. G., Pretorius, W. A., Cillie, G. G., Henzen, M. R., Stander, G. J., and Baillie, R. D., 1969, Anaerobic digestion (review paper), *Water Res.* **3**:385 (I); 459 (II); 545 (III); 623 (IV).

Tomlinson, T. G., and Williams, I. L., 1975, Fungi, in: *Ecological Aspects of Used-Water Treatment*, Vol. I (C. R. Curds and H. A. Hawkes, eds.), pp. 93–152, Academic Press, London.

Unz, R. F., and Farrah, S. R., 1976, Exopolymer production and flocculation by *Zoogloea* MP6, *Appl. Environ. Microbiol.* **31**:623.

Veldkamp, H., 1975, The role of bacteria in energy flow and nutrient cycling, in: *Unifying Concepts in Ecology* (W. H. van Dobben and R. H. Lowe-McConnell, eds.), pp. 44–49, Junk, the Hague.

Wallen, L. L., and Davis, E. N., 1972, Biopolymers of activated sludge, *Environ. Sci. Technol.* **6**:161.

Wallen, L. L., and Rohwedder, W. K., 1974, Poly-$\beta$-hydroxyalkanoate from activated sludge, *Environ. Sci. Technol.* **8**:576.

Walsby, A. E., 1972, Structure and function of gas vacuoles, *Bacteriol. Rev.* **36**:1.

Walsby, A. E., 1974, The identification of gas vacuoles and their abundance in the hypolimnetic bacteria of Arco Lake, Minnesota, *Microb. Ecol.* **1**:51.

Weddle, C. L., and Jenkins, D., 1971, The viability and activity of activated sludge, *Water Res.* **5**:621.

Wolfe, R. S., 1971, Microbial formation of methane, *Adv. Microb. Physiol.* **6**:107.

Wuhrmann, K., 1964, Microbial aspects of water pollution control, *Adv. Appl. Microbiol.* **6**:119.

Zamenhof, S., and Zamenhof, P. J., 1971, Steady-state studies on some factors in microbial evolution, in: *Recent Advances in Microbiology* (A. Pérez-Miravete and D. Peláez, eds.), pp. 17–24, Libreria Internacional S. A., Mexico City.

Zeikus, J. G., and Wolfe, R. S., 1972, *Methanobacterium thermoautotrophicus* sp.n. an anaerobic autotrophic extreme thermophile, *J. Bacteriol.* **109**:707.

# Index